Commercial Energy Assessor's Handbook

Larry Russen, Simon Rees & Stephen Neale

Acknowledgments

Writing any book will inevitably draw on the knowledge and help of people other than the authors, and this book is no different. The authors wish to thank Arjan Buschmann at Pinguin Foods in Kings Lynn, Alister Borthwick of Deepdale Farms, Brancaster in Norfolk, Pat Isbill at the Riverside Restaurant in Kings Lynn, Martin Noyle, who shamelessly steals all of Larry's best ideas and claims them as his own (but, then, Larry treats Martin's ideas in similar fashion!), Tony Herbert, Peter Carr-Seaman, Phil Parnham, Mike Tofts, Bruce Arnold, Hilary Grayson, Mark Sreeves, David Mateus-Flynn, Tim Bull, Kevin Barcock, (particularly for his help with the 'Kings Lynn protocol') and all other members of the training and assessment teams at NHER.

We send our combined thanks to Sophie Brooks at RICS Books for her considerable assistance during the writing of the book; but mostly for her tolerance, patience and forbearance whenever we gave her *another* reason why we weren't in a position to deliver the manuscript on time!

The authors wish to thank Neil Brown, Gary Coulson, Tony Dowsett, Stephen Firth and Bruce Needam at De Montfort University for their assistance in collecting the photographs in Chapters 5 and 6 and Appendix C.

The authors and publishers also wish to thank the following for permission to reproduce copyright material:

   Carbon Trust for Figures 5.4, 5.18.1, 5.23, 5.24, 6.25, 7.2, 7.17, 7.18, 7.19, 7.20
   EarthEnergy Ltd for Figures 5.19.1 and 5.19.2
   H. Guntner (UK) Limited for Figures 6.30.2, 6.31.1 and 6.31.2
   Coolmation Ltd for Figures 6.28.2, 6.29.2 and 6.32
   Viessmann Ltd for Figures 5.8.1, 5.8.2, 5.15.2, 5.15.4, 5.15.5, 5.16.1, 6.33.1, 6.33.2, 6.34.1 and 6.34.2
   De Montfort University
   Gilberts (Blackpool) Ltd for Figures 5.74, 6.1.1, 6.1.2, 6.21.2 and 6.23
   HSE for Table 2.2 and Figure 2.3
   MPBA for Figure 4.18

Published by the Royal Institution of Chartered Surveyors
Surveyor Court
Westwood Business Park
Coventry CV4 8JE
UK

www.ricsbooks.com

No responsibility for loss or damage caused to any person acting or refraining from action as a result of the material included in this publication can be accepted by the author or RICS.

ISBN 978 1 84219 534 5

© Royal Institution of Chartered Surveyors (RICS) January 2010. Copyright in all or part of this publication rests with RICS, and save by prior consent of RICS, no part or parts shall be reproduced by any means electronic, mechanical, photocopying or otherwise, now known or to be devised.

Typeset and printed in Great Britain by Page Bros, Norwich

Contents

Foreword		xi
Preface		xiii
1	**THE COMMERCIAL ENERGY ASSESSOR AND LEGISLATIVE BACKGROUND**	**1**
	Introduction	1
	General energy assessment issues	1
	Non-domestic Energy Assessors	1
	Contribution of commercial buildings to climate change	1
	Historical development of energy issues	1
	European context	1
	The United Kingdom response	2
	Communities and Local Government	2
	National legislation	2
	National calculation methodology (NCM) in the UK	2
	The Simplified Building Energy Model (SBEM)	2
	Other SBEM program options	2
	Different levels of commercial energy assessment – general summary	3
	'Asset' and 'operational' ratings	3
	Building Regulations and SBEM	3
	SBEM 'documents'	4
	Guidance available to CEAs	4
	Air conditioning inspections	5
	Training and assessment	5
	National Occupational Standards (NOS) – Asset Skills	5
	Commercial energy assessor training	5
	Assessment of your knowledge, understanding and competence	5
	End test	6
	Qualification	6
	Quality and auditing	6
	Accreditation schemes	6
	Minimum Requirements For Energy Assessors For Non-Dwellings, CLG, October 2007	7
	Liability	8
	Summary	8
2	**GETTING READY FOR BUSINESS**	**9**
	Introduction	9
	Legal background	9
	Occupiers' Liability	9
	Law of tort	9
	Case study	10
	Law of contract	10
	Privity of contract and third party rights	10
	Health and safety liability	11
	Health and safety legislation	11
	Approved Codes of Practice (ACOP)	11
	Risk assessment	12
	Hazards	12

	Asbestos	12
	Asbestos awareness training	14
	The properties of asbestos	14
	Asbestos risk	14
	Asbestos-related diseases	15
	Emergencies involving asbestos	15
	The management of asbestos in non-domestic premises	15
	Summary	15
3	**PREPARING FOR THE COMMERCIAL ENERGY ASSESSMENT**	**17**
	Introduction	17
	Practical legal definitions and other issues	17
	'Building'	17
	'Conditioning'	17
	Absence of fixed services that provide conditioning	17
	Differences between level 3, 4 and 5 assessment for existing buildings	18
	Exemptions	18
	Enforcement	19
	Organising your practice	19
	Why you need a quality management system	19
	Professional indemnity insurance (PII)	22
	Complaints procedure	22
	Life-long learning	22
	Running and organising your professional practice as a business	22
	Data gatherers	22
	Equipment and measuring practice	23
	Inspection equipment	23
	RICS Code of Measuring Practice	26
	New build construction	27
	Building Regulations	27
	Building Regulations in other parts of the UK	28
	Summary	29
4	**ENERGY AND CONSTRUCTION CORE KNOWLEDGE AND RECOGNITION**	**31**
	Introduction	31
	Energy in construction	31
	General principles	31
	Thermal transmittance (U value)	31
	Effective thermal capacity (K_m value)	31
	Example 4.1.1: Solid one-brick wall, plastered internally	32
	Example 4.1.2: 100mm mediumweight block wall, plastered internally and rendered externally	33
	Example 4.1.3: Timber framed cavity wall	33
	Thermal bridges	34
	Accredited Building Details	35
	Planning and Building Regulations issues	35
	Building Regulations	35
	Conservation areas and listed buildings	36
	Specific construction issues	37
	Identification	37
	Date of construction	37
	Forms of construction	38
	Traditional construction methods	38
	Framed buildings	39
	Fire insulation	40
	Controlling dimensions	41
	Types of external envelope	41
	External walls	41
	Cladding panels	42

	Curtain walling	42
	Profiled cladding systems	43
	Retaining structures	43
	Windows	44
	Doors	46
	Roofs	46
	Floors and ceilings	46
	Building systems	48
	Summary	48

5 HEATING AND HOT WATER SERVICES CORE KNOWLEDGE AND RECOGNITION — 51

Introduction — 51
   The essential task — 51
   Sources of HVAC information — 52
   HVAC system organisation — 52
   HVAC system efficiency — 54
      Equipment seasonal efficiency — 54
      The ECA scheme — 55
   Fuels and emission factors — 55
   Power factor — 56
   Systems not represented in SBEM — 56
Heating systems — 56
   Radiator and convector heating — 57
   Underfloor heating — 59
   Central heating using air — 60
      Systems with only a supply duct distribution system — 60
      Systems with both a supply and extract duct distribution system — 60
   Radiant heating — 61
      High temperature tube radiant heating — 61
      Incandescent radiant heating — 61
      Hot water radiant heating — 62
   Forced convection heating — 62
   Local room heaters — 62
   Destratification fans — 65
Heating sources — 65
   Boiler heat sources — 65
   Boiler flues — 66
   Gas and oil fuelled boilers — 67
   Mineral fuel and biomass boilers — 69
   Medium and high temperature boilers — 70
   Combined heat and power (CHP) — 70
   District heating — 71
   Heat pumps — 72
      Ground source heat pump systems — 72
      Air source heat pumps — 73
HVAC system controls — 73
   Central time control — 73
   Local time control — 73
   Local temperature control — 74
   Weather compensation temperature control — 74
   Optimum start/stop control — 74
Hot water generation — 75
   Hot water demands — 75
   Hot water energy losses — 75
      Distribution losses — 75
      Storage losses — 77
   Hot water generators — 78
      Dedicated hot water boiler systems — 78
      Stand-alone water heaters — 78

	Instantaneous water heaters	78
	Instantaneous combi systems	80
	Heat pump hot water generators	80
Summary		80

6 VENTILATION AND AIR CONDITIONING CORE KNOWLEDGE AND RECOGNITION — 83

	Page
Introduction	83
Ventilation	83
Ventilation openings	83
Natural ventilation	84
Central air conditioning system ventilation	85
Mechanical ventilation	85
Local extract	85
Mixed-mode ventilation	86
Ventilation and air conditioning equipment	86
Fans	87
Filters	88
Dampers	88
Air heat exchangers	89
Air handling units	91
Supply systems with heating cooling and filtration	91
Supply and extract systems with heating, cooling, filtration and recirculation	91
Heat recovery	93
Plate heat exchangers (recuporator)	93
Thermal wheels	93
Run-around coils	94
Heat pipe devices	94
Other heat recovery systems	94
Fan speed control	94
Demand controlled ventilation	95
Specific fan power	95
Air conditioning systems	95
Variable air volume (VAV) systems	96
Single duct VAV systems	97
Dual duct VAV systems	97
Indoor packaged cabinet VAV systems	98
Constant air volume systems	98
Variable fresh air CAV	98
Fixed fresh air CAV	99
Terminal reheat CAV	99
Dual duct CAV systems	100
Multi-zone systems	100
Air-water air conditioning systems	100
Fan coil unit systems	100
Induction units	101
Chilled ceilings	101
Chilled beams	101
Displacement ventilation	102
Water loop heat pumps	102
Refrigerant air conditioning systems	103
Single room cooling systems	103
Split and multi-split systems	104
Variable refrigerant flow systems	105
Air source heat pumps	105
Cooling sources	106
Air cooled chillers	106
Water cooled chillers	107
Chillers with remote condensers	107

	Dry air coolers	108
	Cooling towers	108
Renewable energy systems		109
	Solar energy systems	109
	Photovoltaic systems	110
	Wind turbines	111
Summary		111

7 LIGHTING CORE KNOWLEDGE AND RECOGNITION — 113

Introduction	113
Lighting theory and practice	113
The nature of light	113
Colour temperature	114
Colour rendering	115
Identification of lamps	115
Levels of lighting	115
Building Regulations	115
Lighting glossary	115
Artificial lighting systems and their recognition	117
Types of lamp	117
Recognition of lamp type	117
Tungsten filament (GLS) lamps – incandescent	117
Tungsten halogen (TH) lamps – incandescent	118
Fluorescent 'strip' discharge lamps – (MCF)	118
Compact fluorescent discharge lamps – (CFLs)	120
Low pressure sodium discharge lamps – (SOX)	120
High pressure sodium discharge lamps – (SON)	120
Metal halide discharge lamps – (MBI or HPI or MBI/MH or CDM)	120
High pressure mercury discharge lamps – (MBF)	121
Induction discharge lamps – (QL)	121
Light emitting diodes (LEDs)	121
Identification of fluorescent control gear and ballasts	121
Lamp identification – quick case study	122
Carbon Trust lamp identification information	123
Lighting in SBEM	127
Zone lighting energy	127
Defining lamp type	127
Efficiencies and efficacies of different lamps	127
'Full lighting design carried out'	129
Air extracting luminaires	129
Lighting controls	129
General	129
SBEM light controls	130
Other lighting issues	131
Practical SBEM lighting issues	131
Display lighting	131
Your safe inspection of lamps	131
Damage caused by light	132
Summary	132

8 INSPECTION AND REFLECTION METHODOLOGIES FOR EXISTING BUILDINGS — 133

Introduction	133
First contact with your client	133
The audit trail	133
Initial enquiry, fees and conflicts of interest	133
Client instructions	135
Pre-inspection practice	135
General comment	135
Seller's questionnaire	135

	Desk study	135
	Site inspection – a suggested procedure	136
	Inspection – some general comments	136
	Arrival	136
	Information from the client	137
	Plan of the property	137
	Contents of the plan	138
	Order of inspection	138
	Photographs	140
	Extent of inspection – 'invasive' inspections	140
	Site inspection and reflection notes	140
	Zoning	141
	Two different possible data collection methodologies	141
	After your inspection	142
	Post-inspection enquiries	142
	Summary	142
9	**DATA ENTRY INTO THE PROGRAM**	**143**
	Introduction	143
	Main SBEM tabs	143
	'General'	143
	'Project database'	143
	General comment	143
	Construction for walls	144
	Construction for roofs	144
	Construction for floors	144
	Construction for doors	144
	Glazing	144
	'Geometry'	144
	Global building infiltration	144
	Activity	144
	Building types and activities – an example	146
	Building type and activities – a case study	147
	Choosing the activity	148
	Zones and zoning	148
	Zone height and corners	149
	Envelopes	149
	Adjoining conditions	150
	Specific issues relating to conditioning	150
	Indirect conditioning	152
	Envelope measurement conventions	152
	Thermal capacities within merged areas	153
	External windows and doors	153
	'U values' of vehicular entrance doors	153
	High usage entrance doors	153
	Fins and overhangs	153
	Glazed 'doors'	153
	'Building services'	153
	'Ratings and building navigation'	153
	Summary	153
10	**RATINGS, RECOMMENDATIONS AND REPORTING**	**155**
	Introduction	155
	SBEM results and definitions	155
	Main practical purpose of SBEM	155
	Example of the results page	155
	Different SBEM building types	157
	Other SBEM descriptions	157

Interpretation of the results	157
The Energy Performance Certificate	158
Graphic rating	158
Recommendations	159
How to use the results page	159
Reflection on the EPC and the results generally	159
Audit checks before calculation	159
Audit checks after calculation	160
Recommendations tab	161
The recommendations generated by SBEM	163
Developing your own recommendations	163
The EPC audit	163
Recording comments on your assessment program	164
Reporting to your client	165
General comment	165
Client report	165
Summary	165

APPENDIX A – CASE STUDY 1 – LEVEL 3 'ASSESSMENT' OF THE OLD FORGE, MAIN ROAD, BURNHAM DEEPDALE, NORFOLK PE31 1ET — 167

Introduction	167
Description of the building	167
Photographs	167
Client report	168
Pages from site notes	169
Discussion	171

APPENDIX B – CASE STUDY 2 – LEVEL 3 ASSESSMENT OF THE RIVERSIDE RESTAURANT, KING STREET, KINGS LYNN, NORFOLK PE30 1ET — 173

Introduction	173
Brief description	173
Completed client questionnaire	173
Photographs	190
Site inspection and reflection notes	196
Completed zone plans, geometry data and data reflection notes	233
Zone plan data collection	241
Documents generated	243
Energy Performance Certificate	243
Recommendation Report	245
Main Calculation Output Document	253
Secondary Recommendation Report	254
Client report	262
Discussion	264
Summary	267

APPENDIX C – CASE STUDY 3 – LEVEL 4 HVAC ASSESSMENT OF THE PACE BUILDING, DE MONTFORT UNIVERSITY, LEICESTER LE1 9BH — 269

Introduction	269
Description of the building	269
Building inspection	269
External inspection	269
Boiler room inspection	270
Lobby inspection	271
Theatre rehearsal room	272
Toilets and changing rooms	272
Main rehearsal room	273
Offices	273

      Dance rehearsal rooms   274
      Roof level inspection   274
   Reflection   277

APPENDIX D – PIPE AND DUCT IDENTIFICATION   279

APPENDIX E – DRAWING SYMBOLS AND ABBREVIATIONS   281

APPENDIX F – SURFACE AREA RATIO (AS CITED IN CHAPTER 9, EXTERNAL WINDOWS AND DOORS)   289

APPENDIX G – AREA RATIO COVERED (APPLIED TO ROOFLIGHTS ONLY)   291

APPENDIX H: TRANSMISSION FACTOR CORRECTION (AS CITED IN CHAPTER 9, FINS AND OVERHANGS)   293

   Correction factor   294
   Latitudes   294
   South facing window partial shading correction factors for overhangs, Fo   295
   East or West facing window partial shading correction factors for overhangs, Fo   296
   North facing window partial shading correction factors for overhangs, Fo   297
   South facing window partial shading correction factors for fins, Ff   298
   East or West facing window partial shading correction factors for fins, Ff   299
   North facing window partial shading correction factors for fins, Ff   300

REFERENCES   301
INDEX   305

Foreword

Commercial Energy Assessor's Handbook is an important publication for all surveyors involved in this developing area.

The role of the CEA is new and evolving. The core training and guidance available to practitioners provides the knowledge needed to qualify for the role. However, this book fully equips readers with practical advice for applying this knowledge, and their new skills to the job. With the *Energy Performance of Buildings Directive* and UK *Energy Performance of Buildings* in full flow since October 2008, the CEA role is fundamental and in demand.

The authors of the Handbook: Larry Russen, a Chartered Building Surveyor, Commercial Energy Assessor, trainer, author and assessor with the SAVA and the NHER assessment centres; Dr Simon Rees, a Senior Research Fellow specialising in low energy building design, energy simulation and geothermal systems; and Stephen Neale, a Commercial Energy Assessor, NVQ assessor and verifier who runs his own building consultancy, all provide a wealth of experience to guide surveyors in all aspects of applying their training. Their breadth of knowledge brings comprehensive insight and clear explanation to a complex area.

I am delighted to recommend this handbook to surveyors at all levels who have already trained as, or are considering a role as, a CEA. This book will provide invaluable advice and reference to you all.

Max Crofts
RICS President 2009–10

Preface

This book is primarily directed at those new to commercial energy assessment and Non-domestic Energy Assessors (or Commercial Energy Assessors (CEAs) as we shall refer to them) qualified to levels 3 and 4. However, most of the book – in that it deals with the inspection process and data preparation – is equally applicable to those qualified to National Occupational Standards (NOS) level 5 and those using tools other than iSBEM.

Our intention in writing this book has been to focus on the knowledge, understanding and competence required for the NOS level 3 and 4 CEAs; their role in this new profession, and the processes involved. We are firmly of the opinion that CEAs comprise a profession, with all of the associated benefits, advantages and considerable obligations, particularly legal.

Furthermore, we believe the examples of competence we describe in this book are the minimum acceptable practice; together with the requirements contained in:

- the National Occupational Standards from Asset Skills for Building Energy Assessment (Non-dwellings) on Construction, Sale or Rent;
- the Minimum Requirements for Energy Assessors from CLG;
- the various sets of rules from the different Accreditation Schemes – only 'accredited' individuals ('assessors'), who are members of a recognised 'accreditation scheme' can prepare an Energy Performance Certificate (EPC) and Recommendation Report (RR); and
- any other relevant guidance.

Our focus is not only on preparing non-domestic EPCs and RRs. We consider the other necessary skills and implications of acting in this role.

The book assumes you have some necessary knowledge and understanding, i.e. you are already aware of how buildings are constructed and have some knowledge of typical materials and building services. We include discussion of the energy, construction, service and lighting theoretical knowledge; and the practical skills you will require in the real world. In that regard, we set the new profession in context with other property and related professions.

We refer to many other publications and include what we hope is a helpful but by no means comprehensive bibliography. We recommend you already have a copy of a sister publication, *Domestic Energy Assessor's Handbook* by Phil Parnham and Larry Russen (RICS Books, 2008). We believe this is reasonable as many aspiring CEAs are already DEAs and possess that book. Rather than repeat advice in that book, much of which is relevant, we refer where necessary to the relevant pages.

Since we believe the CEA's role is primarily practical, we include in each chapter photographs and diagrams to illustrate and break up the text. We also include some short case studies (centred around photographs and sketches) to give examples of, and put into context, the skills a CEA requires.

The guidance is specifically for England and Wales in the sense that where reference is made to the various paragraphs and the documents in use, they are those documents in use in those two parts of the UK. CEAs in Scotland and Northern Ireland are advised that paragraph numbers and other references vary, although the advice is still generally applicable.

Energy assessments of level 5 buildings require use of a different methodology (Dynamic Simulation Methodology (DSM)), which is beyond the scope of this guidance. This guidance does not relate to preparation of Display Energy Certificates (DECs). However, the knowledge and understanding about construction, HVAC and lighting will help those practitioners.

The subject of non-domestic energy assessment is extensive and subject to considerable change at the time of writing. There is not universal agreement about how certain data from existing buildings should be interpreted or taken into account in the assessment and entered into software. The book has been written to provide practical advice and tips to energy assessment that we think are important but are not available elsewhere. We do not claim to provide authoritative guidance and we expect some will not agree with our suggestions or advice. Your own Accreditation Scheme Provider should be your authoritative source of guidance and information about both assessment policy and software data entry.

Assessors should always refer to the latest versions of user and technical guidance available from BRE that relates to iSBEM; and should revise their practise, where necessary, whenever BRE issues new versions.

If you can infer any suggestion in the following pages we are at all critical of the BRE iSBEM program, then we apologise now. We are all admirers of the Building

Research Establishment (BRE) and consider it an organisation primus inter pares in the world. It is far easier to critically judge a product than to actually produce it – we believe iSBEM is a good product, and with increased use, reflection and minor amendment by BRE following feedback from users, it will become even better.

Similarly, in our view the Department for Communities and Local Government (CLG) have done the best job possible in the time, and with available resources, to bring the EPBD into force in the UK. Please take any implied criticism of their actions and/or publications in the constructive spirit we intend it.

We also apologise for the inevitable errors and omissions we have incorporated into the text. Those errors are inevitable since:

- to err is to be human (and we hope we are that!);
- the subject area is so vast; and
- changing continuously.

In which regards, readers and practitioners should therefore bear in mind that this book can only be regarded as a general review of iSBEM and all of the attendant information at the date or writing. In view of the current interest in climate change and sustainability they must strive to maintain their knowledge and understanding.

Finally, we would like to say that our hope is that the book will be of value to those undergoing training, those developing their learning and will become a valuable source of information in professional practice.

Larry Russen, Simon Rees & Stephen Neale,
December 2009

About the Authors

Larry Russen

Larry is a Chartered Building Surveyor, Corporate Building Engineer; Commercial Energy Assessor and Home Inspector. He carries out commercial and residential EPCs, building surveys, building pathology, defect diagnosis, acts as a party wall surveyor and advises clients on remedial and new building work. He has appeared in the High Court as an expert witness.

He is a Director of Russen & Turner Chartered Surveyors based in Kings Lynn and BlueBox partners, a national training organisation for property professionals. He is also an assessor and internal verifier with the SAVA and NHER assessment centres in Milton Keynes.

He is the co-author, with Phil Parnham, of *The Domestic Energy Assessor's Handbook* published by RICS Books and the author of the RICS online isurv chapters on residential and commercial energy assessment available at www.isurv.co.uk.

He has a Post Graduate Certificate in Education and regularly teaches on subjects such as energy, surveys and Building Regulations.

Dr Simon Rees

Dr Simon Rees practiced as a Mechanical Building Services Engineer before undertaking a PhD in the field of low energy cooling systems. He has undertaken research in geothermal heat pump systems as a Visiting Assistant Professor at Oklahoma State University and is currently a Senior Research Fellow at the Institute of Energy and Sustainable Development at De Montfort University. He teaches and carries out research in low energy building design, energy simulation and geothermal systems. He is a member of CIBSE and recipient of its Napier Shaw Bronze Medal. He is also a member of the ASHRAE Standards Committee concerned with energy simulation software testing.

Stephen Neale

Stephen Neale BSc(Hons) BA(Hons) MRICS MCMI FInstCPD is a Chartered Building Surveyor who has worked in both the public and private sectors. Stephen has a Certificate in Education and has taught a variety of building surveying and construction subjects at undergraduate level. He is also a qualified asbestos surveyor, commercial energy assessor and NVQ assessor and verifier. Stephen runs his own building consultancy, Hisbis Limited.

1 The Commercial Energy Assessor and Legislative Background

INTRODUCTION

In this first chapter we introduce you to some of the important concepts you will need to understand if you wish to practise as a 'Non-domestic Energy Assessor'. In particular, we address:

- the role of the Non-domestic Energy Assessor;
- climate change and the international, European and national context;
- the NCM and SBEM;
- Building Regulations and different types of 'energy assessment';
- SBEM guidance and documentation;
- National Occupational Standards, training and assessment;
- 'accreditation schemes' and your involvement with them;
- the 'Minimum Requirements' you must comply with when qualified;
- your potential liability as a Non-domestic Energy Assessor.

GENERAL ENERGY ASSESSMENT ISSUES

Non-domestic Energy Assessors

Throughout this book we will refer to a Non-domestic Energy Assessor as a 'Commercial Energy Assessor' (CEA). However, it is to be noted that this is not an entirely accurate acronym, since CEAs can also inspect and report on non-domestic buildings that are 'residential' in nature.

The legislation allows different types, or 'levels', of CEA to practise in the UK: levels 3, 4 and 5. We will explain the differences later.

Contribution of commercial buildings to climate change

Published data confirms that commercial buildings make a significant contribution to carbon dioxide (CO_2) and other 'greenhouse gas' emissions. As you research the subject, you will note that different authors and authorities suggest varying amounts of such emissions. We believe it is reasonable to assume that non-domestic buildings in the UK contribute around 20% of all CO_2 emissions.

Historical development of energy issues

The much quoted Kyoto Protocol, which is available to read in Arabic, Chinese, English, French, Russian and Spanish, comprises 28 Articles and 2 Annexes and was signed by the 'Parties to the United Nations Framework Convention on Climate Change'. The Protocol dates from 11 December 1997.

Energy and climate change issues have been in the news for a number of years. Article 3 of the Kyoto Protocol confirms, among many other matters that:

> 'Parties ... shall ... ensure that their aggregate anthropogenic carbon dioxide equivalent emissions of the greenhouse gases ... do not exceed their assigned amounts ... with a view to reducing their overall emissions of such gases by at least 5 per cent below 1990 levels in the commitment period 2008 to 2012.'

Annex A to the Protocol includes a list of 'greenhouse gases', at the top of which is CO_2. The Copenhagen agreement (October 2009) resulted in few changes in international arrangements to help prevent climate changes.

European context

Initiatives in the EU to improve national energy efficiencies and reduce carbon emissions have resulted in the introduction of the *Energy Performance of Buildings Directive* (EPBD), which came into force on 4 January 2003 (2002/91/EC). In broad terms the EPBD requires EU nations to develop consistent methods of assessing energy performance and to provide certificates that register performance and promote dissemination of energy efficiency information to occupiers and other building users.

In order to implement the requirements of the EPBD every country must have a national calculation methodology (NCM) that allows consistent comparisons between new building designs and national standards, and certification of existing buildings. National calculation methodologies may differ across the EU and may employ more than one calculation method or tool. You should make a careful study of the EPBD to understand the legal context you will practise under.

Paragraph 16 deals with display energy certificates (DECs). DECs are based on a different energy certification methodology. They are 'operational ratings'; whereas energy performance certificates (EPCs) are 'asset ratings' (see later in this chapter).

THE UNITED KINGDOM RESPONSE

Communities and Local Government

The Department for Communities and Local Government (CLG) is the government department responsible for administering the introduction of the EPBD into the UK. CLG has a number of areas of responsibilty, including:

- Building Regulations;
- housing;
- local government;
- planning; and
- sustanable communities

As a CEA you will encounter CLG on a regular basis. We suggest you visit the CLG website at www.communities.gov.uk to familiarise yourself with it and to perhaps add some of the links to your 'favourites' on your PC.

National legislation

The EPBD is not considered to be 'the law' in England and Wales. The main document that frames the law and is the legal basis for your practice is the *Energy Performance of Buildings (Certificates and Inspections) (England and Wales) Regulations 2007* (SI 2007/991).

There are at least four other 'Amendment Regulations' and explanatory documents. We advise you to download all of these documents now for reference purposes. You should have them to hand so you can refer to them during your training, as you read this book, and after you qualify and practise as a CEA.

National calculation methodology (NCM) in the UK

The UK has adopted an approach whereby buildings are compared on the basis of carbon emission rates. This represents a very different approach to previous prescriptive approaches that, for example, relied on specifying minimum levels of insulation.

In the UK the NCM has three elements:

1. the Standard Assessment Procedure (SAP and RDSAP) – for new and existing dwellings;
2. the Simplified Building Energy Model (SBEM) – intended for a significant majority of non-domestic buildings; and
3. approved Dynamic Simulation Models (DSM) – such models may be used for all non-domestic buildings, but in reality will usually only be used for those non-domestic buildings that cannot be assessed using SBEM.

Each NCM is required to take account of factors affecting the energy performance of buildings. Those factors include thermal properties of the fabric, efficiencies of heating, cooling and lighting systems, the 'activities' of the assumed occupiers and the manner they use the equipment implicit in their occupation of the building.

In order to make consistent comparisons between buildings it is essential that calculation methods use agreed assumptions about all of these issues. These standard assumptions amount to a large set of numerical data in a series of databases that can be drawn upon by calculation tools.

The NCM is defined by these databases as well as the mathematical models and calculation methods that are employed. The NCM databases are accordingly common to both the SBEM and DSM.

The Simplified Building Energy Model (SBEM)

The SBEM has been developed by the Building Research Establishment (BRE) on behalf of the CLG and forms the key element of the NCM in England and Wales. The SBEM is defined by a calculation method (a set of equations and algorithms) and the databases of construction and activity information. The SBEM has been implemented in a number of computer applications but is commonly used along with the iSBEM software model developed by the BRE and freely distributed. You should note that in order to make use of the program, you will require Microsoft® Access or Access Runtime on your PC.

BRE originally developed the SBEM for new buildings and it has since been revised to allow assessment of existing buildings. As you learn about SBEM and become more skilled in using the program, you will discover there are a number of anomalies in that regard. The program sometimes seems to be inappropriate for existing buildings. This is a historic issue that may eventually disappear with later revisions. We address some of these issues in this book.

Other SBEM program options

The SBEM has a number of other programs that 'talk' to the main SBEM program. The main program is iSBEM. In this book we generally refer to the generic SBEM program – you can assume such reference includes the other software programs.

Commercially available programs include names such as 'Carbon Checker', 'Design Builder', 'Hevacomp', 'Lifespan' developed by Property Tectonics, 'NHER' and 'SBEM on-line', etc. The majority of these programs include a facility to engage in 3D or similar modelling of the building you are assessing. We consider these other programs and the possibly significant advantages such modelling can bring to the SBEM process in Chapter 9.

Different levels of commercial energy assessment – general summary

There are different levels of commercial energy assessment – levels 3, 4 and 5. The July 2008 guidance from CLG clarifies the Regulations and confirms when you should apply differing levels of assessment.

In very simple terms, the differences and definitions may be described as follows:

- level 3 – buildings with frequently occurring characteristics such as simple heating;
- level 4 – buildings with more complicated heating and cooling systems such as centralised air conditioning systems. In addition, all new buildings are treated as 'level 4' assessments; and
- level 5 – buildings with very complicated heating and/or cooling systems.

At this stage, an important point to note is that the differences between the levels of assessment are generally defined in terms of the services that 'condition' the internal environment, rather than the structure and fabric of the building.

The different levels of energy assessment are mirrored by different types of energy assessor. Thus, level 3 assessors can only sign off energy assessments for level 3 buildings.

'Asset' and 'operational' ratings

An asset rating is a measure of the intrinsic or 'built-in' energy performance of a building. The data input and calculation ignores any impact of the actual occupier. SBEM calculates the asset rating of the building and expresses the result as an asset rating, i.e. an EPC, and other results, as we shall discuss. The EPC is much like the energy performance certificates you can see on the sides of new 'white goods' (see Figure 1.1 for an example).

An operational rating is based on the measured (metered) energy consumptions and therefore incorporates occupier usage. DECs are operational ratings.

Both of these methodologies are based on CO_2 emissions.

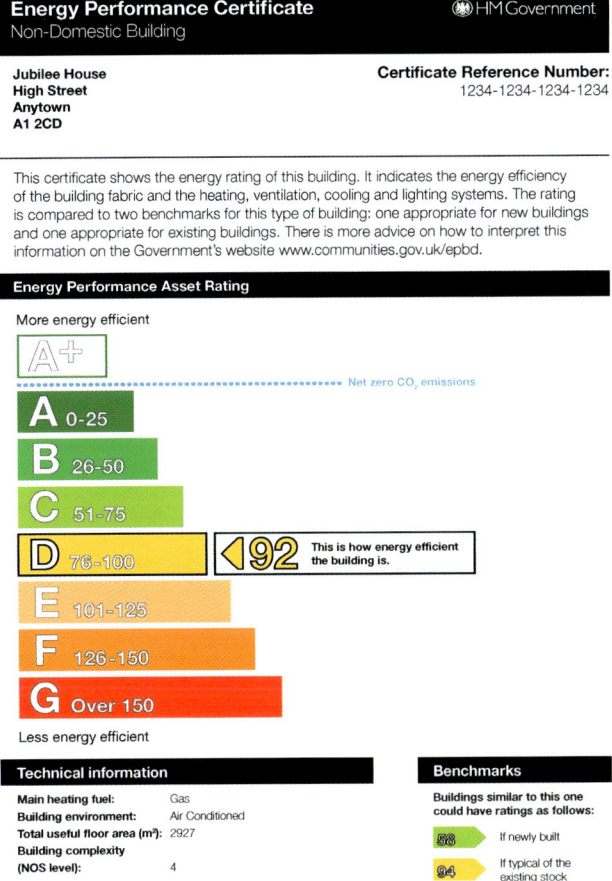

Figure 1.1: Energy Performance Certificate

Building Regulations and SBEM

We consider the Building Regulations for England and Wales in this book. Different regulations apply in Scotland and Northern Ireland. SBEM programs can make assessments for all of the countries in the UK and you must take care to select the Regulations that are appropriate to your project

The Building Regulations are statutory instruments that set standards for the design and construction of building work and seek to ensure that all new construction and significant alterations in buildings and services comply with minimum performance requirements. You can access information and supporting documentation about the Regulations at www.planningportal.gov.uk.

You will discover that a working knowledge of the Regulations is useful when determining the age of construction or when alterations have been made to a building. The Regulations are also used as a means of defining the construction of a building when entering data into a SBEM program.

At the time of writing, the current edition of the Building Regulations is 2000, which has been the subject of several amendments. The current editions of the Approved Documents (ADs) for Part L date from 2006. They comprise four separate subsections:

- L1A New dwellings;
- L1B Existing dwellings;
- L2A New buildings other than dwellings; and
- L2B Existing buildings other than dwellings.

Level 4 practitioners are required to have expertise in assessing and certifying the energy performance of new non-domestic buildings and non-domestic buildings to be significantly extended. They must therefore have a good working knowledge of two of the ADs referred to above, and specifically (with our emphasis):

- ADL2A – 'Conservation of fuel and power in *new* buildings other than dwellings'; and
- ADL2B – 'Conservation of fuel and power in *existing* buildings other than dwellings'.

The SBEM is used for testing the compliance of new buildings with the Building Regulations (Part L2A) and calculation of EPCs for existing buildings (Part L2B). The primary function of the model is to calculate building carbon emission rates given assumptions about occupancy, equipment, etc. The carbon emission data is used in different ways depending on whether a Part L check or EPC rating is required.

If you intend to be a level 3 assessor, you will not require the same level of knowledge as a level 4 assessor. However, you would still be well advised to study both of these documents so you understand how EPCs relate to the Building Regulations.

In addition to the Building Regulations and the ADs to those Regulations, there is a range of 'second tier' documentation relating to commercial energy assessment you should be conversant with. These documents are:

- *Non-domestic Heating, Cooling and Ventilation Compliance Guide*, 2006;
- *Low or Zero Carbon (LZC) Energy Sources: Strategic Guide*, 2006;
- *Accredited Construction Details* (ACD), 2007.

The Building Regulations, ADs and second tier documents are available free from the website indicated above.

SBEM 'documents'

In Chapters 9 and 10, we shall describe how to 'work' the SBEM program. Very briefly, you will enter information ('data') into the program about the external fabric of the building, building services (heating, cooling, hot water, lighting, etc.) and the 'activities' (representing what the intended or actual uses of the building are, e.g. retail, workshop). The software will take the data and generate results that confirm the energy performance of the building. The program compiles some of the results into pdf documents that you can present to your client.

The principal documents are:

- the Energy Performance Certificate (EPC) – an assessment of the asset rating; and
- a Recommendation Report (RR) – a list of suggested improvements to the services and fabric of the building.

Those two documents are the documents 'required' by the EPC Regulations we have referred to. We will discuss these documents and all of the others, including the documents generated for 'new' buildings, later.

Guidance available to CEAs

There is currently very little guidance from the existing professional property bodies such as the Chartered Institution of Building Services Engineers (CIBSE) and RICS, although RICS has an online advice system called RICS isurv. Certain guidance is only available to members of the professional bodies. You should consider joining one of these bodies as a student or associate/affiliate member to have such access.

You must also be conversant with other forms of official, and unofficial, guidance in order to practise safely and to the satisfaction of your clients. The main guidance documents are currently:

- *National Occupational Standards (NOS) for Building Energy Assessment (Non-dwellings) on Construction, Sale or Rent*, Asset Skills, February 2009;
- *Minimum Requirements for Energy Assessors for Non-dwellings*, CLG, October 2007;
- *A User Guide to iSBEM*, BRE, current version;
- *Technical Manual*, BRE SBEM, current version;
- further guidance from CLG, primarily:
  - *Improving the energy efficiency of our buildings – A guide to energy performance certificates for the construction, sale and let of non-dwellings* (2nd edition), CLG, July 2008 (referred to throughout this book as 'the July 2008 CLG guidance'),
  - *Requirements for energy performance certificates (EPCs) when marketing commercial (non-domestic) properties for sale or let*, CLG, October 2008, and
  - *Improving the energy efficiency of our buildings – Local weights and measures guidance for Energy Certificates and air-conditioning inspections for buildings*, CLG, October 2008;
- Guidance and Inspection and Reporting Requirements from the various accreditation schemes – see later; and
- RICS guidance published online – www.isurv.co.uk.

Further guidance will be issued by the various 'stakeholders' in due course. You will need to maintain your knowledge of this guidance so you are always working with current versions.

Air conditioning inspections

The Regulations (SI 2007/991) require that refrigeration and air movement equipment that are part of air-conditioning systems, and their controls; should be regularly inspected as part of the UK response to the EPBD. You will need to confirm whether such an inspection has taken place when you inspect an existing building for an EPC. See our site notes for the case study in Appendix B for further comment.

TRAINING AND ASSESSMENT

National Occupational Standards (NOS) – Asset Skills

The NOS define the knowledge, understanding and competence (skills) that a CEA must possess in order to practise.

The NOS comprise seven 'units' defined as follows:

- Unit 1 – Work in a safe, effective and professional manner
- Unit 2 – Prepare for energy assessments of non-dwellings to produce Regulation 17C calculations (and their equivalent in Scotland and Northern Ireland), Energy Performance Certificates (EPCs), Operational Ratings (ORs), Display Energy Certificates (DECs) and Advisory Reports (ARs)
- Unit 3 – Assess the energy performance of new-build non-dwellings prior to first occupancy using the Simplified Building Energy Model (SBEM)
- Unit 4 – Assess the energy performance of new-build non-dwellings prior to first occupancy using Dynamic Simulation Models (DSMs)
- Unit 5 – Undertake energy inspections of existing non-dwellings with frequently occurring characteristics using the Simplified Building Energy Model (SBEM)
- Unit 6 – Undertake energy inspections of existing non-dwellings using the Simplified Building Energy Model (SBEM)
- Unit 7 – Undertake energy inspections of existing non-dwellings requiring the use of Dynamic Simulation Models (DSMs)

Each unit of the NOS is subdivided into 'elements'. The elements are further subdivided into 'performance criteria' (PC), 'knowledge and understanding' (K&U) and in some instances 'scope'. Thus, for example, element 1.1 deals with contributing to the maintenance of health, safety and security at work; element 3.1 addresses the issue of energy assessments of new-build non-dwellings; and element 5.1 is entitled 'Inspect existing non-dwellings with frequently occurring characteristics', i.e. level 3 buildings.

You must read and understand all those parts of the NOS that are relevant to your work. This is important since you will be initially assessed by an assessment centre against the NOS; and your performance as a CEA will be continuously monitored, once qualified, by the accreditation scheme you are a member of. No partial discussion of the NOS will compare with a good working knowledge of the complete document. The NOS will never be 'set in stone' and will change. It is your responsibilty to maintain your knowledge. Go to www.assetskills.org for more information.

Commercial energy assessor training

You should be aware that 'training' is distinct and separate from the process of your 'assessment'. You engage in training to ensure you have sufficient knowledge and understanding. Your assessment should confirm you have that knowledge and understanding, and competence to practise.

There are a number of different training bodies. Before you pay for a course you should check that their training is sufficient for later assessment purposes. You should read their guidance in full and in particular any reading list. All training tends to be focussed on the syllabus issued by the awarding bodies – see later in this chapter. The subjects you study will depend on the level of assessment, but are likely to include:

- construction;
- heating, cooling, ventilation and hot water services (HWS);
- conventions relating to use of the different software systems that link with SBEM;
- lighting;
- for 'level 4' assessment – 'new-build' construction.

You should spend some considerable time in researching different training schemes. Your training will form the bedrock of your career. You will find it difficult to construct satisfactory foundations on rock that is insubstantial.

Assessment of your knowledge, understanding and competence

This part of the process can usually take place simultaneously with your training, although we would advise most trainees not to begin their assessment process before they have completed some training at least.

The assessment that candidates undertake is 'mapped' against the NOS. At the end of the assessment, the assessment centre must be satisfied you have demonstrated competence against the entire NOS, i.e. you have sufficient knowledge, understanding and competence to practise. When you begin your assessment, you will be assigned to an assessor. By 'assessor' we don't necessarily mean 'energy assessor'; although your assessors will usually be qualified energy assessors. Your assessor will be skilled in assessing candidates against the NOS.

The assessment process varies between different centres. You should ask questions of the centre before you sign up to any scheme. Typical questions you should ask include:

- How long will the entire training and assessment last?
- Does the assessor assigned to the assessment process have occupational competence?
- How does the assessment 'work' and will it suit you (e.g. some assessment centres use a methodology based on a 'face-to-face' process, whilst others prefer to engage in 'online' assessment)?

You can regard your assessment as comprising two generally distinct strands based on:

- 'soft' skills such as ethics, the ability to communicate in a reasonable fashion, health and safety and legal issues associated with being a CEA; and
- 'technical' skills relating to the building and services issues we have referred to above and matters such as the methodology of inspection, data entry and how to provide clients with a report and a service that is fit for purpose.

For the technical part of the assessment process, most centres require that you carry out at least three commercial energy assessments. These assessments, together with other supporting 'evidence', comprise your portfolio. Some centres allow one or more of the assessments to be a 'virtual' EPC assessment, based on given assumed information and photographs, etc. Other centres include an 'observed' or 'accompanied' inspection where an assessor will confirm your on-site competence and practise at a 'real' property.

The process of assessment constitutes a system whereby you will present the assessor with a portfolio of 'evidence'. The assessor will consider that evidence and if it is lacking will provide you with 'feedback'. The feedback; verbal, online or in writing or a combination of all, will seek clarification or further evidence until the assessor is satisfied you have demonstrated 'competence'. Once satisfied, the assessor will 'sign off' your portfolio.

The assessors' work is overseen by an 'internal verifier' who is required to ensure that all assessors work to the same assessment standards and principles. The internal verifier will typically 'randomly sample' the candidates' portfolios to check that the assessors are following the centre's procedures in a consistent manner.

The entire assessment process is overseen by an 'external verifier' who acts on behalf of the awarding body.

End test

The process of qualification normally includes a final test of competence. This is usually a multiple-choice examination with four possible answers to each question. The awarding bodies base the exam on the syllabus and books on the reading list it publishes.

You might be somewhat discouraged and even nervous at the prospect of sitting for an examination. However, there is a simple methodology of answering multiple-choice question assessments as follows:

- read the questions carefully and methodically;
- answer the questions you are reasonably certain you know the answers to – you *will* know many of the answers if you have revised and prepared properly;
- then, answer the questions you are not really sure about:
  ○ start by 'crossing out' the answer you think is probably wrong – you've immediately increased your chances of passing the question from 25% to 33%;
  ○ with luck there will be another answer you 'know', or can guess, is incorrect – cross that one out and your chance of passing this question has gone up from 33% to 50%;
  ○ finally, choose the most probable answer;
- next, answer the questions you haven't got a clue about (there are usually some!), using the same system of elimination; and
- lastly, if you have time, carefully read each question again and check your answers – but take care as your first answer is usually correct, providing you initially read the question accurately.

The 'law of averages' means you will get at least 25% in such a test, unless you're really unlucky! However, by using this system, or something like it, you can dramatically increase your chances. Alternatively, use the system that has worked for you in the past; either way, we wish you good luck.

Qualification

Once you have satisfactorily completed the assessment process, you can currently obtain an approved energy assessment qualification from one of three awarding bodies:

- ABBE – the Awarding Body for the Built Environment (www.abbeqa.co.uk);
- NFOPP – the National Federation of Property Professionals (www.nfopp.co.uk); or
- City and Guilds (www.cityandguilds.com).

QUALITY AND AUDITING

Accreditation schemes

In order to work as a CEA, you must belong to an 'accreditation scheme'. In general, to be an accreditation scheme member, you must:
- be a 'fit and proper' person;
- be qualified (i.e. possessing sufficient skills) to practise; and

- have professional indemnity insurance.

You are permitted to be a member of more than one scheme, subject to acceptance by the scheme. Candidates who apply to any accreditation scheme must either possess an approved qualification following a route such as we have described above; or demonstrate they have sufficient 'prior experiential learning'.

CLG has approved the following accreditation schemes to accredit non-domestic energy assessors and those producing DECs:

- BESCA – www.besca.org.uk;
- BRE – www.bre.co.uk/accreditation;
- CIAT – www.ciat.org.uk;
- CIBSE – www.cibse.org;
- ECMK Ltd – www.ecmk.co.uk;
- Elmhurst – www.elmhurstenergy.co.uk;
- Heating and Ventilation Certificated Associates – www.hicertification.co.uk;
- Knauf – www.knauf.co.uk;
- NAPIT – www.napit.org.uk;
- NES – www.nher.co.uk;
- Northgate – www.northgate-ispublicservices.com;
- Quidos – www.quidos.co.uk;
- RICS – www.rics.org/hips;
- Stroma – www.stroma.com.

If you want more details, visit www.energy-assessors.org.uk.

We wish to make a very important point about the role of accreditation schemes at this stage. It is inevitable you will experience difficulties in the data entry process and/or generation of EPCs, etc. during your career. In such cases, you should consult the documents we have referred to above, this book and discuss the issue(s) with fellow CEAs.

You should develop professional relationships with other CEAs for reasons we discuss in this book. However, you might believe if you admit to difficulties this is somehow a 'loss of face'. We believe it is much better to deal with any issue at the assessment stage rather than later, i.e. when a client makes a complaint. None of us know all of the answers – and in that respect we include we three!

If you cannot solve your problem in the manner we have described above, your ultimate port of call must be your accreditation scheme. One of the roles of a scheme is to provide guidance to CEAs on specific assessment issues. When you have contacted the scheme, remember to record the answer to your query on your file; including the name of the person you spoke to and the date. We discuss why this is important in Chapters 9 and 10.

Minimum Requirements For Energy Assessors For Non-Dwellings, CLG, October 2007

This is an essential publication for you to read as it sets out the absolute minimum requirements for CEAs. The primary aim of the standards is:

'to inform Energy Assessors about the professional approach and standards they need to adopt in order to become accredited and maintain their accreditation for the production of Energy Performance Certificates and accompanying Recommendation Reports.' (paragraph 1)

Paragraph 3 confirms that 'monitoring of the Energy Assessors' performance is an essential part of the Accreditation Scheme'. Accreditation schemes will carry out regular quality audits and checks on your work, as the 'purpose of accreditation schemes is to ensure that consumers and others who rely on Energy Performance Certificates can have confidence in both the certificate and the accompanying recommendations for cost-effective improvement; as well as in the Energy Assessors responsible for producing them' (paragraph 4). This process is additional to your own quality system we have suggested you should adopt, in Chapter 3.

Accreditation scheme monitoring is likely to include:

- regular online compliance checks (say every three months) of an EPC and RR, comparing your result with what the scheme believes should be the 'correct result' and 'appropriate recommendations' for the building;
- an accompanied inspection (say once every year) to check on your site inspection and associated practice – we suggest you pay particular attention to your health and safety procedures;
- evidence to confirm life-long learning and continuing professional development.

You should be aware that if this audit process confirms you are not working competently; you are likely to be required to take remedial action such as issuing an amended report, engaging in more CPD and attending further structured training. Indeed, the *Minimum Requirements* confirm that: 'If a result falls outside the quality standard, the Energy Assessor must reproduce a revised EPC/RR and issue it to the client and make sure the revised edition is lodged on the Register' (paragraph 27).

The following is a brief resume of the main points you should note from this important document, although this is not intended to be a replacement for careful study of the entire document. You must:

- have sufficient knowledge, understanding and competence to carry out EPCs and RRs, and indeed must work only within your competence (paragraph 15);
- carry out assessments with 'reasonable care and skill' (paragraph 11);
- identify and resolve (ideally avoid) any conflict of interest (paragraph 11);
- adopt and abide by the scheme complaints procedure (paragraph 11);
- disclose any financial or personal relationship (paragraph 16) – in practice we think you should confirm this in the EPC and elsewhere as appropriate;

- have 'suitable indemnity cover' (Annex A, paragraph 3, p. 8); and
- remember that 90% of your EPCs must be within + or − 5% of the Buildings Emission Rating and 100% of EPCs must be within + or − 10% of the Buildings Emission Rating that is 'determined by the Scheme Operator's Energy Assessor undertaking quality monitoring' (paragraph 26).

It is possible a revised version of this CLG document will be published before this book appears, following feedback from the accreditation schemes as a result of energy assessments being prepared in 'the real world'. You should ensure you maintain your knowledge of this important document.

Liability

Research in the US, Canada, Australia and the UK suggests 'green' buildings enjoy different operating and market conditions when compared with what might be called 'grey' buildings. The benefits include:

- lower absenteeism in the workforce;
- higher staff morale and motivation;
- fewer headaches at work;
- higher retail sales and productivity;
- lower operational and maintenance costs.

Commercial, as opposed to residential, occupiers place far more importance on these issues and the value of their premises. Green benefits that might affect the value commercial properties include:

- tenants are attracted more quickly;
- such properties are let more readily;
- reduced tenant turnover;
- higher rents, higher yields and higher capital values, some research suggesting significantly higher sale prices of properties of up to 27%.

As early as 2005 RICS stated that 'a link is beginning to emerge between the market value of a building and its green features and related performance' ('Green Value: Green buildings, growing assets', RICS, 2005, p. 3). Stakeholders likely to be interested in such information include:

- chartered surveyors asked to value the property;
- lending institutions, i.e. banks and their underwriters;
- solicitors and accountants acting for sellers and purchasers;
- building owners;
- you – for reasons below.

In the short term, we believe energy issues are more likely to affect larger properties. However, in time the effect could filter down to medium-sized and smaller properties, as public and professional awareness inevitably increases. If your EPC is incorrect and a client or purchaser who can rely on your EPC suffers financial loss, you could be sued successfully – for up to at least 10 years afterwards. We discuss these issues in Chapter 2.

The first successful claim against a DEA occurred as this book was going to press and concerned a purchaser successfully alleging they relied on the (incorrect) EPC prepared for the seller. The amount involved is possibly relatively small when compared with the types of claim that could be made against CEAs on some of the larger commercial properties – 'only' £6,500.

We know some energy assessors are currently offering a cheap and very basic level of service. You should try to ensure *your* fees are sufficient to allow you to do a thorough job and accurately identify the correct rating.

SUMMARY

In this chapter, we have tried to 'set the scene' for you in a reasonably general way so you can understand some of the significant international, national and other issues you need to grasp in your CEA career.

In the next chapter we begin to consider many of those, and other, issues in greater detail.

2 Getting Ready for Business

INTRODUCTION

In this chapter we consider the legal and statutory implications of inspecting commercial buildings. Health and safety legislation applying to building inspections will be identified and some of the common hazards found in commercial buildings will be described. The matters discussed will be used to explain why it is essential to carry out an effective risk assesment when undertaking a commercial energy assessment. The main topics considered are:

- occupiers' liability – illustrating how the liabilities of the owner or occupier can affect the CEA;
- contract and tort – these legal principles are the basis of a CEA's business relationship with their client and are also the reason why they can be sued;
- health and safety liability – legislation; Approved Codes of Practice; risk assessment; asbestos and occupiers' activities.

LEGAL BACKGROUND

Occupiers' Liability

The liability of an occupier or person having control over any premises is covered by the *Occupiers' Liability Act* 1957 and the *Occupiers' Liability Act* 1984. For simplicity, whenever reference is made to an occupier in this section it shall include 'a person having control over premises'. The Acts describe the 'common duty of care' that an occupier owes to any visitor. The 1957 Act dealt with lawful visitors and the 1984 Act extended the duty to all other visitors, including trespassers.

The common duty of care is a duty to see that visitors will be reasonably safe when using the premises, but it is not imposed on an occupier where a visitor willingly accepts risks. The Acts have effect 'in place of the rules of the common law' and as such they only result in civil action. In some cases the duty of care can be discharged by giving an appropriate and reasonable warning of danger. However, the Acts make it clear that an occupier should expect children to be less careful than adults and this would extend to their ability to heed any warning that may be given.

You should be aware of the circumstances created by the *Occupiers' Liability Act* by taking heed of any warnings given by an occupier or other person who could be thought as having some control over the premises, such as owners, managing agents or caretakers. Your acceptance of risk is a matter covered by health and safety legislation, but the implication of the *Occupiers' Liability Act* is one of the factors to be considered when making a risk assessment: you will be deemed to be solely responsible when accepting risks willingly. However, an occupier may retain some liability as an employer in their own capacity, or as your client. For example, the *Work at Height Regulations* 2005 place duties on any person who controls the work of others, where there is a risk of a fall causing an injury.

LAW OF TORT

The Law of Tort is part of the English civil law system that makes it possible for a person to seek a remedy for various types of loss. For example, a loss may be as a result of personal injury, damage to a property, financial loss or emotional distress. A tort has been described as a 'civil wrong', which affects one party and is caused by the deliberate or negligent action of others.

Setting aside any deliberate act of tort, which should be beyond the contemplation of a professional person, the matter of negligence, or how to avoid it, requires some consideration. The principle of negligence is when a person is liable for another's loss due to his or her acts or omissions for which the results should have been reasonably foreseeable. Although the principle of negligence is relatively simple, the premise of reasonableness and the chain of causation have been tested many times in litigation. A key requirement in any negligence case is that the defendant must owe a legal duty of care to the claimant, or plaintiff.

One legal test of whether a duty of care exists between two parties was based on the 'neighbour principle', which was first defined in the famous case of *Donoghue v Stevenson* in 1932 (a case involving a snail in a bottle of ginger beer). This principle is useful for understanding the possibility of negligence in practice: reasonable care must be taken to avoid any acts or omissions you can reasonably foresee which could injure your neighbour. Your neighbour, in this case, is anyone who may be so closely and directly affected by your action that you should have reasonably foreseen that they would have been affected.

We believe it is likely CEAs will be sued for negligence. This has already happened to DEAs. To summarise, circumstances that will give rise to a possible claim are as follows:

- a claimant must demonstrate you owe them a 'duty of care';
- you must have breached that duty; and
- damages must flow from that breach.

In property cases, judges tend to assess any damages based on a concept called 'diminution in value', i.e. the extent to which the value of the property has been adversely affected by the alleged negligent act. In Chapter 1 we discussed the possible effect EPCs could have on the value of properties, i.e. there is evidence suggesting properties with high ratings sell for more than those with low ratings. Let us assume such an effect is generally accepted by the courts with an example.

Case study

The CEA prepares an EPC and RR for a seller, the client. An error is made by the CEA; say incorrect data entry or failure to identify the correct HVAC system. The rating is 71 – 'C'. A purchaser buys the building. They later sell the building, say in 9 years ('required documents' are valid for 10 years). At that time, the second purchaser commissions another EPC. They discover the correct rating is 89 – 'D'. The first purchaser (not the first CEA's client, but nevertheless somebody that the CEA could 'reasonably foresee' might rely on the EPC when it was prepared) could claim they can only sell their property for less than it is worth (a grade 'D' building being worth less than a grade 'C'). The difference in value (the diminution) would be the level of damages.

Law of contract

A contract may be formed between two parties verbally or in writing or simply implied by their conduct. Whilst contracts are made easily in everyday life, such as buying a cup of coffee, the legal principles can be complex. There are five fundamental features of a contract: offer; acceptance; consideration; intention; and capacity. All five features must be present to form a binding contract. The verbal request for a cup of coffee in a cafe is an offer; a simple verbal response to the customer by the person at the counter is acceptance; the payment of money would be consideration; the commercial setting of a cafe that sells food and drink would demonstrate an intention to form a legal relationship and providing both parties are sane adults (capacity) a binding contract is formed.

However, all of the elements have been tested in the courts and various rules have been established for each. For example, one of the rules for acceptance is that it must be communicated to the person making the offer, not a third party. So, the reference to a 'verbal response to the customer' in the cafe illustration is specific and deliberate. Although simple contracts can be easily formed in the way described, some transactions demand the use of a formal contract. Leases for buildings or land, hire purchase agreements and the terms and conditions for a credit card are examples of formal contracts.

The law of contract and tort is too extensive to cover in detail in this book. However, it should be evident that the conditions for forming a legally binding contract will occur in the process of producing an EPC. It should also be evident that the work of data gathering, building inspection and making recommendations all provide plenty of opportunities for acts of negligence. It is therefore vital that you define the scope of your contract by using a clear and detailed set of terms and conditions of engagement. These documents need to spell out what you will and will not do in the course of producing an EPC and RR. They should also identify the information needed to carry out the work; the person responsible for providing the information and what will be done if the information is not provided on time. However, it should be noted that it is not possible to exclude liability for negligence in any terms of business. Nevertheless, a comprehensive set of terms and conditions of engagement can set the scene for what is reasonably foreseeable and can therefore reduce your risk of being found liable in negligence.

The timing of when the terms and conditions of engagement are provided to the client or customer is a critical factor in whether they will be effective. Ideally, the terms should be available at the time the contract is formed. Obtaining a written signature and date on a copy of the terms can be a useful safeguard.

Some professions now require their members to agree these terms, or conditions, of engagement to protect the public, and their members. Indeed, the NOS (element 2.1) require that CEAs:

> 'explain to clients the terms and conditions and fee structure under which you will undertake an energy assessment ... explain to clients the limitations and constraints of the planned energy assessment ...[and] ... confirm to clients the terms, conditions and arrangements that have been agreed.'

On balance, we believe you can do this best in writing.

Privity of contract and third party rights

It is a long held doctrine of contract law that only parties to the contract can make claims against it. There are some exceptions to the doctrine. For example, the person who instructs an agent to appoint you can sue you even though they did not make the contract. An assignment of a contractual right can also create an exception to the doctrine.

Both of these examples are likely to affect your appointment as a CEA and you should therefore be aware that you could be contractually liable to people other than your immediate client. Investors and prospective tenants may rely on your certificate and report. You should also be aware of the *Contracts (Rights of Third Parties) Act* 1999 which makes provision for third parties to enforce contractual terms. Specific criteria must be fulfilled for the Act to apply, including the identification of the third party in the contract. You should be aware of these provisions to ensure that you do not unwittingly become part of a contract with others.

HEALTH AND SAFETY LIABILITY

Health and safety legislation

Health and safety obligations have been imposed upon both employers and employees for over 30 years as a consequence of the *Health and Safety at Work etc Act* 1974. One duty set out in the Act is for employers to ensure, so far as reasonably practicable, that a safe working evironment is provided for employees. Your working environment as a CEA extends to the buildings that you inspect. Any person or organisation that employs you should therefore fulfil this duty and take account of your safety when you make an inspection. However, if you are an employee you will also have duties under the Act to:

- take reasonable care for the health and safety of yourself and others affected by your acts or omissions;
- cooperate with your employer and others to enable them to fulfil their legal obligations.

The *Health and Safety at Work etc Act* 1974 is an enabling Act. This type of legislation allows the Secretary of State to introduce further regulations without going through Parliament to pass another Act. This power has been used extensively to introduce a wide range of health and safety legislation, including the *Management of Health and Safety at Work Regulations* 1999, sometimes referred to as the 'Management Regulations'.

The Management Regulations impose a specific requirement on all employers, including the self-employed, to carry out risk assessments. These assessments must be 'suitable and sufficient' and should consider the 'risks to the health and safety' of the employees, or self-employed person, to which they are exposed whilst they are at work. The Regulations also require the employer, or self-employed person, to assess the risks to health and safety of persons not in their employment where it is in connection with work carried out by them (*Management of Health and Safety at Work, Approved Code of Practice and Guidance L21* (2nd edition), HSE Books, 2000). The duties set out in the Regulations overlap with other regulations. Compliance with the related regulations is usually sufficient to demonstrate compliance with the Management Regulations.

You will need to follow your employer's health and safety procedures and those of the building's occupier. For example, some circumstances may require you to have a permit to work, which will alert you to the hazards that exist and impose specific obligations on your activity.

Figure 2.1: Permit to work sign

The implications of these statutory requirements on your potential liability for negligence should also be considered.

Approved Codes of Practice (ACOP)

Approved Codes of Practice (ACOP) have been published for many health and safety regulations. These documents have a special legal status. Although the main purpose of an ACOP is to provide guidance on how to achieve the requirements of the regulations, they are not mandatory. However, in the event of a prosecution a defendant would have to demonstrate to the court how they satisfied the regulations, if they have not followed the ACOP. In view of the importance of an ACOP, you should possess, or have access to, relevant ACOPs, particularly the ACOP for the latest version of the *Management of Health and Safety at Work Regulations*.

Risk assessment

A risk assessment must be recorded in writing to satisfy the *Management of Health and Safety at Work Regulations*, where there are more than four employees in an organisation. Where there are four or less employees a risk assessment is still required, but it need not be in writing. It is our view that you should make a written record of your risk assessment, regardless of the size of your organisation, because it may be helpful to refer to it in the event of a claim. On another occasion you may need to use your risk assessment as evidence to demonstrate your competence to your Accreditation Scheme as part of a monitoring audit.

There is no standard method for undertaking a risk assessment, but it is important to follow a process. The statutory requirement for a 'suitable and sufficient' assessment should be considered when a risk assessment procedure is designed. A lot of guidance material is provided online by the Health and Safety Executive (HSE) including a suggested five-step risk assessment process (www.hse.gov.uk/risk/fivesteps.htm):

- identify the hazards;
- decide who might be harmed and how;
- evaluate the risks and decide on precaution;
- record your findings and implement them;
- review your assessment and update if necessary.

The first stage of hazard identification is common to most risk assessment guidance. You can start this stage of the process at the point of instruction, before doing the inspection. It is important to appreciate that others will have a statutory duty of care for your health and safety, such as employers and building occupiers. You would do well to identify the duty holders and ask them for information about any known hazards or safety procedures. These enquiries may be made using a written questionnaire or via a telephone discussion. The response will inform you about the amount of time and detail that may be required in completing the risk assessment and inspection on site. For example, if a duty holder cannot be identified or contacted, or you do not have confidence in the attitude of a duty holder, the likelihood of unknown hazards being found on site will be increased.

Hazards

It is helpful to distinguish between what is meant by a 'hazard' and a 'risk' when conducting a risk assessment:

- *hazard*: a hazard is anything that may cause harm;
- *risk*: a risk is the chance, or likelihood, that somebody could be harmed by the hazard.

A hazard could be an 'activity' or the presence of a hazardous substance. You will need to consider the risk of your own activities, such as using a ladder, and the activities of the occupants of the building being assessed. In some circumstances the latter will be the most significant part of the risk assessment. For example, the activities in an industrial workshop may include welding, high-voltage electrical testing, abrasive wheel grinding and fork-lift truck operations. The materials used and stored by an occupant should also be considered. Examples include flammable materials and toxic chemicals.

A hazard may also be the building, or part of the building, being inspected. There could be the potential to be trapped in the building or falling through a weak part of the structure, such as a defective floor or a lightweight roof. The presence of an asbestos-containing material (ACM) may also present a hazard.

Your assessment of risk should consider the severity of a hazard, or the potential consequences it will cause, and its probability, or likelihood that it will cause harm. Simple risk estimators can be found in various health and safety books which can assist in this assessment. For example, consider Table 2.1 and the use of fork-lift trucks in an industrial building. The potential severity of being hit by a truck or its load is clearly fatal and would be assigned a rating of 5 on the horizontal axis. However, the likelihood of an accident may be considered as unlikely if there are designated walkways and the trucks use audible alarms and flashing lights. This would be assigned a rating of 2 on the vertical axis, resulting in a total rating of 10 and an assessment of the hazard as a medium risk.

Whenever a risk is assessed as high or medium you should take some action to avoid or reduce the risk. In this example, you may consider it reasonable to only inspect the areas affected when the fork lift operations are idle. In other circumstances when the movements of the trucks are unpredictable or you consider they have inadequate warning devices, you may assign a higher likelihood rating of 3, 4 or 5, which would result in a high risk category. In this case you may consider the only appropriate action is to conduct your inspection outside of normal working hours.

Asbestos

Asbestos is a naturally occurring mineral that has been used in many building products, particularly mechanical and electrical services (see Table 2.2 and Figure 2.2).

You will want to inspect plant rooms and service ducts in order to collect data, but a risk assessment must be carried out before this is done. The health risks associated with asbestos are now widely acknowledged and the use of ACMs has been phased out over the past 30 years (see Table 2.3) but ACMs remain in many existing buildings.

Table 2.1: Simple risk estimator

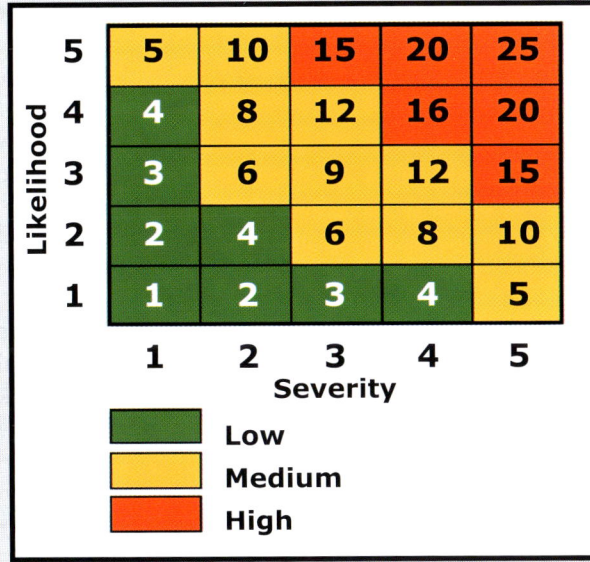

Severity
1. Minor
2. Treatment required
3. Absence from work > 3 days
4. Serious injury
5. Death

Likelihood
1. Very unlikely
2. Unlikely
3. Likely
4. Fairly likely
5. Highly likely

Table 2.2: Typical locations for the most common asbestos-containing materials

Air handling systems	Flooring materials
Lagging Gaskets Anti-vibration gaiters	Floor tiles Linoleum and paper backing Lining to suspended floor
Boiler, vessels and pipework	**Interior walls/panels**
Lagging on boiler, pipework, calorifier, etc. Debris associated with damaged lagging Paper lining under non-asbestos pipe lagging Gasket in pipe and vessel joints Rope seal on boiler access hatch and between cast iron boiler sections Paper lining inside steel boiler casing Boiler flues	Loose asbestos inside partition walls Partition walls Panels beneath windows Panel linings to lift shafts Panelling to vertical and horizontal sides of beams Panel behind electrical equipment Panel on access hatch to service riser Panel lining service riser and floor Heater cupboard around domestic boiler Panel behind/under heater Panel on, or inside, fire door Bath panel
Ceilings	**Roof and exterior walls**
Spray coating to ceiling, walls, beams/columns Loose asbestos in ceiling and wall cavities Tiles, slats, canopies and firebreaks above ceilings Texture coatings and paints	Roof sheets and tiles Guttering and drainpipes Wall cladding Soffit boards Spandrel panels (apron wall) Roofing felt Coating to metal wall cladding
Domestic appliances	**Other**
Gaskets, rope seals and panels on domestic boilers 'Caposil' insulating blocks, panels, paper, string, etc. in domestic heaters String seals on radiators	Fire blanket Water tank Brake/clutch lining (e.g. lift motors)

Source: *A Comprehensive Guide to Managing Asbestos in Premises,* HSE Books, 2002

The chance of you encountering ACMs in a building built in or before 1999 is relatively high. Some materials used in the fabric of buildings, such as old types of sprayed fibre fire insulation on steel frames (see Chapter 4), are particularly hazardous. However, the highest risk presented by asbestos to you is likely to be where it has been used in building services.

Figure 2.2: Insulation lining on a boiler casing containing asbestos

Asbestos awareness training

The ACOP for the *Control of Asbestos Regulations* 2006 recommends asbestos awareness training for those who are 'liable to disturb asbestos while carrying out their normal everyday work'. A list in the ACOP of people who fall into this category includes 'architects, building surveyors and other such professionals'. There is a clear implication that a CEA should undertake asbestos awareness training.

The recommended topics for training stated in the ACOP are:

- the properties of asbestos and its effects on health, including the increased risk of lung cancer for asbestos workers who smoke;
- the types, uses and likely occurrence of asbestos and ACMs in buildings and plant;
- the general procedures to be followed to deal with an emergency, for example an uncontrolled release of asbestos dust into the workplace; and
- how to avoid the risks from asbestos; for example, for building work, no employee should carry out work which disturbs the fabric of a building unless the employer has confirmed that ACMs are not present.

The properties of asbestos

The three types of asbestos most often found in buildings are those commonly referred to as blue, brown and white asbestos. The colours are not evident from a visual inspection of a material, nor can a positive identification of asbestos be made without a laboratory test. Materials are suspected of containing asbestos based on the date of construction; the purpose of the building product (e.g. fire resistance); its location and general appearance.

There are two main forms of asbestos:

- amphibole, categorises five types of asbestos, which have long straight fibres – crocidolite (blue), amosite (brown), fibrous actinolite, fibrous tremolite and fibrous anthophyllite;
- serpentine has curly fibres, of which chrysotile (white) is the main type.

Both forms of asbestos are hazardous, but amphibole is considered to be more dangerous than serpentine.

Table 2.3: Key dates

Key dates that chart the withdrawal of asbestos from use in the UK	
1971	use of blue asbestos stopped
1974	use of sprayed asbestos stopped
1984	use of asbestos in Artex products stopped
1985	use of brown asbestos stopped
1992	use of asbestos in bitumen products stopped
1992	use of amphibole prohibited by law
1999	use of chrysotile prohibited by law

Asbestos risk

Asbestos only poses a risk to health if its fibres are inhaled. In many cases ACMs are safe if left undisturbed, but any disturbance that causes fibres to become airborne will create a risk. In view of the potential danger of airborne asbestos fibres, any work on ACMs that can cause a disturbance, such as drilling, cutting, sanding and removal work, is strictly regulated. You need to be able to identify potential ACMs in order to avoid accidentally damaging the material and causing a release of fibres (see Figure 2.3).

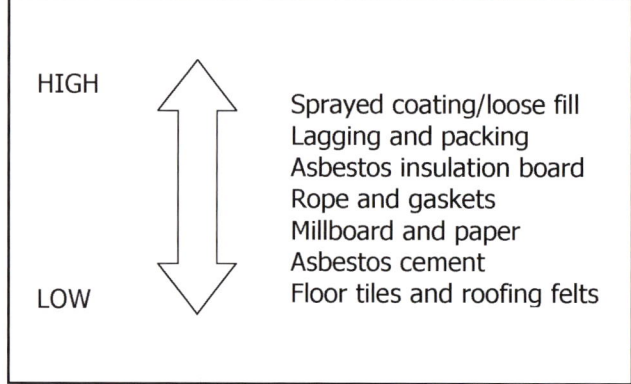

Source: *Introduction to Asbestos Essentials*, HSE, 2001

Figure 2.3: Potential for fibre release

The HSE claim that annual deaths from asbestos-related disease in the UK exceed 3,500.

Asbestos-related diseases

There are three fatal asbestos-related diseases:

- *asbestosis* – ill health and respiratory problems due to scarring of the lung tissue by fibres;
- *lung cancer* – a tumour develops in the lung, much like cancers caused by smoking. The likelihood of contracting lung cancer is therefore vastly increased for an asbestos worker who smokes. There can also be a latency period of 15 to 30 years;
- *mesothelioma* – this is a tumour that develops around the lung and in the chest cavity. The tumour has been specifically linked to asbestos exposure, whereas in other circumstances it is a relatively rare form.

All three are reportable diseases under the *Reporting of Injuries, Diseases and Dangerous Occurrences Regulations* 1995 (RIDDOR).

Emergencies involving asbestos

Where there has been an unplanned release of asbestos action should be taken to ensure that the situation does not get worse. For example, the activity that has caused the release should be stopped. Anyone who may be affected by the release should be informed. A licensed contractor and analyst should be employed to check and clean the affected area thoroughly before its normal use is resumed. Any contaminated clothing should be removed in a controlled environment and disposed of as contaminated waste. The person wearing the clothing should wash and decontaminate themselves in a controlled washing and changing area. A note should be made on the medical or personal records of any employee who was not wearing respiratory protective equipment when they were exposed to the asbestos. This note should be kept indefinitely.

The management of asbestos in non-domestic premises

The legislation and guidance for non-domestic premises requires that in each building there should be a duty holder to ensure that any asbestos in the premises is identified and managed. Information about the location of the asbestos and the procedure for managing it should be made available to you when working in the premises.

You should therefore consider ACMs as part of your risk assessment, for which an initial step would be to contact the duty holder and ask for information about the location and condition of asbestos in the building to be made available.

SUMMARY

In this chapter we have discussed the main legislation that can affect your liability. We have identified people who may have some responsibilty for your safety, but even in these circumstances remember you can still be liable.

3 Preparing for the Commercial Energy Assessment

INTRODUCTION

In this chapter, we discuss more of the issues you need to consider before leaving the office. Briefly, these are:

- discussion of some important practical legal definitions;
- the different levels of commercial energy assessment;
- possible ways to organise your 'office' and internal systems;
- methods of ensuring you have sufficient capacity to meet client demands, by having access to available staff;
- inspection equipment and measuring practice;
- important aspects of the Building Regulations for 'new builds'.

PRACTICAL LEGAL DEFINITIONS AND OTHER ISSUES

'Building'

In the Regulations, a building is defined as:

> 'a roofed construction having walls, for which energy is used to condition the indoor climate.' (reg. 2)

Thus, a building that is open-sided, e.g. a cart-shed, would not require an assessment.

'Conditioning'

The July 2008 CLG guidance states that:

> 'services that are considered to condition the indoor climate are the following fixed services: heating, mechanical ventilation or air-conditioning. Although the provision of hot water is a fixed building service, it does not "condition the indoor environment" and would not therefore be a trigger for an EPC. The same argument applies to electric lighting.'

Accreditation schemes appear to accept this definition of conditioning.

Absence of fixed services that provide conditioning

The Regulations state that buildings must use energy to condition the indoor climate if an EPC is required. The July 2008 CLG guidance attempts to clarify this statement and confirms:

- 'Where a building is expected to have heating, mechanical ventilation or air conditioning installed, it will require an EPC based on the assumed fit out in accordance with the requirements in Part L of the Building Regulations' (p. 7);
- 'if there is no intention of having fixed services and no ability to include fixed services to condition the indoor climate, then an EPC will not be required' (p. 8);
- 'if a building is to be let with fixed services, the EPC for the building should reflect the fixed services actually installed' (p. 8);
- 'if a building is to be let without fixed services, but there is an intention that fixed services will be installed, the EPC should be based on the building's use class under the planning legislation. This applies whether fixed services have ever been installed previously in the building, or whether the building is newly constructed on a "shell and core" basis. For the purposes of producing the EPC, the activity within the building should be specified in line with business activity typical of the use class and the most energy intensive fit-out adopted in line with Part L of the Building Regulations in force when the building was built' (p. 8).

There is currently considerable confusion regarding this conditioning requirement. Some commercial agents have allegedly advised their clients to remove fixed services that provided conditioning in order to avoid the need for an EPC and RR. We believe that such action may demonstrate an incorrect understanding of the law, and runs the risk of legal action by the authorities and clients, for the following reasons:

- the 'spirit' of the Regulations and the EPBD is to record, reflect and then manage and try to reduce, CO_2 emissions;

- it can reasonably be anticipated that most commercial buildings will require some form of conditioning for health and safety reasons, even if fixed services are not currently installed.

The law has yet to be clarified in the courts. However, if you can reasonably foresee that fixed services that provide conditioning will not be provided, it would seem appropriate that an EPC and RR will not be required. In practice though, and in our experience, most solicitors require production of the required documents upon exchange of contracts and/or completion of the sale or lease. In addition, a number of commercial occupiers are becoming increasingly aware of the practical importance of such documents to their use of the building and the value of the property.

On balance therefore, you would be best advised to assume an EPC and RR are required unless the building is exempt – see later under 'exemptions'.

Differences between level 3, 4 and 5 assessment for existing buildings

At present, the July 2008 CLG guidance states:

'Level 3 and level 4 buildings may both be assessed using SBEM. Currently the only distinction between these two levels is on the basis of HVAC systems. A level 3 building includes frequently occurring characteristics such as simple heating systems, simple natural ventilation and small comfort cooling systems. Level 3 does not require the candidate to demonstrate competence in new build.

Frequently occurring characteristics are defined in the NOS in terms of HVAC, fabric and lighting. **(i), (ii) and (iii)** cover HVAC:
(i) simple heating systems (Boiler Systems <100kw)
(ii) simple natural ventilation
(iii) small comfort cooling systems (up to 12kw).' (p. 40)

Some accreditation schemes have confirmed that if a boiler system is 'simple' and CEAs believe they are competent to enter the data, they can ignore the CLG definition of the differences between levels 3 and 4. Thus, in March 2009, NES (one of the accreditation schemes) issued clarifying guidance including confirmation that 'their' CEAs should treat any buildings incorporating MTHW or HTHW (medium/high temperature) systems as a level 4 assessment

In practice, you should take advice from your accreditation scheme and bear in mind that if you assess buildings outside your competence, you may run the risk of your professional indemnity insurers refusing to provide cover in the event of a claim for professional negligence.

The July 2008 CLG guidance is silent on the difference between level 4 and 5 assessments. The *BRE User Guide* confirms it is difficult to use SBEM if the building has 'properties that vary non-linearly over periods of the order of an hour … not a universal rule' and lists certain features you cannot currently represent in SBEM including light transfer between highly glazed internal spaces such as atria or light-wells. Most buildings with large areas of glazing to any atrium are unlikely to respond well to SBEM, because of the resultant relatively rapid movement of air convection currents behind glazing in buildings when compared with other types of external fabric.

If you enter data into the program, it will not 'flag up' when that data is based on a level of assessment that is different to the level you intended; i.e. there is no 'fail safe' mechanism in SBEM. You could easily enter data for a level 4 assessment at level 3 and generate a result. Your judgment is therefore crucial in this regard. If you are uncertain you should discuss the issue with your accreditation scheme and follow their guidance.

Exemptions

In addition to the 'absence of fixed services' issue referred to above, certain other properties do not require an EPC and RR. You are likely to be questioned closely by clients, solicitors and others regarding exemptions, and should be sure of your facts. We include a brief consideration in Table 3.1.

The July 2008 CLG guidance provides clarification of what constitutes a building with 'low energy demand'. This description includes buildings, or parts of buildings designed to be used separately, whose purpose is to accommodate industrial activities in spaces where the air is not conditioned. Activities that would be covered include foundries, forging and other hot processes, chemical process, food and drinks packaging, heavy engineering and storage and warehouses where, in each case, the air in the space is not fully heated or cooled. Whilst not fully heated or cooled these cases may have some local conditioning appliances such as plaque or air heaters or air conditioners to serve people at work stations or refuges dispersed amongst and not separated from the industrial activities.

The intention here is to include buildings that are conditioned primarily, or entirely, by the industrial process that is taking place in the space. You will need to judge each case on the individual merits of each building.

A non-residential agricultural building with low energy demand includes a building such as a greenhouse.

We have included a case study of an exempt building in Appendix A.

As a practising CEA you will have to deal with potential clients who do not want to commission an EPC. If the building(s) is/are exempt, you may find you have a friend for life! However, you must be sure of

Table 3.1: EPC exemptions

Exemption	Discussion	Regulation
Buildings used primarily or solely as places of worship	Note the word 'primarily', i.e. at least 50%	SI 2007/991, reg. 4(1)
Temporary buildings with a planned time of use of two years or less	The intention would appear to be to capture temporary accommodation on construction sites	SI 2007/991, reg. 4(1)
Industrial sites, workshops and non-residential agricultural buildings with low energy demand	Examples are described in the July 2008 CLG guidance. You will also find the description in ADL2A, para. 16	SI 2007/991, reg. 4(1)
Stand-alone buildings with a total useful floor area of less than 50m² which are not dwellings	A small mid-terrace shop, of say 10m² floor area needs an EPC	SI 2007/991, reg. 4(1).
Where the construction of the building has not been completed	Incomplete buildings do not require an EPC or RR	SI 2007/991, reg. 4(2)
When the building will be demolished an EPC need not be provided by the owner	Specific rules apply to this exemption – see the Regulation and July 2008 CLG guidance, suggesting you confirm the owner has made an application for planning permission. A mere intention is unlikely to be sufficient. You should record details of the application on your file.	SI 2007/991, reg. 7(2)

your facts and opinions. At the time of writing, the relevant authorities (see later in this chapter) are taking very little enforcement action against occupiers and owners who fail to commission an EPC when they sell. However, solicitors tend to insist they see an EPC before they complete a contract for sale, unless the building(s) is/are exempt of course.

You must therefore ensure you are certain of your reasons when you confirm that an EPC and RR are not required, for a number of reasons:

- if the local Trading Standards Department take enforcement action, they will serve your client with a penalty notice;
- obtaining the EPC and the RR could cause a delay in the sale with resultant loss of time, and money;
- once you have prepared the EPC and the RR, the purchaser may view the results with some concern if they are unfavourable, and could argue 'had they known the building was so inefficient' they would not have proceeded at the agreed price;
- in extreme instances, the resultant delay and EPC results could give the purchaser time to reflect and withdraw from the sale – your client is unlikely to be impressed!

Enforcement

Regulation 38(1) of the 2007 Regulations (SI 2007/991) states that 'every local weights and measures authority is an enforcement authority for the purposes' of the Regulations.

At the time of writing we know some Trading Standards Officers (TSOs) are aware that many owners of commercial buildings and some commercial estate agents are ignoring parts of the legislation. This means they are neither complying with the spirit of the EPBD, nor ensuring purchasers are aware of the energy performance of the building. A study by one of the accreditation schemes in May–June 2009 revealed that only around 10% of commercial properties on the market had an EPC available. Time will tell whether TSOs eventually take coordinated enforcement action across the country.

ORGANISING YOUR PRACTICE

Why you need a quality management system

Every CEA should have a quality management system (QMS) to help them practise to the best of their abilities and provide a quality service. Some early quality systems were criticised as being focussed more on the process rather than the result, and generating too much paper. You must ensure your QMS is the complete opposite.

An effective quality system should:

- achieve better efficiency and consistency in the service you provide;
- reduce errors, some potentially costly;
- increase client satisfaction;
- reflect on your 'target market' and improve your penetration into that market, and new markets;
- manage expansion in your business since new team members will have your systems to follow;
- continuously improve your services and systems.

The system you establish should constitute a framework you use to manage your key business processes. We include some examples in the appendices and at Figure 3.1.

You should regularly review and carry out formal audits of your system. The best method of auditing is to take a randomly selected completed job file and measure your performance against the standards in your quality manual and systems. Figure 3.1 is an example of a completed audit sheet.

RT2.D0678   SAT COMMERCIAL EPC INSPECTION AUDIT SHEET

Braxton Hicks Plc – Exeter office
Quality management system – commercial EPC / RR audit

| CEA name: *Martin Carr-Noyle* | Reviewer: *Larry Russen* | Date: *060609* |

Note: copy double-sided
Consult Code of practice for specific guidance. Please note this process is intended to fulfill two purposes:
- Monitor past / current performance and competence against our quality standards; and
- Prompt us to individually reflect and take appropriate personal action, and alter our QMS systems

| AUDIT | File no: *007* | Address: *Pretend Property, High St* | Postcode: *EX1 N1T* |

Performance required	Satisfactory	Notes	Action(s) required
1. Desktop audit (pre-inspection) checks made and recorded	**Yes** / No / Not seen	Satisfactory	None
2. Satisfactory behaviour towards seller / agent	**Yes** / No / Not seen	—	None
3. Health & safety risk assessment completed and recorded	Yes / **No** / Not seen	Erected step ladder upside down	Training
4. Site notes completed contemporaneously	**Yes** / No / Not seen	—	None
5. Accurately identified walls, roofs, floors, doors and windows	**Yes** / No / Not seen	Used measuring rod for ceiling	None
6. Accurately identified heating / cooling and controls	**Yes** / No / Not seen	Carefully removed cover to controls	None
7. Tried to identify lamp type including Watts / ballasts / make	Yes / **No** / **Not seen**	Unable to access due to customers	None
8. Binoculars for roof, flues, destratification fans etc	**Yes** / No / Not seen	For destrat fans	None
9. Sufficient photographs taken	**Yes** / No / Not seen	—	None
10. Full floor plans with check measurements / sections	**Yes** / No / Not seen	—	None
11. Site notes complete at end of inspection	Yes / **No** / Not seen	Failed to complete info about lamps	Check this aspect next month. MCN to reflect

Statement of continuing professional competence
I confirm that I observed the CEA inspect a property for the purpose of a commercial EPC / RR on the above date at the above address. During my observation I saw the CEA make a visual inspection of relevant aspects of the property in a safe and methodical manner. I also witnessed the CEA make observations and measurements appropriate to the inspection. Upon the basis of my observations and subject to further actions identified, I believe the CEA is **competent** / ~~not competent~~ [delete and circle as required] to undertake this type of inspection.

Signed..Dated *6th June 2009*

| RT2.D0676 | Page 1 of 2 | Quality Management Record Form | June 2009 |

Figure 3.1: Commercial EPC audit

RT2.D0678   SAT COMMERCIAL EPC INSPECTION AUDIT SHEET

ACTION – see below for results

Action(s) required	By whom?	Due date	Date completed	Sign off
1. Read ladder instructions and LR to re-train	MC-N + LR	070609	070609	B
2. MC-N to reflect on need for full copy of site notes before leave site	MC-N	130609	130609	MC-N
3. Further check of site note practice	LR	070709	070709	B
4.				

Reviewer to copy both sides, place original on file KG6 and pass to first person for action by due date. That person to action, confirm actions etc below and pass to next person if required. When action(s) complete, last person to pass form back to reviewer for filing on KG6.

1. I trained Martin in how to set up his ladder in accordance with our health and safety (H+S) policy. He read the instructions and signed his action off on the H+S file.

2. I realise why Larry noted I had not completed my site notes on lighting. This was because of time pressure + the unit was closing – client had to leave. I took good photos as a record.
   MC – Noyle

   M C Noyle

3. I brought forward Martin's next audit by one month. He demonstrated good practice and fully completed his site notes – see audit sheet for file 008.

RT2.D0676   Page 1 of 2   Quality Management Record Form   June 2009

Figure 3.1: Commercial EPC audit (cont.)

If you are a sole practitioner, do not allow that fact to give you an excuse not to engage in a quality or audit process. If you are sufficiently reflective and tough with yourself you can audit your own work. However, this can allow you to develop bad practice and two suggestions you should consider are:

- ask a friend, or your partner to complete your audits – this is not ideal though as although they can be entirely objective, your auditor ought to understand SBEM; so, a better approach is
- agree that another CEA sole practitioner will audit your systems and you will audit their work.

In the latter instance you will need to trust each other; but there are wider advantages to such an arrangement. You can provide holiday cover for one another and 'bounce ideas' off each other if either have a professional question or issue causing stress.

Professional indemnity insurance (PII)

Annex A of CLG's *Minimum Requirements for Energy Assessors for Non-dwellings*, entitled 'The Role of the Scheme Operator' includes para. 3, which confirms one of the accreditation schemes' roles as 'Ensuring that members of the scheme have in force suitable indemnity cover'. At the time of writing we believe most schemes offer PII for level 3 assessments.

You require PII to protect yourself and/or your practice against claims of negligence. We have discussed this issue in more detail in Chapter 2. We suggest you take advice from an experienced PII broker even if you 'only' practice at level 3.

Complaints procedure

The CLG *Minimum Requirements* specify you or your employers must have a written statement 'setting out in unambiguous terms and with definable milestones the procedures that [you will follow] ... in the case of a complaint' (para. 35). You should have such a policy and moreover should adhere to it if you receive a complaint.

Many of the accreditation schemes specify what you should include in your complaints procedure.

Life-long learning

You should be aware of the need for every professional to engage in the important concept of life-long learning, or continuing professional development (CPD). The world is an ever-changing place, and energy assessment is no different – the only constant thing in life is change! For that reason, the CLG *Minimum Requirements* (para. 11) specifies that you must remain 'current' in your knowledge through CPD.

You should ensure you comply with this requirement based on a 'life-long reflective learning cycle' including continuous *appraisal* of your strengths and weaknesses, *planning* to address any weaknesses, *development* of required skills you identify and *reflection*. You should record all of this process. Your accreditation scheme may require a copy for monitoring. Your PII insurance company may require reassurance on the subject. Worst case – a judge might want to see it!

You must truly embrace this important concept and accept that you learn throughout your entire career as a CEA – if you're not learning, or believe you are when you're not, you may find professional arrogance can easily lead, like pride, to a fall.

Running and organising your professional practice as a business

You should organise your work so it makes a profit, whether self-employed or not. You should write a business plan with time-specific goals and objectives.

In addition to these issues, you also need to have systems for matters such as data protection, document storage (e.g. use good CDs as cheap versions do not last), discrimination, public liability insurance and equality issues. You will find more information in the *DEA Handbook*.

You should ask advice from an accountant and/or similar professional. You can seek help at Business Link and the local Chamber of Commerce.

Data gatherers

The July 2008 CLG guidance document confirms (p. 16, 3.4) that a 'team of people can work on gathering the information for an energy assessment as long as they are working under the direction of an Accredited Energy Assessor', under strict conditions including:

- any data gatherers (DGs) must be 'fit and proper', including having appropriate technical ability;
- you are responsible for the EPC, RR and all the DGs actions, data and output as though you had undertaken their data gathering yourself.

CLG's requirements are onerous (correctly in our view), potentially costly and can be very consuming in terms of your time if you do not make your arrangements for data gathering in an appropriate manner. Subsequent guidance from CLG (May 2009 sent direct to all accreditation schemes) has clarified the use of DGs. Thus, CLG has stated you must not involve DGs in the assessment of level 3 buildings. This is rather a shame since it will make it more difficult to train new CEAs and introduce them to the profession.

You should therefore only use DGs for level 4 and 5 assessments. You would be foolish to take on just anybody; and should send them on a recognised course.

You should inform your professional insurance company and accreditation scheme. You will thereby demonstrate your good management and understanding of risks. You are likely to be required to provide this information anyway, particularly as those organisations grow to understand the risks involved in using DGs. In addition, you should require in your written contract with them that they maintain their life-long learning; whether they are self-employed or directly employed. You should seek legal advice before you enter into any contract.

You should ensure your new team member knows about your standard of work, acquaint them with your quality systems and ensure they comply with them. You should audit them regularly when they begin work for you, and should only reduce (but never entirely cease) audits when you are happy.

You must not sign off reports where only your DG has inspected the property and entered the data (or you may have entered the data) without you visiting the property. The July 2008 CLG guidance confirmed this was possible. However, by May 2009 CLG confirmed this practice was no longer permissible due to concerns about the quality of EPCs. You must inspect every property for which you issue an EPC.

We foresee the following possible approach as a reasonable basis upon which to proceed:

- at first, your DG accompanies you on inspections to learn how to do the job your way and observes as you enter data and reflect on the final results;
- they then take more responsibility, e.g. they inspect internally while you inspect externally and they enter some of the data;
- finally the DG could take over responsibility for data entry;
- however, you will still need to regularly audit a certain percentage of their work on site and their data entry – CLG guidance suggests you should be able to provide evidence of this process.

Indeed, you must supervise the DG's work at all relevant stages and have an audit trail to prove you have done so. We further suggest you always consider carrying out a desktop study before they go to the property to decide if your DG is competent for the commission, certainly in the early stages of your relationship.

We believe use of DGs is likely to be a major issue and there is a significant possibility they could put your PII at risk if you use them inappropriately. We are more comfortable with a concept of 'data enterers' than data gatherers. This is because it is easier to supervise data entry into the program, rather than the collection of data on site – correct identification of the fabric, insulation and services in a building can be very difficult. In some buildings it is likely that only an assessor with many years experience will be able to confidently identify all the required data.

If you manage a DG well they could make a positive contribution to your business. However, you let a DG loose at your peril if you do not engage and manage their services properly.

EQUIPMENT AND MEASURING PRACTICE

Inspection equipment

You will need an inspection 'kit' to function properly as a CEA. There is no 'official' list of equipment. We have developed what we consider is a minimum list based on our experience, other publications and current practice by other professionals (see Table 3.2).

A fundamental point to remember about all equipment is you should always read the manufacturer's instructions, for reasons of:

- health and safety (particularly for items such as ladders, etc.); and
- ensuring you get the best out of the item – we know some professionals using their tools incorrectly 30 years after they bought them because they cannot be bothered to read how to use them properly!

Table 3.2: CEA minimum equipment

EQUIPMENT	USE (examples only)	COMMENT
Data collection equipment, e.g. site inspection notes, pens, coloured highlighters, A4 and A3 clipboards, recording equipment	Record information gathered on site in accordance with this and associated guidance	Most properties, even the apparently 'simple' ones, will require collection and evaluation of a *significant* quantity of information
Binoculars	Externally – for chimneys, roofs and other upper building elements including walls Internally – for heat sources, lamps, de-stratification fans, otherwise inaccessible construction details	Should be used regardless of eye-sight as you will be surprised how much you can miss if you do not use them. Best vantage points may be far away. Some properties may not require their use, e.g. low-rise buildings
4m 'surveyor's' ladder and short step ladder only for short duration, light work when it is not reasonably practicable to use other, safer access equipment, (e.g. binoculars)	For wall construction at low level, flat roofs (particularly to view cooling systems on the roof), roof voids, service installations such as cooling systems attached to walls, or lamps	Carry out a risk assessment in accordance with the *Work at Height Regulations* 2005, and take appropriate action For further information see www.hse.gov.uk/falls/ladders.htm

Table 3.2: CEA minimum equipment (cont.)

EQUIPMENT	USE (examples only)	COMMENT
Camera with spare batteries	Record details of construction, materials and services	Should be used in all cases. Particularly helpful for complicated construction and also recording restrictions to the inspection. Must only be used as supplementary to the written site notes
Measuring devices, e.g. 30m tape with claw, retractable tape or timber rule and quality laser measuring device (not sonic)	Calculate floor areas, envelope areas, heights and lengths, wall and insulation thicknesses, lamp sizes	Use only metric measurements Take care with laser devices around work people and animals – the factory cat will find the red spot of light fascinating, but will not appreciate the light shining in her eyes Be aware also that laser devices sometimes refuse to work properly outside and can be 'fooled' when bouncing off obstacles between you and the surface you are measuring to
Spirit level and protractor	Confirm pitch (slope) of roofs and walls	Use the protractor by eye – a reasonably accurate calculation will usually be satisfactory
Torch and reserve, spare batteries and bulb	Roof voids, plant and boiler rooms, riser ducts and cupboards and other poorly lit areas	Ensure you have a torch with sufficient strength. Your beam will sometimes have to extend significant distances
Pocket mirror	Voids with very restricted access, at eaves level within loft spaces to view insulation	You can combine a mirror with use of the torch. The type on a short arm used by car mechanics is particularly useful
Triangular key for meter boxes	Opening service meter boxes	Essential for electricity and gas meters
Range of screwdrivers	Opening up otherwise inaccessible service ducts	Only appropriate on infrequent occasions and where this can be easily accomplished and without any health and safety issues arising
Compass	Identify the property's orientation	Combine with local knowledge of prevailing winds to consider 'condition' issues – see Chapter 10 Stand away from metal objects that might confuse the tool
Hard hat	Protect against falling debris or materials and when working in confined spaces	In some instances a hat can restrict vision
Steel toe capped safety boots and shoes	In all circumstances where there are sharp protrusions, slippery surfaces	Heavy boots can be tiring to wear, so use the lighter safety shoes once you have inspected those areas where you must use boots
High visibility jacket	Where vehicle traffic is a hazard, e.g. industrial estates, in buildings where fork-lifts are in operation	Some versions are waterproof. Advertise your firm's name
Safety glasses/goggles/sunglasses	Close proximity to certain industrial processes, in strong sunlight, driving	Sunglasses are essential in strong sunlight
Clothing for different weather conditions – e.g. waterproof coat, footwear, sun hat, shorts, etc.	Adverse weather conditions	Sometimes an overlooked issue. Weather conditions can have a significant and potentially adverse affect on your performance. If you are cold, hot, wet because you do not have the correct clothes, you might restrict your inspection to the detriment of collecting the required data
Overshoes	Wet and/or muddy conditions	The client is unlikely to appreciate your excellent safety shoes/boots when they carry mud onto the expensive boardroom carpet. We know of a number of jobs that have become 'difficult' through lack of common sense and courtesy
Overalls	Manufacturing and warehouse buildings; but possibly not in smart offices	You will arrive dressed in your business suit – inappropriate for the dark and possibly dirty spaces a CEA must sometimes enter

Table 3.2: CEA minimum equipment (cont.)

EQUIPMENT	USE (examples only)	COMMENT
Coverall with elastic hood, wrists and legs; and sticky flap over zip	Areas where there is a higher risk of ACMs than usual, e.g. roof voids, boiler and plant rooms	Must comply with EN ISO 13982 Type 5 for protection against fibres such as asbestos. Particularly for use where issues such as absence of an asbestos register, older buildings or other matter puts you on notice you must 'raise your game' and take additional precautions – see Chapter 2 for more information
Disposable masks (FFP3)	Loft spaces and areas where the possibility of mineral fibre particles is high, e.g. as above and areas above suspended ceilings, ducts and risers	Essential to protect your personal safety
Heavy duty gloves	Protection against dirt/debris/oil-based fuels/sharp edges/contamination (e.g. in plant and boiler rooms)/lamps	You are likely to come into contact with some sharp edges, etc.
Latex gloves	Protection against dirt/debris during more delicate work	Important for your personal safety and hygiene
Hand cleaning 'wipes'	Unoccupied buildings where no water connected or available	Important for your personal safety and hygiene
First aid kit complying with minimum HSE requirements	Minor cuts and bruises, indeed any incidence of personal injury	Important for your personal safety. You should learn how to use the kit – attend a first aid course
Ear defenders/plugs	Workshops and industrial buildings where you encounter excessive noise levels	In wearing this equipment, your hearing will be reduced and this can be a danger so beware of occupiers' activities and your surroundings
Personal alarm/mobile phone	Calling for help if injured or threatened Calling ahead to confirm late arrival	Essential to protect your personal safety and client relations
Bag or container for all/most of the above	Provide protection to all equipment, particularly the delicate tools	A stout plastic or wooden tool box is ideal
Food and/or drink, e.g. thermos with hot tea or coffee	Not strictly 'equipment', but frankly sometimes necessary to maintain morale!	Some inspections are long and arduous. Lack of sustenance can reduce your concentration and in extreme circumstances your health and safety

In some buildings your client may insist you wear other equipment, e.g. a hair net in food preparation areas.

You might wish to consider other equipment you could use to provide your clients with an 'added value' service (see Table 3.3). All of the equipment in Table 3.3 involves or infers some type of 'invasive' inspection. You should only use equipment in the list by agreeing you will do so in your conditions of engagement. Some accreditation schemes may require that you do not engage in invasive inspections. If you use additional equipment you might possibly extend your liability (see later when we discuss moisture meters). You should certainly consider whether you charge additional fees for such extra work.

Table 3.3: CEA possible additional equipment

EQUIPMENT	USE (examples only)	COMMENT
'Merlin' laser gauge	Confirm thickness of glass/air gap in sealed unit and whether glass is coated	Will not confirm whether air space is filled with gas or air or vacuum. Is sometimes 'fooled' when glass has pattern, dimples, or incorporates laminate
Thermo-graphic survey equipment	To confirm presence of insulation	Expensive equipment, very useful in older buildings
'Mirror on a stick'	As above for 'pocket mirror', and to investigate lamps from below	The 'stick' is extendable and facilitates access into areas otherwise inaccessible. However, sometimes awkward to use and requires strong torch
Boro-scope	In areas behind cladding, wall cavities, or flat roofs	May also require drill hole to gain access into the void. Limitations include how much can you infer from such an invasive inspection in, say, one location only. However, can be very useful to confirm information you may have doubts about, e.g. insulation or thermal bridges
Safe systems of working at height, e.g. 'cherry-picker'	Services and construction at high level where other equipment such as binoculars are inadequate	Will require training if you use the equipment personally – best to hire in on ad hoc basis
Light meter	Confirm lighting (lux) levels	Use with care, can be 'fooled' by natural daylight. Some models can be grossly inaccurate. Must take into account age of lamps

We have sometimes been asked by candidates in training whether they should use a moisture meter during an energy inspection to confirm the presence of dampness, i.e. water penetration and condensation.

Building surveyors and other property professionals use moisture meters to check and confirm the condition of building elements in a survey report. Moisture meters can be difficult to use and interpret and can be misleading if used incorrectly.

We have deliberately omitted a moisture meter from the lists of possible equipment. We believe if you use a moisture meter, and certainly if your client knows you use one and/or if you confirm this in your report, you will extend your liability. This is because a client might then reasonably conclude you have considered this important issue for the entire building. They could suggest they relied on that 'fact' later.

You are an *energy assessor*. You are instructed to report on the energy performance of the building, not on the condition. However, you *will* need to consider the condition of certain parts of the building's fabric. We strongly recommend you rely on a visual inspection and your other senses such as smell and touch when you consider damp. We suggest you leave the building's condition to the building surveyor. If you are a surveyor who is also an energy assessor, ensure you separate the two different roles.

Similarly, if you are a services engineer, adopt the same practice and do not get drawn into giving advice, solicited or otherwise, about the condition of the boiler or air conditioning plant, or whatever. Indeed, there is some justification for arguing you should not use a light meter for such reasons.

RICS Code of Measuring Practice

This is a guidance note published by RICS. In the July 2008 CLG guidance (p. 39), CLG confirms:

'The **total useful floor area** is the total area of all enclosed spaces measured to the internal face of the external walls, that is to say it is the gross floor area as measured in accordance with the guidance issued to surveyors by the RICS. In this convention:

a. the area of sloping surfaces such as staircases, galleries, raked auditoria, and tiered terraces should be taken as their area on the plan; and

b. areas that are not enclosed such as open floors, covered ways and balconies are excluded.'

You need to understand the definition of 'total useful floor area' (TUFA). RICS 'core definition' of gross internal area (GIA) is 'the area of a building measured to the internal face of the perimeter walls at each floor level'.

You should read this RICS document. Further definitions from the document are helpful; thus, GIA (and by implication TUFA) includes:

- areas occupied by internal walls and partitions (2.1 of RICS definition); and
- columns, piers, chimney breasts, stairwells, lift-wells, other internal projections, vertical ducts (2.2 of RICS definition).

We deal with some more of the detail of measuring practice in Chapter 8.

NEW BUILD CONSTRUCTION

Building Regulations

You need to understand some other important concepts in the Building Regulations if you intend to be a level 4 assessor and certify new non-dwellings. We do not have space to deal with all of the issues, so provide a general comment only.

Part L of the Building Regulations in England and Wales, 'Conservation of fuel and power' requires that:

> 'L1 Reasonable provision shall be made for the conservation of fuel and power in buildings by:
>
> (a) limiting heat gains and losses:
>
>   (i) through thermal elements and other parts of the building fabric; and
>
>   (ii) from pipes, ducts and vessels used for space heating, space cooling and hot water services;
>
> (b) providing and commissioning energy efficient fixed building services with effective controls; and
>
> (c) providing to the owner sufficient information about the building, the fixed building services and their maintenance requirements so that the building can be operated in such a manner as to use no more fuel and power than is reasonable in the circumstances.'

ADL2A (referred to in Chapter 1) deals with new 'non-dwellings' and extensions to non-dwellings if the TUFA of the extension is greater than 100m² and more than 25% of the existing building; and 'fit out works' if they are part of the new building.

ADL2A sets out five criteria a new building must comply with. These are:

- *Criterion 1 – achieving an acceptable building CO$_2$ emission rate (BER)*
  - the BER must not be greater than, i.e. less than or equal to, the TER (TER is a rating based on a 'notional' building but with improvements above the CO$_2$ emissions allowed in the 2002 Regulations plus improvements by including low carbon technologies),
  - calculations must be made using an appropriate NCM,
  - low carbon technology is encouraged by the AD;

- *Criterion 2 – limits on design flexibility.* This is achieved by ensuring there are 'limiting' standards below which standards should not be allowed to fall, although designers can vary their designs, e.g.:
  - minimum 'U values' for the fabric of the building (we have reproduced Table 4 from ADL2A to the Building Regulations below as an example, see Table 3.4),
  - maximum air permeability of 10m³/h/m² @ 50 Pa,
  - separate control zones for services, e.g. the default condition for central plant should be 'off',
  - heating and cooling systems should not operate at the same time,
  - use of energy meters – e.g. all low carbon technologies must be separately monitored and any building greater than 1,000m² must have automatic metering,
  - for HVAC systems follow guidance in the *Non-Domestic Heating, Cooling and Ventilation Compliance Guide* (NDHCVCG),
  - ductwork should be 'reasonably' air tight,
  - limit specific fan powers,
  - insulate pipes and ducts in accordance with NDHCVCG,
  - lighting efficacies to be not less than certain specified limits, and
  - rules about lighting controls;

- *Criterion 3 – limiting the effects of solar gains in summer,* by
  - sizing and situating windows appropriately,
  - providing protection to windows, e.g. shading,
  - using the thermal capacity of the building, and 'night ventilation';

- *Criterion 4 – quality of construction and commissioning,* achieved by, e.g.:
  - builder having an appropriate system of site inspection,
  - air pressure testing – unless the building is smaller than 500m² TUFA when air permeability is assumed at 15 m³/h/m² @ 50 Pa – if the building fails the pressure test, remedial measures will be required and another test carried out,
  - pressure testing of ductwork, with similar provisions for remedial work if required;

- *Criterion 5 – providing information,* by e.g.:
  - giving the owner a logbook with information about how to manage the building and services:
    – details of services,
    – information about controls,
    – how to use them,
    – maintenance requirements.

Table 3.4: Table 4 ADL2A –

Table 4 **Limiting U-value standards (W/m²K)**

Element	(a) Area-weighted average	(b) For any individual element
Wall	0.35	0.70
Floor	0.25	0.70
Roof	0.25	0.35
Windows[1], roof windows, rooflights[2] and curtain walling	2.2	3.3
Pedestrian doors	2.2	3.0
Vehicle access and similar large doors	1.5	4.0
High usage entrance doors	6.0	6.0
Roof ventilators (inc. smoke vents)	6.0	6.0

Notes:

1. Excluding *display windows* and similar glazing. There is no limit on design flexibility for these exclusions but their impact on CO_2 emissions must be taken into account in calculations.

2. The U-values for roof windows and rooflights in this table are based on the U-value having been assessed with the roof window or rooflight in the vertical position. If a particular unit has been assessed in a plane other than the vertical, the standards given in this Approved Document should be modified by making an adjustment that is dependent on the slope of the unit following the guidance given in BR 443.

We discuss many of the rules and how to achieve them in later chapters. You will not achieve the BER if you consistently apply the minimum requirements in every case, so you will need to make further adjustments.

For new build assessments you will need to carry out a two stage assessment, 'as designed' and 'as built'. For the first stage, you will require a considerable quantity of information from a number of sources. Information should include:

- fully dimensioned 'proposed' drawings including plan (bird's eye) views of each floor, sections through complicated constructions and elevations;
- detailed specification information about the:
  - building fabric,
  - HVAC,
  - hot water services,
  - lighting systems and controls;
- information about the proposed building's use.

Once you have the information from the building and service designers you will need to carefully consider the drawings, etc. (You can skip ahead at this point to other chapters if you want to know more about how to deal with issues relating to the fabric and services of the building.) You will need to enter the following data into SBEM:

- 'U values' of each part of the building fabric, nowadays normally using a software program and including confirmation of:
  - general losses through floors, walls, roofs, doors and glazing,
  - repeating thermal bridges, e.g. steel or timber framework,
  - non-repeating thermal bridges, e.g. corners, edges of floors, eaves, reveals, etc. around windows and doors,
- 'Km values', i.e. effective thermal capacities in kJ/m²K of the different parts of the fabric;
- ventilation losses, i.e. a measurement of air leaking in and out of the building. through gaps in the fabric;
- efficiencies, etc. of the heating, ventilation, hot water systems and air-conditioning systems. You will:
  - obtain this information from the HVAC engineers,
  - use the 'second tier' documents where appropriate;
- full details of the lighting systems and controls:
  - you obtain this information from the lighting engineers.

When you have entered details of the proposed building into SBEM, you will often find it does not comply with the requirements, i.e. BER > TER. At that stage you should consider how the designers can best achieve the target emissions as they are likely to ask for your advice. They will then provide you with alternative design information and you can enter the new data into SBEM and confirm whether the revised data generates a compliant design.

When the building has been constructed, you will receive information about results of testing and commissioning of the building and services. If the building as constructed does not comply, you will provide advice to the design team and contractor/developer on how best to achieve compliance. They will carry out remedial work and provide you with revised test data for you to enter and, hopefully, achieve a compliant building.

Throughout this process you will not necessarily be required to visit the building, even after construction – you will rely on information provided by others. You will though find it can be a real challenge to assemble all of the data you require once construction is complete, particularly if there is no project manager or person in overall charge of the design.

Building Regulations in other parts of the UK

In Scotland, the Building Standards require lower levels of carbon emissions than previously, regular inspection of air conditioning and energy certificate provision to occupiers. For more information, go to www.sbsa.gov.uk.

In Northern Ireland, the Building Regulations (NI) 2000 Part F – 'Conservation of fuel and power' came into force in November 2006 and match the requirements in England and Wales. For full details, you need to consider the two technical booklets:

Preparing for the Commercial Energy Assessment

- DFP Technical Booklet F1: 2006 – *Conservation of fuel and power in dwellings* (www.dfpni.gov.uk/tb_f1_mp_v6.pdf); and
- DFP Technical Booklet F2: 2006 – *Conservation of fuel and power in buildings other than dwellings* (www.dfpni.gov.uk/tb_f2_v15-2.pdf).

SUMMARY

You will not practise in a theoretical vacuum defined only by SBEM. You must operate within the law and in accordance with the statutory constraints that flow from Kyoto via the EPBD and down to each sovereign part of the EU. We have drawn on past experience to provide you with some guidance on how we believe you can practise successfully as a CEA.

With that in mind, in these first three chapters we have set the scene for you. In the next chapter we shall begin to consider the buildings you will inspect.

4 Energy and Construction Core Knowledge and Recognition

INTRODUCTION

In this chapter we consider the fabric of the building in the context of the information you have to gather and use in an SBEM assessment. We first consider energy principles, as applied to the building fabric, and then we describe common forms of building construction used in commercial buildings. The matters we cover are:

- the terminology used to define and describe energy;
- the principles of thermal transmittance and thermal capacity;
- the identification of thermal bridges and the use of Accredited Construction Details;
- an overview of the Building Regulations;
- methods of establishing the date of construction;
- the distinction between domestic-style construction and framed buildings;
- a description of typical materials used for different building elements with some advice for avoiding mistaken identification and a warning about health hazards.

ENERGY IN CONSTRUCTION

General principles

The process of energy transfer in a building is extremely complex. For the sake of clarity 'energy' will now be referred to as 'heat energy' or simply 'heat', although it is recognised that this is not always held to be the correct scientific terminology. 'Heat' and 'heat loss' have been the commonly used terms in building construction for many years. The process of heat transfer operates by energy moving from substances which have a high temperature to those that have a lower temperature.

Hence the notion of 'heat loss' from buildings is associated with winter conditions where heat is transferred from the warm internal space to the cold environment outside. However, this is only one part of the story because heat is also transferred from hot external conditions to the colder conditions inside a building in summertime. It is this process of heat transfer through the building fabric, and the interaction with internal heat sources and solar heat gains, that has a direct influence on the demands for heating and cooling in a building.

Thermal transmittance (U value)

A 'U value', or *thermal transmittance*, represents one of the essential heat transfer characteristics of an element of building fabric. The heat loss through an element is proportional to its U value. The 'U value' is significant in energy assessment because the thermal transmittance through the building fabric is used in the calculation to determine the heating or cooling energy demands of each zone.

The concept of a 'U value' is based on the principle of steady-state heat losses and gains. U values are specified with the units W/m^2K. The U value literally quantifies the number of watts of heat that will be transmitted through one square metre of fabric for each degree of temperature difference between the inside and outside of the building. The U value is the quantity that best defines the insulating properties of the fabric – lower values mean better insulation. The U value of an element is determined, in simple terms, by taking the reciprocal of the sum of thermal resistances of all of the material layers in the element.

The recommended methodologies for calculating U values are described in the *iSBEM User Guide* (BRE, *A User Guide to iSBEM*, CLG, 2009) and we will mention the principle of thermal bridges later in this chapter.

Effective thermal capacity (Km value)

The thermal capacity of the materials used to construct a building affect the way the building responds to the dynamic movement of heat energy caused by varying temperature differences on either side of the building's fabric. The effective thermal capacity also governs the response to internal and solar heat gains and is important in overheating calculations. It also governs the response to the intermittent operation of a HVAC system. The effect occurs across both external and internal elements, i.e. walls, floors, roofs and partitions.

Whereas the U value has been seen to represent the heat flow through a material, the K_m value represents the amount of heat which is absorbed and temporarily stored by the material and released at different times.

The SBEM energy assessment process accounts for the thermal response of building materials by assigning a Kappa (K_m) value to building materials in the project database. The Kappa value was referred to as a Cm value in early versions of SBEM programs. The K_m value is stated in units of kJ/m²K and is determined by calculating the contribution each layer of construction makes towards the thermal capacity of an element. The calculation starts from the layer of construction closest to the internal space and adds together the values for each layer until:

- the total thickness of the layers reaches 100mm; or
- the mid-point of the construction thickness is reached; or
- an insulating layer is reached.

An insulating layer is defined as having a thermal *conductivity* value of less than 0.08W/mK (recognised insulating materials have much lower values than this). Cavities are included in the calculation if their conductivity is greater than 0.08 W/mK.

Figure 4.1: Example Kappa calculation

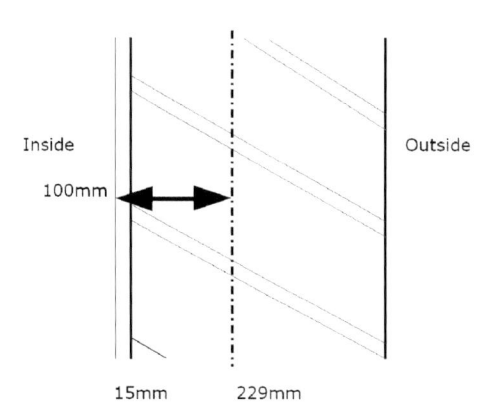

Example 4.1.1: Solid one-brick wall, plastered internally

	A	B	C	A × B × C
Element layer	Density	Thickness	Specific heat capacity	K_m value
	kg/m³	m	kJ/kg K	kJ/m²K
Plaster	1120	0.015	0.960	16.13
External brickwork	1700	0.085	0.800	115.60
Totals		0.100		**131.73**

In this example the value is determined from when the total thickness of layers reaches 100mm from the inside face of the finished wall.

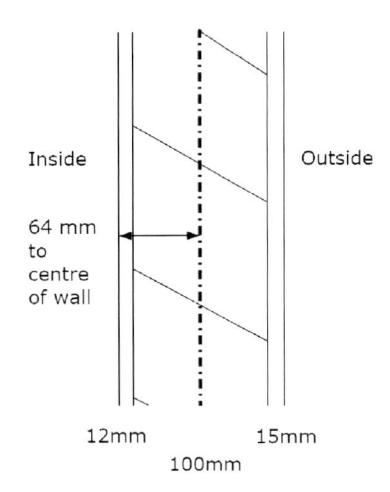

Energy and Construction Core Knowledge and Recognition

Example 4.1.2: 100mm mediumweight block wall, plastered internally and rendered externally

	A	B	C	A × B × C
Element layer	Density	Thickness	Specific heat capacity	Km value
	kg/m³	m	kJ/kg K	kJ/m²K
Plaster	1120	0.012	0.960	12.90
Mediumweight block	1350	0.052	0.840	58.97
Totals		0.064		**71.87**

In this example the value is determined from the mid-point of the construction thickness.

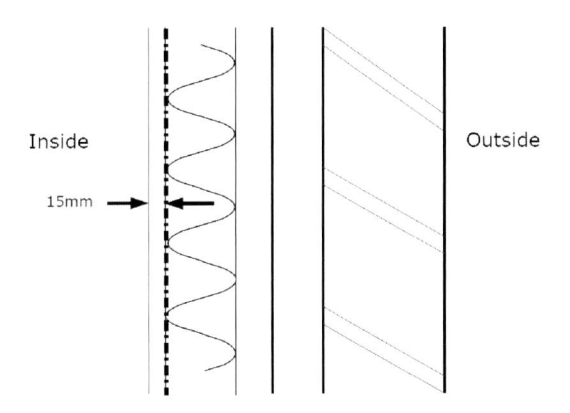

Example 4.1.3: Timber framed cavity wall

	A	B	C	A × B × C
Element layer	Density	Thickness	Specific heat capacity	Km value
	kg/m³	m	kJ/kg K	kJ/m²K
Plasterboard	950	0.015	0.840	**11.97**

In this example the value is determined from the thickness of the internal plasterboard only because the next layer is insulation.

Density and specific heat capacity values were taken from *Environmental Design CIBSE Guide A*.

There are differences between the Km values in these examples and the values for similar construction methods on the SBEM database. This is due to slight variations in the data used. The results are representative of the value range for light, medium and heavyweight construction.

33

The thermal capacity of a building and building materials is sometimes referred to as 'thermal mass'. The thermal response of different buildings is often illustrated by referring to the differences between 'heavyweight' and 'lightweight' forms of construction. In simple terms, this is because the scientific definition of thermal capacity involves the density and thickness of the material as well as its specific heat capacity. So, the greater the density and thickness (mass) of the material, the more energy is required to raise the temperature of that material. In this sense thermal mass is related to physical mass. However, as you will see below, this is not the whole story.

The quantity Kappa is called the 'effective' thermal capacity because it is recognised that not all the fabric makes a contribution to the dynamic storage and release of heat. The inner layers of the fabric make the greatest contribution to the thermal mass and have the greatest effect on dynamic behaviour. This is the reason for the rules relating to what layers are included in the calculation. The position of the insulation layer is very significant. For example, a fabric that has an outer masonary layer but a timber inner frame filled with insulation will have a relatively low Kappa value as little heat can be stored in the inner plasterboard layer. The insulation prevents heat inside from reaching the masonry. Fabrics can have the same U value but very different Kappa values.

The amount of heat energy used to raise the temperature of a building's fabric also relates to the speed at which it responds to changing thermal conditions in both summer and winter. In summertime a building with a high thermal capacity will respond slowly to the increase and decrease in the external temperature during a 24-hour period, or diurnal cycle. On the other hand, a building with a low thermal capacity will respond rapidly to the daily temperature cycle; that is the internal space will heat up and cool down quickly. The contrast in thermal response will be the same in wintertime, but it will then be the building's response to the heat produced by the heating systems, not the external conditions.

The overall heating or cooling energy demands depend on the transmission of heat through the fabric (governed by the U values as you would expect) but, when considered over a whole month, also depend on Kappa. The energy demands due to transmission are offset by internal heat gains that are retained in the building fabric and this depends partly on Kappa. In winter, energy demands can be reduced if some of the internal heat gains can be retained and put to good use. More heat can be retained if the fabric has high effective thermal capacity. However, if the fabric has a low U value some of the heat retained in the fabric will be eventually transmitted outside rather than back inside. In other words, both the U value and the Kappa value are important to heating and cooling energy demands. This sort of process also depends on whether the HVAC system switches on and off or is on continually. The calculation method used by SBEM (based on various EU standard methods) takes this all into account when it calculates the monthly zone heating and cooling demands.

The thermal capacity of an entire building affects its response and energy demands. You must therefore record the type of materials used for both the internal and external fabric. In other words, the thermal capacity of intermediate floors and internal partitions are important too. Some SBEM programs include a time-saving procedure where similar adjoining zones can be merged. For example, two adjoining cellular offices (i.e. the same activity) with identical HVAC services, lighting and orientation could be entered as a single zone. Ideally when you do this, the dimensions and materials used for the dividing partitions must also form part of the data entry in order to account for their thermal capacity. However, the *iSBEM User Guide* suggests you only need to do this if the construction is 'heavy' – e.g. if lightweight partitions are present you can omit them and this omission 'should not cause any significant effects on the calculation' (section 3.3).

Thermal bridges

A thermal bridge is 'part of a structure which has a lower thermal insulation value than the structure which surrounds it' (Garratt and Nowak, 1991). Thermal bridging will therefore occur in buildings where two materials with different thermal *conductivity* values are connected. A typical example is a one-piece steel or concrete lintel built into an insulated wall. The lintel will have a higher value of thermal *conductivity* than the insulated wall and the one-piece design will mean there is a path for heat *transmittance* between the inside and outside of the building. Thermal bridges can also occur in well-insulated buildings, if the detailed design or on-site construction results in discontinuous insulation.

A typical place for this to occur is at the junction between an insulated external wall and an insulated solid ground floor. The problem is avoided in modern construction by installing a vertical upstand of insulation material at the junction of the wall and floor. A thermal bridge can also be formed at corners, even when the connecting materials are identical. This is caused by the flow of heat energy becoming concentrated at corners (NHER, OCDEA *Technical Bulletin* Issue 19). The unit of measurement used for thermal bridges is similar to the U value except that it refers to linear metres and not area. The unit is identified by the Greek symbol psi (Ψ) and the value as W/mK. The heat loss through a thermal bridge is calculated by multiplying the Ψ value by the length of the thermal bridge.

SBEM programs consider two types of thermal bridge: *repeating thermal bridges* and *non-repeating thermal bridges*. The mortar joints in lightweight

Repeating thermal bridges

Masonry joints   Timber / metal frames

Figure 4.2.1: Repeating thermal bridges

Non-repeating thermal bridges

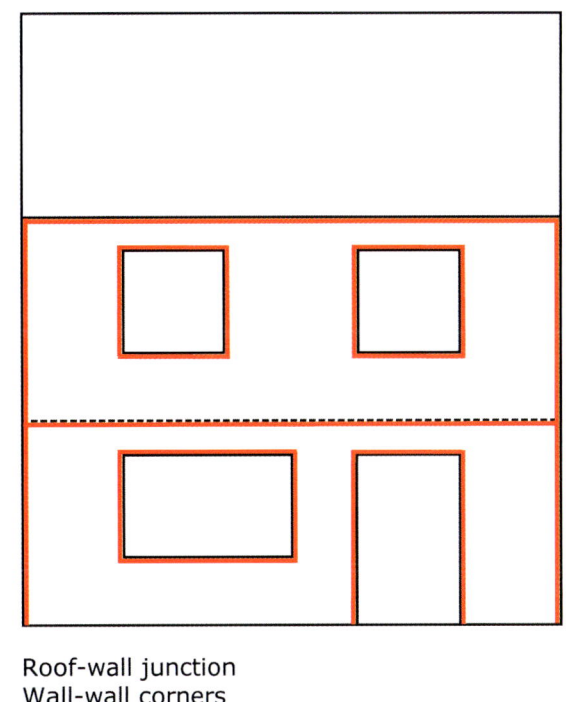

Roof-wall junction
Wall-wall corners
Wall-floor junction
Window and door surrounds

Figure 4.2.2: Non-repeating thermal bridges

blockwork are an example of repeating thermal bridges. The joints are considered to be repeating because the mortar, which will have a higher thermal conductivity value than the blocks, affects the whole of the wall area. Another example is the individual timbers in timber frame construction. On the other hand, the construction of an opening in a wall is dependent on the design of a building and its location is specific to that design. The thermal bridging associated with the perimeter of openings is therefore considered to be non-repeating.

Repeating thermal bridges are included in the U value of building elements in SBEM programs, whereas you must decide on whether to accept default values or enter your own values for non-repeating thermal bridges.

Accredited Building Details

An alternative method for indicating non-repeating thermal bridges on the program is to select from a list of Accredited Building Details. This method should only be used when there is conclusive evidence that Accredited Details have been used in the building being assessed. The details are part of a Government approved scheme for building designers and developers to demonstrate compliance with specific parts of the Building Regulations. They are also an attempt to prevent heat loss that is due to poor construction quality, such as the example of discontinuous wall insulation mentioned previously. The scheme has only been operating since 2007 (CLG, 2007) and the details approved so far are only for domestic style construction. However, you should note that the scheme is intended to apply to all types of building and it is possible to build small-scale, commercial buildings using domestic style construction. Where an Accredited Detail is used in a building it is taken to have a known thermal performance. The values representing the performance of the detail will be used in the SBEM calculation, if the detail is selected by the assessor. Accredited Building Details are not currently considered for buildings that involve metal cladding.

PLANNING AND BUILDING REGULATIONS ISSUES

Building Regulations

You will find that a working knowledge of the Regulations can be useful when determining the age of construction or when alterations have been made to a building. The Regulations are also used as a means of defining the construction of a building when entering data into a SBEM program.

The first set of Building Regulations in England and Wales were published in 1965 and introduced in 1966. Regulatory control of building work before these dates was achieved by a system of locally administered bye-laws. Public Health Acts also applied to some aspects of construction prior to the Building Regulations, such as setting standards for the density of development and the provision of drainage. Legislation that applied to the use of buildings also had an indirect affect on the way they were constructed in the past. For example, although there was no regulatory requirement for how heating or cooling was achieved in buildings, the *Office Shops and Railways Premises Act* 1963 set temperatures for working conditions which influenced the selection of heating, cooling and fabric thermal insulation for a building.

The relevance of the Building Regulations to commercial energy assessment concerns the performance standards set for the fabric and services of buildings. In the following chapters we will describe how SBEM programs relate to the regulations. For example, you will see that one of the data entry requirements is to compile a project database, naming each type of building element used at a property. One method of defining these elements is to use an 'inference procedure' that references the element by the date of the Building Regulations that would have applied to the building when it was constructed or altered.

The Building Regulations 1965 set out requirements for the thermal insulation of dwellings, but it was not until 1979 that insulation standards for non-residential buildings were introduced. This does not mean that a commercial building built before this date was not insulated. It has already been stated that other legislation influenced the design of buildings and building services. In addition, the *Thermal Insulation (Industrial Buildings) Act* 1957 set a statutory requirement for the roofs of industrial buildings. It should also be noted that SBEM programs infer different thermal performance values for uninsulated construction elements and those for Building Regulation standards set between 1965 and 1981.

Since the 1985 Building Regulations the control of energy efficiency has been covered by Part L, 'The Conservation of Fuel and Power'. This edition of the Regulations was introduced under a new legislative approach that was set out in the *Building Act* 1984. The new Regulations were less detailed and concentrated on describing the performance standard that was required. Compliance with the Regulations could now be achieved in a number of ways, one of which was to follow the guidance from a set of 'Approved Documents' that were published to accompany the Regulations. At this point it is important for you to note that the Approved Documents are not the Regulations, they are guidance on just one method of satisfying the Regulations.

In 1985, compliance with Part L was still relatively simplistic, concentrating on limiting heat losses through the building fabric and from hot water storage vessels, pipes and ducts. The Regulations also required the output of space heating and hot water systems to be controlled 'appropriately'. This could be achieved by the use of basic room temperature controls, weather compensation controls, time clocks and optimisers. However, regulations as a whole were only being applied to the construction of new buildings and large refurbishment projects. Retrofitted replacements that were done for the sake of repair or maintenance did not always satisfy the standards.

It was not until the 2002 amendment of Part L of the Building Regulations 2000 that work to existing buildings became more tightly regulated. This was achieved by making the regulations apply to 'controlled fittings and services, and thermal elements', which include: windows, boilers, mechanical ventilation, air-conditioning plant and lighting.

There are now four Approved Documents that accompany the Building Regulations 2000, as amended (see Chapter 1).

The 2006 Approved Documents provide one method for demonstrating compliance with Part L, based on the calculation of carbon emissions. However, earlier versions of the documents provided two or three alternative methods. The 'elemental' approach to demonstrating compliance is still an option in ADL2B. However, rarely will you find information to tell you which of these methods was used. One of the methods accepted in the past was a whole building calculation that allowed a trade off between parts of the building having relatively poor fabric insulation with parts that had an improved performance. For this reason you should not make assumptions that the fabric in a building has a specific transmittance value (U value). When you select a transmittance value on an SBEM program to define an element of the construction that has been built at a particular time you will merely be inferring the performance value that is stated; hence the term 'inference procedure'.

Conservation areas and listed buildings

Conservation areas are areas of special architectural or historic interest that can be designated by local authorities. English Heritage also has designation powers in London, subject to it consulting with the local Borough Council and obtaining the consent of the Secretary of State. The Secretary of State can also designate conservation areas in some circumstances. The effect of a conservation area designation is that the relevant local authority has extra control over demolition, specific aspects of building development and the protection of trees. Demolition and building development may still be permitted in a conservation area, but it is likely to be subject to conditions that retain buildings, or parts of buildings and restrict the types of materials that can be used in order to preserve the character of the area.

Similar controls may be applied to individual buildings through a process of listing. A listed building designation can only be made by the Secretary of State, but any individual or organisation can recommend a building for listing by submitting an application to English Heritage. Generally a building must be more than 30 years old to be eligible for listing, but the older a property is the chance of it being listed increases. Most buildings built between 1700 and 1840 are listed and any building that was built before 1700 and is unaltered will be listed. Listed buildings are classified in three grades: Grade 1 (usually written with Roman Numerals as Grade I) buildings of exceptional interest; Grade 2 star (Grade II*) particularly important buildings of more than special interest; Grade 2 (Grade II) buildings that are nationally important and of special interest. Listed building consent is required for any demolition, alteration or extension of these buildings.

The additional statutory controls applied to buildings in a conservation area and listed buildings will clearly affect what work can be done to them to improve their energy performance. The controls do not mean the work cannot be done, but that formal consent will be required to do the work. You should therefore establish whether a building being assessed has conservation area or listed building status. The local authority will hold this information, which is often published on their websites. You are then mainly concerned with how the building's status affects the EPC's RR (see Chapter 10).

Some SBEM programs have a simple tick box to indicate a building's special conservation status, which may automatically remove recommendations that are considered to be inappropriate for the building's circumstances. However, you should still check through the recommendations with the building's status in mind, before the report is finalised. It is probably better to make a comment against any relevant recommendations, rather than remove it because there remains the chance that statutory consent may be given for the improvement work. You may also feel certain that a local authority will refuse consent for the work today, but there is the possibility that policy could change during the potential 10-year lifespan of an EPC.

SPECIFIC CONSTRUCTION ISSUES

Identification

In this chapter we will describe the main forms of construction found in non-residential buildings and comment on some of the practical issues encountered when recording and using the data. You will need to be able to identify both the external and internal elements of construction used in a building because of the zoning methodology of the SBEM assessment (see Chapter 9).

The construction methods used in non-residential buildings can vary from traditional forms to complex, system-built assemblies. Level 3, 4 and 5 energy assessments could involve any form of construction, although level 3 assessments are more likely to involve traditional construction methods. As with all aspects of the commercial energy assessment, you will gather evidence of the method of construction from the inspection and documentary evidence. The NOS for the CEA qualification require an assessor to be able to identify the form of construction visually and from plans and specifications.

Date of construction

The date of construction is important to establish for a SBEM calculation. However, non-residential buildings tend to be altered frequently and you will need to determine the date and extent of any alterations. The style of construction, type of materials used and the extent of deterioration can provide clues about when a building was built or altered. Enquiries made with owners, occupiers and managing agents can also provide information. In some situations a person with specific responsibility for a building, such as a facilities manager or caretaker, could be available to ask. A friendly chat with anyone at the building can be very worthwhile; a concierge who has worked at a building for several years is likely to know to something about its history. However, visual and verbal evidence alone can be unreliable. You should always attempt to obtain documentary evidence to validate other sources of information. We include a list of some of these documents in Chapter 8. You should also take care to authenticate and validate any documents seen. For example, planning permission may have been granted, but the work may not have been carried out as proposed.

The architectural style of a building can indicate the period in which it was built. For example, the strong, imperial appearance of Edwardian buildings can be distinguished from the simplified, modernist look which followed it (see Figures 4.3.1 and 4.3.2).

Architectural styles are often connected to historical trends in the arts or technology. This can still be seen in modern buildings, where environmental concerns have led to the use of sustainable materials and green roofs. It must be stressed that style is only an indicator of age and once again you must be cautious, not to be misled. In the past 20 years many buildings have been completely redeveloped, but have retained the facade of the original building (see Figure 4.4). These buildings, including the external walls, should conform to the building standards in force at the time of the redevelopment.

Figure 4.4: Image of retained facade

Figures 4.3.1 and 4.3.2: Contrasting building styles

These three categories provide a useful comparison of how the structures of buildings are arranged and how materials are used.

Traditional construction methods

Traditional forms of construction are typically those where the external envelope performs the function of supporting the building and is also the protective fabric. Structural masonry is a typical example of this form of construction.

Traditional building methods comprise of those commonly found in dwellings. These include: solid masonry walls; cavity walls; domestic scale timber/metal frame walls; suspended timber floors, solid, ground-bearing floors; beam and pot suspended floors; cut-timber roofs; truss-framed roofs; double-glazed windows. For identification purposes, all of these methods are dealt with at length in many construction technology books, including the *Domestic Energy Assessor's Handbook* (2008). Whilst many inspection skills are the same for domestic and commercial energy assessment, you must remember that the purpose of the inspection and data collection is for an SBEM calculation, not RdSAP.

Forms of construction

You will be concerned mainly with establishing the thermal performance of a building, not how it is supported. However, some understanding of the structural arrangement of the building can help with the identification of the thermal elements.

ABBE, one of the awarding bodies for the CEA qualification, categorise non-residential buildings into three types:

- traditional forms of construction;
- low-rise, wide span (portal frame); and
- multi-storey.

Masonry walls in commercial buildings are more likely to be of a larger scale than in domestic properties. A characteristic of tall masonry walls is they are thicker at the bottom than at the top (see Figure 4.5). This will increase the floor area of the rooms on the upper floors if the decrease in width is made on the inside of the wall.

Figure 4.6: Multi-storey masonry construction showing smaller windows on upper floors

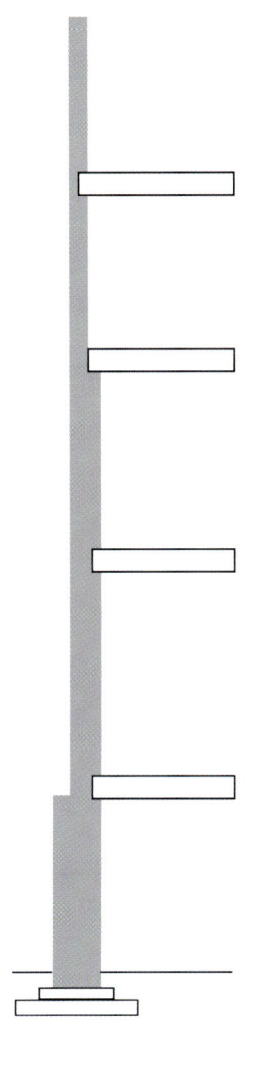

Figure 4.5: Section through multi-storey masonry construction showing thinner walls and lower storey heights on upper floors

There is also an optimum height at which brick walls can be built before they become too slender and for this reason the storey heights of a tall masonry building become less as the height of the building increases. This is often evident from the smaller windows on the upper floors (see Figure 4.6). You should be alert to these construction features because internal floor areas and storey heights are important data for a SBEM assessment.

Framed buildings

In other types of building the structural support is generally provided by a frame, onto which the fabric of the building is attached. Masonry is often used for the external fabric and internal partitions of framed buildings, but in this case the masonry only supports itself (see Figure 4.7).

Structural frames for non-residential buildings are generally built from either steel or reinforced concrete. Concrete frames can be either from in-situ or pre-cast concrete. In-situ concrete is poured on-site into temporary moulds, known as formwork, whereas pre-cast concrete products are produced off-site. Concrete alone has insufficient strength to be used as a frame and must always be reinforced with steel. However, reinforced concrete provides a stiff form of construction and in most cases a frame made from the material is a simple arrangement of columns and beams.

Structural steel frames are often more complex for two reasons. First, they require some form of bracing to prevent movement. The bracing could be in the form of additional diagonal members (see Figure 4.8) or provided by the infill panel between beams and columns. Secondly, steel loses strength at relatively low temperatures. A building fire could cause a steel frame to distort and collapse.

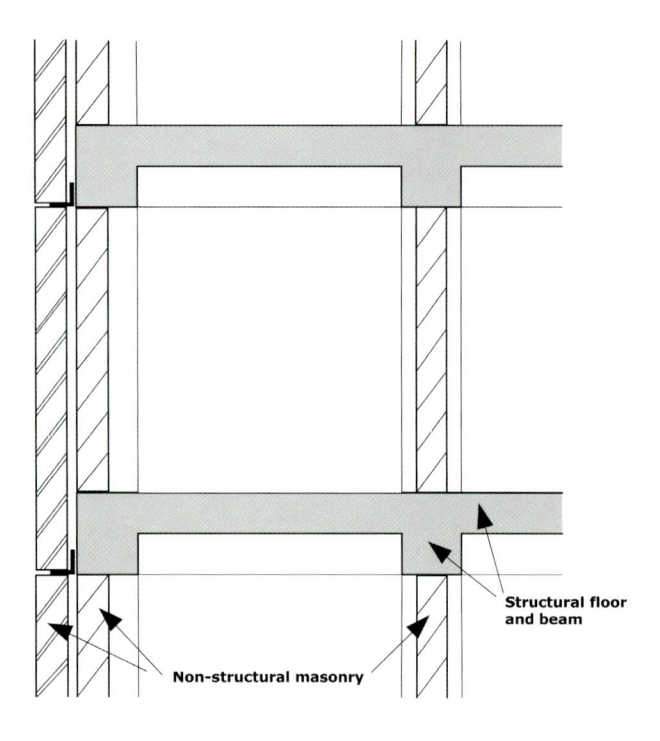

Figure 4.7: Non-structural masonry

Figure 4.8: Bracing

Fire insulation

The inherent weakness of a steel structure in a fire is overcome by adding fire protective materials to vulnerable parts of a frame. Concrete casings have been used in the past, but this practice was quickly discontinued because a reinforced concrete frame would produce the same effect more economically. In recent years intumescent paint has been used for fire protection of steel and timber. The high temperature from a fire causes the paint to expand and provides a protective layer to the surface it is covering.

The most common forms of fire protection on structural steel frames are blown-fibre coatings (see Figure 4.9.1) or light-weight casings (see Figure 4.9.2) both of which need to be treated with caution. These forms of protection involve materials with good

Figure 4.9.1: Blown-fibre insulation

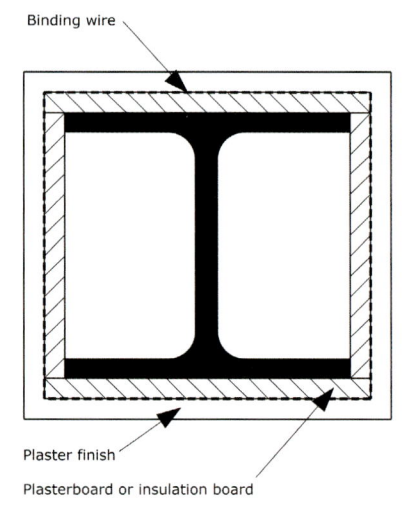

Figure 4.9.2: Frame casing

insulation properties. However, in this case the materials are only being used to insulate the steel against the effects of fire, not to reduce heat losses. In addition, asbestos containing materials (ACMs) have been used for both types of fire proofing. Asbestos in blown-fibre applications is particularly hazardous because there is a tendency for it to contain a high asbestos content. In addition, this material is extremely friable and prone to damage that could easily release fibres into the air. Insulation boards have also contained asbestos in the past, but the asbestos in these products is generally present in relatively low concentrations and bound tightly into the material. Nevertheless, should you encounter either material you must avoid disturbing it and follow the guidance of the Approved Code of Practice (ACOP) for the Control of Asbestos, described in Chapter 2.

Controlling dimensions

A common characteristic of framed buildings is they are built to a dimensional grid. This means there are horizontal and vertical controlling dimensions, such as the distance between columns. This feature allows standard-sized, prefabricated materials to be used for both the structure and fabric of a building, which could result in a reduction in construction time and costs. You can use the controlling dimensions of a framed building to your advantage when measuring it for an energy assessment. A record of the grid dimensions on which the building has been built can serve as a useful check for other dimensions and for calculating the floor area.

A controlling dimension can also be used to estimate the size of materials and components that are inaccessible. For example, a measurement of the distance between two portal frames at ground level could help to estimate the size and arrangement of rooflights when access to the roof is not available.

TYPES OF EXTERNAL ENVELOPE

External walls

The external walls of framed buildings have been constructed using traditional masonry techniques, such as solid brickwork and cavity walls. An external wall, or part of the wall, may be built to fill in between the structural frame or cover it, depending on the material and construction method used for the frame. For example, concrete frames are often left exposed, whereas until recently steel framed buildings were nearly always fully enclosed by the external fabric. Most of the early framed buildings were relatively large and prestigious. Stone, terracotta and faience were all used on the external walls of these buildings, but generally only as a facing built onto a backing of solid masonry.

The external wall construction of framed buildings built in the last 50 years can be more difficult to identify from a visual inspection alone. A wall that looks like a brick wall from the outside could be thin brick 'slips' adhered to a solid concrete panel. Alternatively, it could be a half-brick (102.5mm) skin of brickwork in front of a timber or metal-framed wall. This could be the case even when a 'solid' brickwork pattern, such as English or Flemish bond, is visible, if 'snap headers' have been used. You should therefore take specific care identifying these types of wall. Normal inspection practices of recording the materials used, brick patterns and measuring wall thicknesses should be continued, but supplemented with additional investigation. For example, some time could be spent looking for areas where the wall construction may be exposed from damage or where services have been installed. Background research of documentary evidence could also be useful.

Where different construction methods are present in a single wall, they may have different thermal performance values. For example, if a cavity wall has been used to fill-in between an exposed concrete frame, the wall will have a lower thermal transmittance value than the concrete. You may therefore consider whether to enter the two methods of construction separately in an SBEM program (see Appendix B, case study 2). However, this can be a time-consuming operation and may not affect the output significantly. In cases where you consider the contribution of different materials might be significant, such as in Figure 4.10, you could make an estimate of the percentages of the wall and frame.

Figure 4.10: Cavity wall and solid reinforced concrete frame in one elevation

In the circumstances where inference values are used to define the thermal performance of construction elements, these are based on the Building Regulations at the time the building was built. Theoretically the whole building should achieve compliance and it should not be necessary to discriminate between different materials used in one wall. However, thermal bridging, discussed earlier in this chapter, is likely to be a problem with the form of construction used in this case, and adjustments may be required in the SBEM calculation.

Cladding panels

Cladding panels are another popular method of constructing external walls on framed buildings. These panels can be of pre-cast concrete or a lightweight frame construction. The panels are often made up of composite layers. In its simplest form this could be an external waterproof layer, a middle layer of insulation and an internal finishing layer. Cladding panels are prefabricated off-site and should therefore achieve a better build quality and level of performance than can be achieved by construction on-site. The panels are commonly hung from and fixed to the perimeter of the structural frame (see Figure 4.11 (c)) or the cantilevered edge of the floors.

Curtain walling

Curtain walling is another method used for the external fabric of framed buildings. This is a frame assembly, infilled with glazing or composite boards, that forms a complete envelope around the structural frame. The curtain walling frame is usually a lightweight metal, which supports itself and the infill material, but does not provide structural support to the building. There are two features of curtain walling that distinguish it from other types of external wall: first, it envelops the whole of the structural frame; and secondly, the curtain wall assembly alone is the external wall (see Figure 4.11 (a)). These distinctions are necessary because sometimes curtain walling elements are used as infill panels, where the structural frame is exposed on the outside of the building. On other occasions, a curtain wall assembly will be used to cover the whole of the structural frame, but there will be additional layers on the inside face of the wall. This form of construction is known as rainscreen cladding (see Figure 4.11 (b)).

It should be evident that both infill panels and rainscreen cladding will have different thermal properties than a curtain walling assembly used on its own. The matter is complicated further because the database of construction types in some SBEM programs does not accurately distinguish between

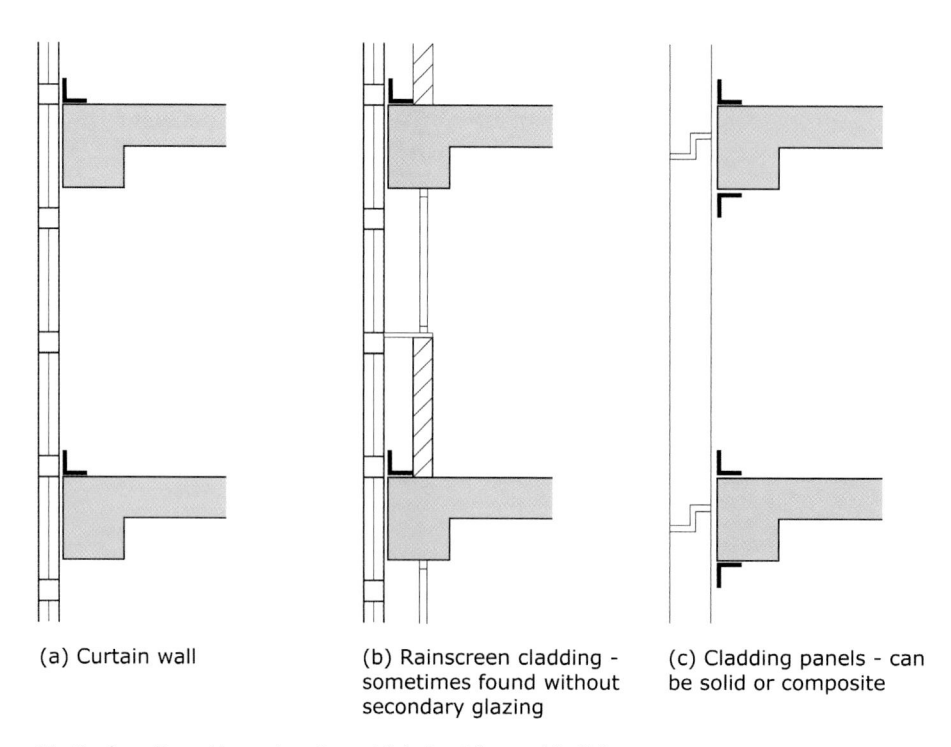

(a) Curtain wall

(b) Rainscreen cladding - sometimes found without secondary glazing

(c) Cladding panels - can be solid or composite

Vertical sections through external fabric of framed buildings

Figure 4.11: Different forms of enclosure for framed buildings

these different forms of wall. For example, data might be available for rainscreen cladding, but as a traditional style comprised of solid brickwork with timber weatherboarding on the external surface. In both cases you will have to check the U and Kappa values of the construction method you select.

Profiled cladding systems

A popular choice of construction for the external walls of low-rise, medium and wide span buildings is profiled sheet cladding. The profile in this case refers to the cross-sectional shape of the cladding sheet. Early forms of cladding had a simple, sinusoidal, corrugated profile and were made from asbestos cement. A non-asbestos version of this sheeting is still available. Sometimes this type of material is referred to as NT, or 'New Technology', which may be seen imprinted on the surface of the sheets. However, nowadays the material is only generally used for agricultural buildings and maintenance.

Modern profiled sheet cladding is made from steel or metal alloy and can be obtained in a variety of profiles, which are often a rectangular shape. Metal sheets are generally delivered with a decorative finish that offers protection against deterioration. From the early 1970s to the late 1980s plastic coated profiled metal cladding was used commonly, but it fell out of favour because of a series of failures where the coating blistered and peeled off the sheets. Resin based coatings have been used in recent years.

Profiled cladding alone is a thin material and when it is made of metal it has a high thermal transmittance value. Wall and roof cladding is therefore installed as some form of insulated system. There are two main types of system: one where the layers are installed separately on-site and the other which is supplied as pre-formed panels that encapsulate an insulation material. Profiled cladding is often used for the external and internal layers of an installed system, with mineral fibre or foam insulation between them. This method of construction can help you to determine that a system has been used, as opposed to a single sheet, because different shapes of profile are often used for the external and internal layers (see Figure 4.12). Alternatively, the internal layer of a cladding system may be formed from flat surfaced plasterboard or insulation board. The Metal Cladding Manufacturer's Association provides comprehensive details of different cladding types and construction methods used (www.mcrma.co.uk).

Fibreglass profiled cladding may also be encountered on occasions, but it is not commonplace because it tends to be degraded easily by ultraviolet light. All forms of profiled cladding are vulnerable to impact damage and low-rise, medium and wide-span buildings usually have masonry construction to a height of about 2 metres at ground level. You must record measurements and details of both the masonry

Figure 4.12.1: Uninsulated cladding

Figure 4.12.2: Insulated cladding

Figure 4.12: Profiled cladding

and cladding for the data used by SBEM programs because the materials will have different thermal performance values.

Retaining structures

Retaining walls may be encountered in basement construction of traditional and framed buildings. The main function of a retaining wall is to resist the lateral forces of the soil and water in the ground adjoining the basement. A retaining wall may also support part of

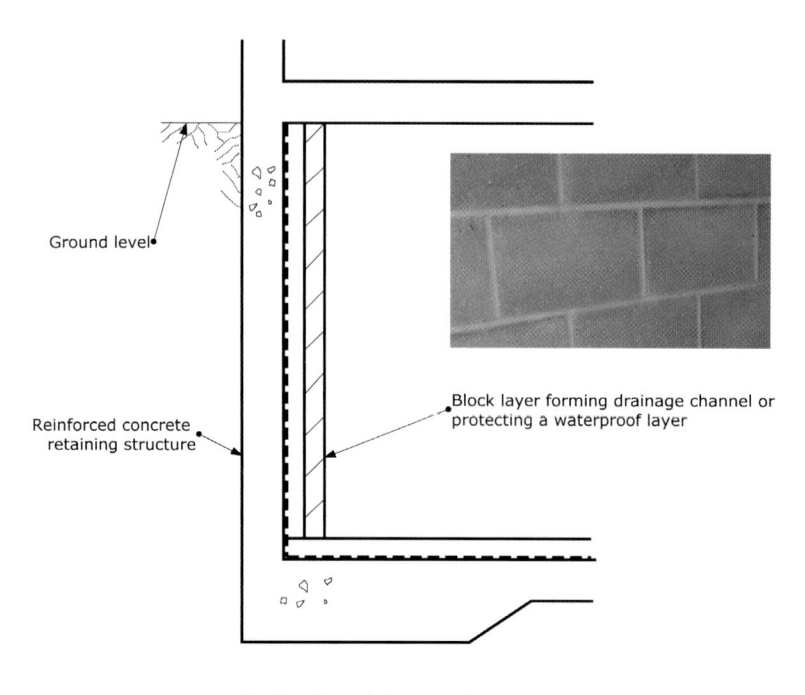

Fig 4.13: Section through basement construction

the building above it. The strength required to resist the lateral forces in older, traditionally built buildings was achieved by using thick masonry walls. You should note that where these walls are thickened on the inside of a building, the floor area of a basement level is likely to be less than the area of the floor above it.

Reinforced concrete is used to achieve the required strength in most basement construction. In both old and new buildings you are therefore more likely to be recording a solid construction method with a high thermal capacity for basement wall construction. Where the adjoining condition is below the external ground level it will be recorded as 'underground'. Care is required to establish the position of the external ground level in relation to the internal floor level of the basement in case only part of a basement wall is below ground level – see case study 2 in Appendix B.

You must also be wary of not being misled by the appearance or feel of the internal surface of an external basement wall. Retaining walls are often multi-layered and when a waterproofing layer is applied to the inside of a wall a further skin of brick or block can be added to provide protection against physical damage and to prevent water pressure pushing the waterproofing layer off the wall. Where a single skin of masonry is built on the inside of the waterproofing layer it will be laid in stretcher bond (see Figure 4.13). You should not confuse this with cavity wall or lightweight construction; there will be something more to the wall in order to fulfil its retaining function.

On other occasions the inside surface of an external basement wall could be lined with a lightweight boarding. This is often for aesthetic purposes, but may be to provide a drainage cavity for ground water penetrating the external part of the wall. Once again you should not mistake this as the main form of construction for the wall.

Windows

Windows, including rooflights, are considered as part of the SBEM calculation in terms of thermal and light transmission. Like other parts of the external fabric of a building the thermal transmittance through windows is considered for both heat loss and gain. However, heat gain is a more significant feature of a window's thermal performance. In addition, the amount of natural light transmitted through a building's windows will affect how much artificial lighting is used, which in turn will affect the CO_2 emissions.

SBEM calculations use the U value for the whole of a window or rooflight, including the frame. You need to record information about the materials and construction of both a window's frame and glazing for the calculation to be made. The materials used for the frame and whether it is constructed with a thermal break should be established if possible. Relevant information about the glazing includes:

- number and thickness of the panes and gaps;
- type of gas in any gap; and
- whether the glass is coated.

This amount of specific detail is more likely to come from documentary evidence because it is difficult to determine visually. Glass analysis equipment is available that can detect glass coatings (see Figure 4.14)

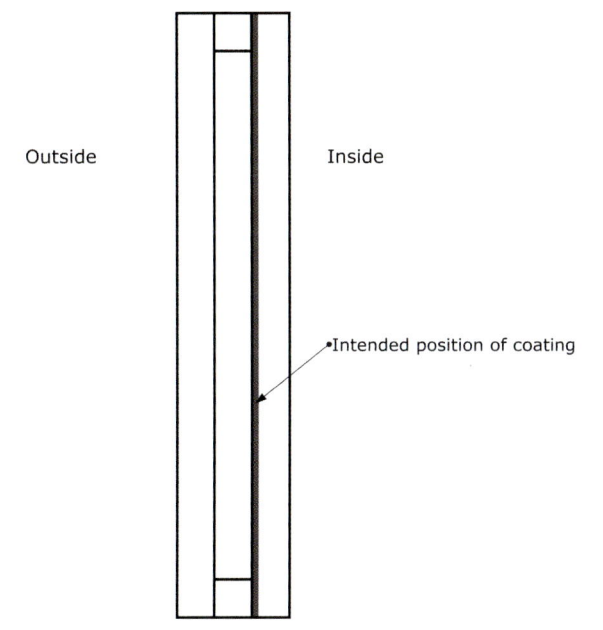

Energy and Construction Core Knowledge and Recognition

Figure 4.14: Intended position of coating on sealed double glazing unit

Figure 4.15: Glazing film

and measure the thickness of multi-layered glazing. Equipment that can identify the gas fill in glazing units is highly specialised and currently beyond reasonable affordability as part of a CEA's toolkit.

Glass is coated as a separate stage of the manufacturing process. The coating is applied onto the surface of the glass and is not homogenous. This makes the coating vulnerable to damage and it should therefore be on a surface within the gap of multiple glass units. For optimum thermal performance, Pilkington recommends the coating is on the third surface of a double glazed unit, counted from the outside to the inside.

The *iSBEM User Guide* states in the 'Glazing Database', 'Uncoated, clear' refers to ordinary clear glass which has no low emissivity coating and no tint, 'Reflectance, low-emissivity' refers to glazing in which at least one glass pane has a low-emissivity coating (such as 'Pilkington K' glass or 'Optitherm' glass), and 'Tinted' refers to glazing where at least one pane is colour-tinted. It should be noted that tinted and low-emissivity glass produce different heat output results on SBEM calculations. It is therefore important to make the correct data entry.

Many older buildings with single glazed windows have had a plastic film coating applied to the glass (see Figure 4.15). Often this will have been done in an attempt to reduce heat gains or glare, but sometimes it will have been done to improve safety or security. The film is often strongly tinted or reflective and it is therefore easy to identify. It is also vulnerable to damage, particularly from window cleaning, because the film is on an exposed surface. The film then becomes evident from the scratches that can be seen.

A final check that can be made for retro-fitted film is to look closely at the junction between the glass and the window frame where a thin line of uncovered glass is likely to be visible at the cut edge of the film. Currently the SBEM database does not include solar transmittance values for coated single glazing, which is the most likely arrangement where a film has been used. The *iSBEM User Guide* recommends the use of ISO 9050, EN410, EN 13363-2 and ISO 15099, when solar transmittance values have to be calculated.

In addition to recording details about the glazing it is important to take accurate measurements of windows and rooflights to determine the proportion of glass to other parts of the fabric. Three data entry defaults have to be considered: Surface Area Ratio; Area Ratio Covered (rooflights only); and Transmission Factor (see Chapter 9 and Appendices F, G and H for examples).

Figure 4.16: External doors with different glazed areas

Doors

SBEM programs categorise doors as either personnel or vehicle access doors. Only external doors are considered for assessment (see Figure 4.16). Other than the category of a door you shall only be concerned with recording its size (including the frame), proportion of glazing and whether or not it is insulated. A door with more than 50% glazing is recorded as a window and you must therefore record more detail.

Roofs

Two categories of roof are considered in the SBEM assessment: flat and pitched. A roof with a gradient of 10 degrees or less is considered to be flat. Steep roof surfaces, such as those in mansard roofs, are treated as walls if their gradient exceeds 70 degrees. Flat roof descriptions in SBEM programs distinguish between lightweight and heavyweight construction. A flat roof with timber roof joists and a board deck or profiled metal deck construction, typically found on low-rise, wide-span buildings, are examples of lightweight roofs.

Heavyweight flat roofs would include in-situ concrete or pre-cast beam and pot/block construction. As a rule of thumb, lightweight structures are normally covered with lightweight fabric. Metal waterproof coverings on flat roofs, such as copper, lead and zinc are generally laid on lightweight board, decking. However, asphalt and bituminous felt waterproofing could be used on either light or heavyweight construction.

Floors and ceilings

Floor construction is categorised on SBEM programs as either 'solid ground floor' or 'suspended floor'. Suspended floors can be located at the lowest floor level or as an intermediate floor between different levels of the building. The reference to 'solid ground floor' can be better described, for avoidance of doubt, as a 'ground-bearing solid floor' because intermediate floors can also be of a solid form of construction, such as concrete. The calculation of U values for floors that are in contact with the ground is complex and subject to its own methodology. Should you need to calculate a value yourself, you should use an appropriate software tool and ensure that you have valid information for the calculation. The Building Regulation's Approved Document L2B is a useful reference source for the differences between standards for new and replacement elements and limiting U-value standards.

When a suspended floor construction is located at the lowest floor level, care must be taken to consider the condition of the adjoining space. For example, some modular building systems sit on supports that raise them above the external ground surface. The adjoining space in this case would be selected as the 'exterior'. On the other hand traditional suspended timber floor construction at ground level is built over a ventilated void, which can sometimes be a relatively deep space. Over-site concrete, gravel or hardcore is generally laid under these spaces. In these circumstances the over-site construction becomes the floor level for measuring storey heights and the floor construction described as solid. The condition of the adjoining space is described as 'underground', which is the condition under the over-site construction. The depth of the void under the floor can sometimes be estimated from the difference between the external ground level and the internal floor level.

On other occasions it may be possible to measure the void through an access trap or in a pipe duct. In circumstances when a floor void cannot be measured it will be necessary to make an allowance for an estimated dimension. This dimension may be recommended as an assessment protocol by your accreditation scheme.

You should be concerned with the heat transmittance and thermal capacity of a floor when identifying it during your inspection. The issue of transmittance is mainly relevant to the lowest ground floor level and floors adjoining unheated or strongly ventilated spaces. This is a matter of determining whether or not the floor is insulated. Floor insulation has been required by the Building Regulations in these situations and the respective U value will be assigned to the floor when inference procedures are followed for the SBEM data entry. Thermal capacity is a matter that is relevant to all floors and this is largely a matter of distinguishing lightweight, timber and metal floors from dense, concrete floors. You should take care on both counts.

For example, wooden floors are often laid over solid floors for sound reduction or aesthetic purposes. On the other hand, when a visual inspection reveals the underside of intermediate floors constructed from profiled metal sheets these are generally only forming part of a composite construction that is covered with concrete. Some commercial buildings have *raised access floors* that accommodate cabling and building services (see Figure 4.17).

Sometimes the floor void serves as an air plenum for the ventilation system (see Chapter 6). The floors are formed by a platform of individual panels supported on a grid of blocks or metal struts. The depth of the floor void could be a few centimetres or several hundred millimetres. It is therefore desirable to inspect under the floor to look for building services; check the structural floor construction and establish the zone height, but this should be done with caution. Raised floor panels are notoriously difficult to replace tightly or evenly after being lifted. A metal faced panel, in particular, may need to make physical contact with all of its supports and the adjoining panels in order to maintain electrical earth continuity. You may therefore consider it appropriate to only inspect the floor void when the owner or occupier arranges for the panel to be lifted and replaced.

The inspection of the underside of *intermediate floors* can be similarly complicated by the presence of *suspended ceilings*. There are two main types of suspended ceiling: exposed grid or lay-in tile; concealed grid or flush. Lightweight fibrous ceiling tiles are commonly used for both types of ceiling. Heavier, rigid tiles made from fibrous plaster or cement are sometimes used in exposed-grid ceilings. On both counts, you should remember the potential for ACMs being used in older forms of these ceilings and take due precautions to avoid disturbing them. Flush ceilings are sometimes formed with rigid timber or metal frames lined with boarding. This form of suspended ceiling is common in corridors or spaces with short floor spans. Once again, in older properties, the board used may be an ACM, particularly if the joints of the board, or fixing screws, are visible. You should also consider whether the ceiling is intended to serve a fire-resistance function, in which case the use of an ACM is more probable.

The void above suspended ceilings may contain building services and be relatively shallow or very deep. The void above flush ceilings is often inaccessible, unless an access hatch is available. It may be possible to move light fittings, but this is not advisable for health and safety reasons, unless it is undertaken by an electrician. The tiles of lay-in ceilings can often be lifted with care, but they may be fitted with a concealed, spring-clip which may cause damage to the tile during lifting. In addition to the potential hazards already mentioned, the inspection of a suspended ceiling void is likely to require a separate risk assessment as required by the *Work at Height Regulations* (see Chapter 2).

Figure 4.17: Raised floor

Building systems

Non-traditional forms of system-built construction used for dwellings, such as Cornish and Airey, are unlikely to be encountered in commercial properties. On the other hand, most framed buildings are likely to involve some form of construction system, in the sense that parts of them will have been manufactured off-site. However, there are forms of proprietary, whole-building systems that have been used for non-residential construction. Clasp buildings were a popular choice of public organisations such as education authorities and the health service, between the 1960s and 1980s. The buildings were often intended to provide temporary accommodation, perhaps during the refurbishment of a permanent building or to cope with periodic increases in the population. Despite their original intention, some of the buildings continue to be used today and the system is still available on the market. The Clasp and similar systems provided by other manufacturers were built from a timber or metal structure enclosed with lightweight timber panels.

Other proprietary forms of temporary building used for non-residential uses, such as Portakabin and Terrapin, have used a modular system, where the whole building is built off-site. Many manufacturers of temporary buildings now also provide permanent building systems. You will need to be able to distinguish between the two because different building standards apply to them. Permanent, modular and portable buildings will have had to comply with the Building Regulations at the time of construction, whereas the Regulations applying to temporary buildings will vary according to percentage and date of manufacture of modules, panels and sub-assemblies used to construct the building. This is illustrated on the flowchart in Figure 4.18 to show compliance, produced by the Modular and Portable Building Association (MPBA).

For an assessment of permanent modular and portable buildings built to comply with the Building Regulations you can use the inference values on SBEM, but you will need to select appropriate thermal performance values and definitions for older, temporary buildings that were not covered by the regulations and have been extended into permanent use.

SUMMARY

In this chapter we have discussed the principles of thermal behaviour relating to the building fabric in the context of an SBEM assessment. We have also described common forms of building construction used in commercial buildings. An overview of the Building Regulations has also been made. In the next chapter we begin to consider how buildings are heated, cooled and otherwise 'conditioned'.

Energy and Construction Core Knowledge and Recognition

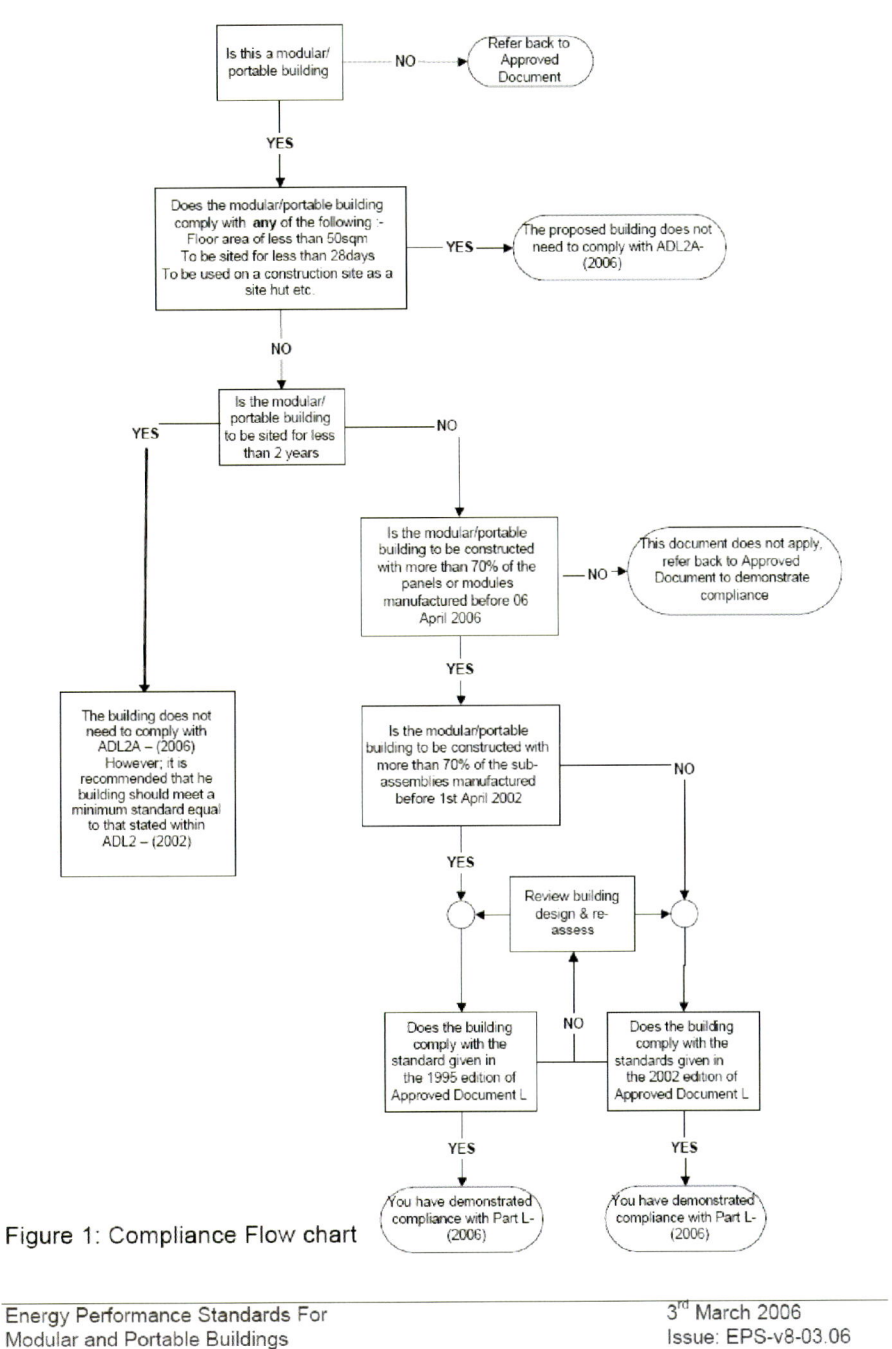

Figure 4.18: MPBA flowchart (Reproduced with the kind permission of the Modular and Portable Building Association)

5 Heating and Hot Water Services Core Knowledge and Recognition

INTRODUCTION

In the next two chapters we aim to provide the core knowledge required to undertake assessment of building heating, ventilation and air conditioning (HVAC) systems. This chapter focuses on heating and hot water services. Ventilation and cooling systems are dealt with in Chapter 6. In both chapters we try to give you enough information and practical guidance for you to:

- understand the basic operating principles of commonly occurring HVAC systems;
- identify key system components and distinguish between systems; and
- collect the data you need to carry out SBEM calculations.

In the introductory part of this chapter we deal with subjects that are of relevance whether your interest is in assessing buildings with simple heating systems or complex air conditioning systems:

- the objectives of HVAC assessment;
- sources of information useful in HVAC assessment;
- the organisation of HVAC systems in buildings;
- system efficiencies;
- fuels and electrical power factor;
- dealing with HVAC systems not available in SBEM.

The major topics dealt with later in this chapter are:

- types of heating system;
- heat sources; and
- hot water systems.

We aim to provide core knowledge concerning the assessment process and data preparation and so this should be of interest to those working at all levels and with a variety of approved software. We have not deliberately separated material according to NOS Level 3 and Level 4.

The heating technologies we discuss in this chapter are confined to those that can be represented in SBEM calculations and are relevant to those qualified to all levels. Those qualified to NOS Level 3 are able to assess properties with boiler installations up to a total capacity of 100 kW and cooling systems up to 12 kW (your Accreditation Scheme should be able to advise on the interpretation of these particular capacity limits). This chapter deals with boiler systems of this size range, which are likely to be classed as low temperature hot water (LTHW) systems; but also larger medium and high temperature (MTHW and HTHW) boiler systems. All cooling systems are discussed in Chapter 6 along with renewable energy systems.

The essential task

In many ways HVAC is at the heart of the EPC calculation process and so the importance of proper inspection and assessment of buildings' HVAC systems cannot be overstated. The complexity and variety of HVAC systems found in commercial buildings is one of the things that make the assessment process very different, indeed so much more difficult, to that of domestic buildings.

Building HVAC systems – with their interconnected room terminals, pipes, ducts and central equipment – are often well concealed if not invisible to the casual occupant. Assessing an existing building for EPC purposes requires wider appreciation of any HVAC installation; beyond what can normally be seen. Fortunately this does not mean having to inspect every pipe and duct. The essential task is to establish the logical connections between each room and the system it is served by, the heating and cooling sources being used and their fuel. You must build up a representation of the building and the HVAC systems that defines how rooms, systems and heating/cooling sources are logically related. This task can be summarised as follows:

- each room must be identified with a heating and/or cooling *system type*;
- each system must be associated with a heating or cooling *source*; and
- each heating or cooling source must be associated with a *fuel type*.

In rooms with local room heaters, for example, there should be no need to look beyond the room to determine this information. In larger buildings with centralised HVAC systems more steps in the inspection process are required before all these associations can be made. In general, if we drew the logical links between rooms, systems and heating and cooling sources we find something of a tree structure. This is illustrated in Figure 5.1.

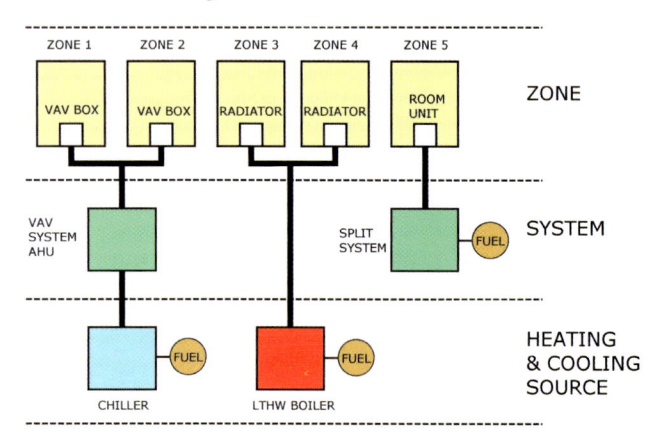

Figure 5.1: The elements of HVAC systems and their interconnections

This 'picture' is not immediately obvious except in the very smallest buildings but requires inspection of a number and variety of spaces before the correct links can be established. There is not a 'correct' way to go about this as it is possible to work from top-down or bottom-up to find the same information. Nevertheless, every room, system, source and fuel has to be associated to complete the assessment.

Sources of HVAC information

Arriving at a conclusion about what type of system is being used to heat and cool a particular room and establishing what heating/cooling source it is associated with very often requires some detective work.

Identification of the type of system is complicated by factors such as:

- similar types of room air outlet are used by many different types of air conditioning system;
- room terminal equipment is often concealed by suspended ceiling tiles;
- distribution ducts are often hidden in risers and ceiling voids; and
- where there are multiple air handling units, which zone they are associated with is not always obvious from visual inspection.

Very often a number of pieces of evidence will need to be collected from a number of rooms before certain possibilities can be rejected and firm conclusions can be drawn. Sufficient core knowledge and some logical deduction should, however, enable you to reach the correct conclusion.

You will also need to develop some skill in dealing with a wide range of types and sources of information. These include:

- contractors' drawings;
- operating and maintenance manuals;
- building log books;
- schematic drawings;
- manufacturers product information and labelling; and
- facility management staff.

Like a good detective, it sometimes pays to be sceptical about the evidence you find. It also pays to seek corroborating evidence. For example, contractors' drawings can be very helpful but it is also possible that the building has been altered since construction. Although it is useful to collect as much information about the building from the client before your visit and it is invaluable to have a member of staff accompanying you on the inspection, it is sometimes surprising how little clients and staff know about their building.

Drawings of building HVAC systems are rather different to architectural drawings and you may have to develop some additional skills to make good use of them. Schematic drawings can be particularly valuable and they are often clearly displayed in boiler rooms and other plant rooms. To be able to make use of them does require some understanding of the common symbols and abbreviations used. For this reason we have included Appendix E on this topic.

HVAC system organisation

It is helpful for an assessor to understand how HVAC systems are organised and distributed in buildings. Understanding how systems differ in their organisation helps when deciding how rooms are connected to the building's systems and how these systems are connected to their heating and cooling sources. We can identify broadly three ways in which HVAC systems are organised:

- centralised systems;
- distributed (or partially centralised) systems; and
- local systems.

Large buildings and building complexes usually have centrally located heating and cooling sources that are shared with, and distribute energy to, the zone HVAC systems in other parts of the building. In this arrangement hot and chilled water is commonly generated at a central location and distributed to HVAC equipment throughout the building. Air is, in turn, distributed to each room via duct systems.

Centralised systems can be thought of as being a hierarchy of three types of equipment:

(i) central plant;
(iii) air handling equipment; and
(iii) room terminal equipment.

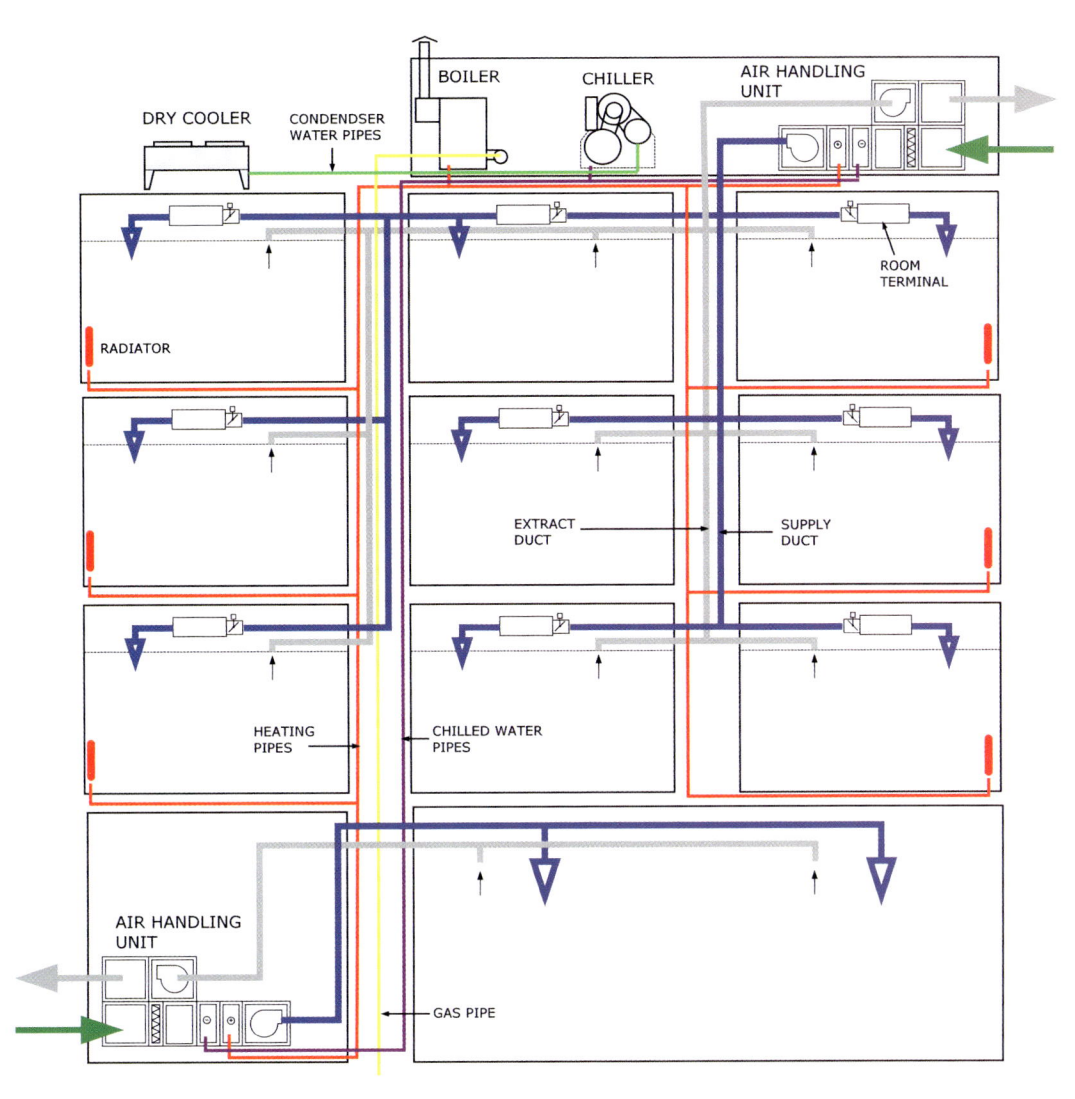

Figure 5.2: A diagram of a multi-storey building with a centralised air conditioning system

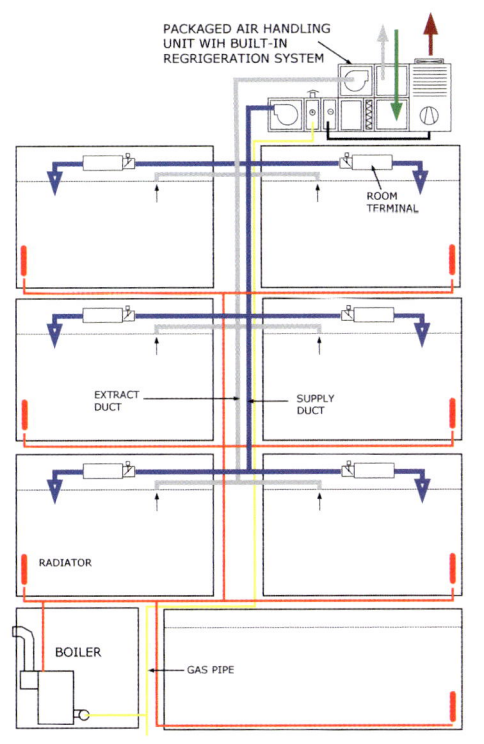

Figure 5.3: A diagram of a multi-storey building with a decentralised air conditioning system

In other words, the heating and cooling plant is centralised but the air handling equipment is distributed about the building. This form of system organisation is illustrated in Figure 5.2.

In distributed systems (sometimes called partially centralised systems) sources of heating and cooling are combined with each air handling unit. This usually means some refrigeration equipment and either gas or electric heating equipment – often on the roof – is built into the air handling unit or located immediately adjacent to it. In this arrangement each part of the building has 'packaged' air handling equipment often on the roof with its own air distribution ducts to individual rooms. This form of system organisation is illustrated in Figure 5.3.

There are a wide variety of types of heating and cooling equipment that can be thought of as local. This includes room heaters but also local cooling systems like split systems and unitary (packaged) heat pumps. This means that there is no connection to systems in other rooms.

In the case of local heating and cooling systems inspection of the room and its heater or cooling unit will enable you to decide the type of system and also the fuel. In distributed systems it will be necessary to

53

inspect each room but then also inspect the air handling equipment before deciding the system type and fuel source. In centralised systems it may be necessary to inspect the rooms and air handling plant to decide what type of system is being used. It will however be necessary to go one step further and find the central plant before deciding on the fuel type.

HVAC system efficiency

The relationship between the heating and cooling energy required in each zone, and that consumed by the HVAC systems can be defined in terms of seasonal energy efficiencies. In general, overall HVAC system efficiency depends principally on:

- heating or cooling system type;
- pipe and duct thermal losses;
- duct leakage;
- the effectiveness of the system controls;
- the effectiveness of monitoring measures;
- heating or cooling generator seasonal efficiency; and
- fan and pump energy demands.

These are taken account of in the SBEM according to your choice of system type and the heating or cooling generator seasonal efficiencies but also a number of other factors dependent on the information you provide.

Equipment seasonal efficiency

The efficiency data for the cooling generation equipment that you can directly specify in SBEM is the 'seasonal energy efficiency ratio' (SEER) and the nominal 'energy efficiency ratio' (EER). The EER is the nominal efficiency ratio value when the chiller, for example, is running at full load. The SEER is a weighted average of the efficiencies at different load conditions and will be a lower value than the EER. The *Non-Domestic Heating, Cooling and Ventilation Compliance Guide* (NDHCVCG) gives some rules for calculating the SEER where there are multiple chillers and different load profiles.

You should note that SEER and EER in some manufacturer's literature are defined rather differently and have units rather than being ratios. Values will be much higher (3.41 times larger) than the data used in the SBEM. You must be very careful when you obtain manufacturer's information in this regard, as otherwise you could enter fundamentally incorrect data.

The efficiency data for the heat generation equipment that you can directly enter is the 'effective heat generating seasonal efficiency'. This is well defined in the NDHCVCG. Heat generator efficiency is simply the ratio of useful heat output to heat input. The heat generating seasonal efficiency represents the average efficiency over the operating season given the fact that efficiency may be lower when equipment such as boilers do not operate at full capacity.

The NDHCVCG states minimum efficiencies expected for various equipment to comply with the 2006 Building Regulations Part L. It also states a range of efficiency credits that can be taken into account according to equipment having certain features (improved control features for example). Each credit is a percentage increment that can be added to the nominal efficiency. These credits can be taken account of in defining the efficiency of heat generating equipment in existing buildings in SBEM EPC calculations just as it can for new buildings (provided you have evidence of the relevant equipment features).

When you take any heat generating efficiency credits into account you can find the 'effective heat generating seasonal efficiency'. In other words:

**effective heat generating seasonal efficiency =
heat generator seasonal efficiency +
heating efficiency credits**

This is the value you enter into your SBEM software when you define a heating source. Your accreditation scheme may have particular guidance as to when to take efficiency credits into account in existing buildings. Unless you have clear evidence that the equipment does have the features associated with a particular credit you should not increment the seasonal efficiency.

SBEM software provides default values for EER and SEER in the case of cooling sources and effective heat generating seasonal efficiency in the case of heating sources. These values have been derived so that they represent the performance of typical systems in the UK commercial building stock. These values do not represent the minimum standards in the 2006 Building Regulations Part L2 as set out in the NDHCVCG and so when assessing buildings built or refurbished under these regulations you may reasonably override the defaults accordingly.

The bulk of the energy used to provide heating, cooling and hot water is calculated according to the energy consumed directly by the heating and cooling sources. However, other energy demands should rightly be associated with building HVAC systems. For example, many systems use pumps, fans or both and the energy used to run these can be significant. The number of pumps and fans varies between one system type and another. This energy is always electrical energy and is known as 'auxiliary energy'. You do not need to take this energy into account when providing efficiency data such as SEER values as auxiliary energy is calculated automatically. (We will discuss other parameters that do affect auxiliary energy later.)

It is acknowledged that buildings can operate with better annual efficiency if energy is carefully monitored and compared with energy targets. Such targets can be intelligently set from analysis of previous energy consumption data. Energy managers can use this information to actively manage building operation and diagnose problems. Energy meter systems and specialist analysis software can be used to do this in an automatic and rigorous manner. The benefits of doing this 'targeting and monitoring' can be acknowledged in an energy assessment and you can specify such a system is in use in your approved software. Such energy monitoring hardware and management systems may be used in larger organisations where a particular member of staff has responsibility for energy management. You should be able to simply verify if this is the case with your client.

The ECA scheme

One important distinction that is made regarding heating and cooling source efficiency is according to whether the equipment qualifies for enhanced capital allowances (ECAs). It can be worthwhile checking if this is the case as the minimum efficiencies that can be used in the SBEM calculation can be noticeably higher. For example, the default EER for a water cooled chiller is 2.5 but rises to 4.12 if it qualifies for ECAs. Full details of the ECA scheme can be found at www.eca.gov.uk/etl.

How do you know if equipment qualifies? First, the equipment must fall within certain categories. Those of relevance are:

- air-to-air energy recovery;
- automatic monitoring and targeting (AMT);
- boiler equipment;
- heat pumps for space heating;
- heating ventilation and air conditioning zone controls;
- lighting;
- refrigeration equipment (e.g. chillers);
- solar thermal systems;
- warm air and radiant heaters; and
- combined heat and power (CHP).

Most products appear on the schemes energy technology product list (ETPL) and you can search for these using the online database. CHP equipment is not on this list but could still qualify (see www.chpqa.com for more information). The other and most convenient way to see if equipment is listed is to see if it carries the badge shown in Figure 5.4.

Figure 5.4: Label found on equipment on the ECA scheme energy technology product list (Reproduced with the kind permission of Carbon Trust)

Fuels and emission factors

We have already noted that the relationship between heating, cooling and hot water demands and energy is dependent on the efficiency of the HVAC systems. Ultimately the predicted energy consumption has to be expressed as a CO_2 emission rate. The relationship between energy consumption and carbon emission rates is entirely, and very simply, determined by the fuel type you assign and its emission factor.

The fuel CO_2 emission factors are a simple ratio that indicates how many kilograms of CO_2 are released for every kilowatt hour (kWh) of energy consumed. These ratios vary considerably and so the type of fuel you assign to a system, and its heating source or cooling source, can have a significant effect on the asset rating. The agreed emission factors are reproduced from ADL2A 2006 in Table 5.1. You will see that that grid supplied electricity has an emission factor noticeably greater than all the other fuels. Of the fossil fuels that are defined, natural gas has the lowest emission factor and you will see that bio fuels have emission factors approximately one-tenth that of fossil fuels. This will give you an appreciation of the significance of choosing whether a system uses gas as opposed to electricity (e.g. as the fuel used to heat hot water). Grid supplied electricity is always assumed in SBEM calculations to be the fuel associated with auxiliary energy use (e.g. fans and pumps).

Table 5.1: Fuels and their CO_2 emission factors (data from ADL2A 2006, Table 2). Identifying storage facilities are also indicated

Fuel	CO_2 Emission Factor ($kgCO_2$/kWh)	Fuel delivery or storage type
Grid supplied electricity	0.422	Grid connection and building power distribution system
Natural gas	0.194	Grid connection and pipe distribution system – pipes colour coded yellow
Oil	0.265	Liquid tank (possibly buried). Fuel pipe connections colour coded brown
LPG	0.234	High pressure storage tank
Biogas	0.025	
Coal	0.291	Bunker, hopper or silo
Anthracite	0.317	
Smokeless fuel (inc. coke)	0.392	
Dual fuel (mineral + wood)	0.187	
Biomass	0.025	
Waste heat	0.018	From industrial process

The fuels associated with cooling systems are almost always grid electricity. The choice of fuel in heat sources is naturally very wide ranging.

When you inspect a property some of the first clues to identification of fuel types include fuel storage facilities, solid fuel feeds to boilers and fuel supply pipe connections to equipment. We have identified some of these features in the last column of Table 5.1. Features of particular equipment that will help you identify fuel type will be dealt with later as different system types and equipment are discussed. Fuels other than natural gas or grid supplied electricity require on-site storage of some type and this can be one important thing that can be determined from walking around the site at the start of the inspection. Very often a building owner or occupier will be well aware of their main heating and hot water fuel type and so you may be able to collect this information from your client prior to any visit. It may be helpful to view fuel delivery information in the case of mineral fuels if it is not clear exactly which type is being delivered.

Power factor

One of the parameters that effect the calculation of carbon emissions from all the building's electrical systems is the building's power factor. Ideally, electrical system current should alternate exactly in phase with the voltage. However, in most installations this is not quite true and the current generally lags behind the voltage. This happens because of the reactive properties of electric induction motors and the disturbances to the power supply due to things like computer power supplies and some types of fluorescent lights. The results of this are that additional current is drawn from the grid, efficiency is reduced and so overall carbon emission rates are higher.

Low power factors are often accepted in domestic and small commercial buildings. In large buildings and particularly industrial premises – where there are often many large motors – distortion in the power distribution system can be more of a problem and so energy companies can impose additional charges if the measured power factor is poor. Power factor correction equipment (essentially collections of large capacitors) can be used to reduce this problem and avoid such charges. It is likely that only buildings with this sort of equipment will achieve power factors better than 0.9 (values less than 0.9 are the default assumption in SBEM). Power factor correction equipment such as that illustrated in Figure 5.5 is usually found near the building's incoming power supply. You may be able to get information about the site power factor if your client can show you billing information.

Systems not represented in SBEM

A large number of HVAC system types and heating and cooling sources can be explicitly represented in SBEM

Figure 5.5: Example of power factor correction equipment. More modern equipment may have a digital display

calculations. The list of systems is not exhaustive and you may find situations where systems do not clearly fall into a recognised category. In many cases it is possible to represent HVAC systems that are similar to, but not exactly the same as, those listed in SBEM software by means of a work-around. This will mean selecting the system type most similar but changing the fuel type, efficiency and other variables to provide a reasonable calculation of the carbon emission rate (this is what matters in the end). An example of a work-around like this is shown in our Level 4 building case study in Appendix C. Some work-arounds are also described in the *iSBEM User Guide*.

If you are not clear of the most appropriate approximation to the system in question you should seek the advice of your accreditation scheme. In every case you should document what you have done and be prepared to justify what you have chosen to do.

HEATING SYSTEMS

In many buildings the energy required for space heating is the largest source of demand and probably the most significant source of carbon emissions. It is vitally important that you are able to identify which type of heating system is used in each zone and which heat source is associated with that system. Identifying the heat source associated with each zone – be it a local source or elsewhere in the building – generally enables the fuel type to be identified.

The various heating technologies – although their characteristics seem to be very different – can be found in nearly all building types. It is important, then, that you are able to deal with a very wide range of heating systems whether qualified to National Occupational Standard Level 3, 4 or 5. Similarly, every assessor should be able to identify all types of hot water generation system as, although scale may vary, the different types of system can be associated with any building type and size.

Heating systems differ in how heat is delivered to each room or zone and can be categorised as:

- central heating using water or air:
  - radiators,
  - convectors,
  - underfloor heating,
  - warm air heating;
- high temperature radiant heating:
  - flued radiant heating,
  - unflued radiant heating,
  - multi-flue radiant heating;
- forced air heating:
  - flued convection heating,
  - unflued convection heating;
- individual room heaters.

Approved software tools can be used to assess a number of heating systems in each category and we will discuss these in the following sections.

RADIATOR AND CONVECTOR HEATING

Heating systems that use radiator panels to deliver heat to each room are relatively common and share many characteristics with domestic heating systems. Radiators emit heat by a combination of radiation and convection heat transfer processes. They have sufficient surface area to emit a large part of their heat by long-wave radiation (infrared heat) but also generate plumes of warm air that convect heat into the room. Some types of radiator have a significant area of concealed fins and produce as much heat (or most of their heat) by convection as they do by radiation.

A wider variety of radiator types are in use in commercial buildings than in domestic buildings. A number of types of radiator are shown in Figure 5.6. These range from cast iron column to extruded aluminium designs. There can be good reasons to conceal radiators in some rooms and this clearly makes

Figure 5.6.1: A common steel panel radiator

Figure 5.6.2: A cast iron column radiator

Figure 5.6.3: A modern panel radiator

Figure 5.6: Different types of radiator

Figure 5.6.4: This radiator has an additional front panel to reduce the surface temperature and grilles to prevent fingers, etc. getting trapped

identification a little harder. Radiators may be concealed for aesthetic reasons in historic buildings, for example, or to protect occupants from burning themselves against the radiator in changing rooms, schools and medical facilities. In such cases closer inspection is worthwhile as it is also possible that the heat emitter is a convector rather than a radiator.

Central heating systems that use hot water to distribute heat to each zone may also use convector heat emitters. These types of heat emitter consist, essentially, of long horizontal pipes with a large number of closely spaced vertical fins. Heat is emitted by buoyant air flow around the pipes and through the fins. Convector heat emitters can be found in a variety of forms depending on how the finned element is enclosed. The simplest forms have the fins quite visible and are simply fixed near the floor at the room perimeter (Figure 5.7.1). Trench heating is a form of convection heating system that has the finned element (or pair of elements) concealed below a grille in the floor as illustrated in Figure 5.7.2.

In non-domestic buildings you will often find building services concealed in purpose made enclosures or casings that form a continuous feature below window sills. An example is shown in Figure 5.7.3. You will need to take some care to inspect this type of feature as a variety of air conditioning terminals can also be concealed in this way. If an air conditioning terminal is installed under a window and enclosed in this way you will find a linear grille to discharge the air up the window. If it is simply a convective heater that is concealed there may be a similar linear grille but often something simpler or just a slot facing into the room. Convective heating like this is often called 'perimeter heating' or 'sill line' heating (Figure 5.7.4).

Figure 5.7.1: A simple low level convector heater

Figure 5.7.2: A trench heating convector

Figure 5.7.3: This type of sill enclosure, with an upward facing grille, can conceal a variety of heating equipment or air conditioning terminals. In this case there is a convector heater inside

Figure 5.7.4: Perimeter heating installed in an office building. The panels hide all the pipes and valves as well as the convection heating elements (Reproduced with the kind permission of Gilberts (Blackpool) Ltd)

Figure 5.7: Different types of convection heating

Central heating systems that use either convector or radiator heat emitters can be used in zones with natural ventilation or a variety of mechanical ventilation and cooling systems. Examples include rooms that have central variable air volume air-conditioning (discussed in Chapter 6) or where small room air conditioners have been retrofit. Although rooms with radiators and opening windows are an extremely common heating and ventilation design, it would be a risk to assume that because a room has a radiator it has no mechanical ventilation system or cooling system.

UNDERFLOOR HEATING

Distributing heat to rooms by embedding either pipes or electrical elements in floors is another technology that is used in both domestic and commercial buildings. These types of heating system are well suited to spaces with moderate heat losses and open spaces in which it is difficult to position radiators – atria and foyers for example. Such systems provide thermal comfort to occupants by radiating heat upwards and forming a warm zone relatively close to the floor – such systems can be effective in tall spaces like atria.

The commonest form of underfloor heating consists of serpentine coils of plastic pipes fixed to insulating panels and covered by the floor screed layer. Although underfloor heating pipes are usually embedded in a floor screed layer it is also possible to incorporate similar plastic pipes into wood sub-flooring panels. Consequently, you should not assume because the floor is of wood construction that there cannot be any underfloor heating. You should also be aware that underfloor heating can be retrofit using these systems.

Underfloor heating systems are intrinsically difficult to identify in existing buildings. One reason to investigate the possibility of an underfloor heating system being used in a particular room might be the absence of other visible forms of heat emitter (bearing in mind that many air systems may provide a heating function). All warm water underfloor heating systems need a series of pipes connected in parallel to supply sufficient heat to anything but the very smallest room. These multiple plastic pipes have to be connected into the main warm water circulation system at some point. This is commonly done at a pipe and valve arrangement called a 'manifold' not far from the edge of the floor – as illustrated in Figure 5.8.1.

These manifolds are very often hidden behind metal or purpose-made panels in the walls of the room being heated or in an adjacent room or cupboard. The manifold and its enclosure (Figure 5.8.2) is probably the most visible indicator of the presence of an underfloor heating system.

Underfloor heating systems should be traceable to a low temperature heating source such as a boiler. Where a boiler provides heat to both radiators and/or convectors it is necessary for the water distributed to the underfloor heating to be on a separate lower

Figure 5.8.1: Underfloor heating pipes and a distribution manifold at the edge of the room shown during installation (Reproduced with the kind permission of Viessmann Ltd)

Figure 5.8.2: The pipe and valve manifolds of an underfloor heating system can be concealed in several ways but this type of steel cabinet is common (Reproduced with the kind permission of Viessmann Ltd)

Figure 5.8: Underfloor heating

temperature circuit. This requires a separate circuit with its own pump and control valve and you might be able to see this next to the boiler. You may also find a buffer tank.

Central heating using air

Buildings can be heated by distributing warm air as well as by distributing hot water. The types of central heating system that use air to distribute heat vary depending on the complexity and size of the building and whether the intention is also to provide a fresh air supply. The basic configurations can be summarised as follows.

Systems with only a supply duct distribution system

In this form of system, warm air is distributed from the heating equipment through a duct system to diffuser outlets in each room. Where multiple rooms are heated air may leave each room via grilles through the doors or walls to be drawn back towards the heating equipment. It is not necessary to have a extract duct system and so only one fan is needed. This arrangement (Figure 5.9.1) is similar to a domestic warm air central heating system and is only suitable for a small number of rooms.

Large spaces in industrial and retail facilities can be heated using warm air distribution systems. In these applications warm air may be distributed from a single heater unit throughout the space via a supply duct system. Air may then be drawn back into the heater at a single point, for example near the roof at the back of a supermarket.

Systems with both a supply and extract duct distribution system

In the second form of system (Figure 5.9.2) warm air is distributed from the heating equipment through a duct system to diffuser outlets in each room and air is extracted from the room through a second system of grilles and ducts. Larger numbers of rooms, possibly on more than one storey of the building, may be heated using this sort of central heating. The main heating equipment may be situated inside large single spaces but also on the roof or on the ground outside.

The fan and heat exchanger equipment in larger warm air heating systems may not be very different to central air conditioning systems. For this reason, even if you don't intend to assess buildings with central air conditioning equipment (i.e. Level 4 or 5), it is still worthwhile developing your knowledge of larger ventilation equipment such as air handling units (see Chapter 6).

The essential components of a warm air central heating system are a fan, a heat exchanger and most likely a set of filters. (Various types of heat exchanger that can be used to heat the air are discussed in the next chapter.) These components can be packaged in a wide range of equipment types or simply assembled as components of the duct system. A wide range of heating sources can be used to heat the air. These include LTHW, heat pumps, electric resistance heating as well as gas and oil.

In all SBEM calculations it is assumed that a sufficient quantity of fresh air is supplied to every zone (according to the activity specified) and you can specify this to be either natural or mechanical ventilation. You should note, at this point, that 'Central heating using air' is the one type of heating system that is always assumed to provide ventilation by mechanical means. In other words, it is assumed that fresh air is drawn into the system and distributed via the duct system along with the heat required to offset the zone heat losses. In practice you will find that fresh air for ventilation may be introduced into a building with air central heating in a number of ways. As fresh air is always accounted for in the SBEM calculation you do not need to be overly concerned with resolving the way in which fresh air is introduced to the system.

Probably the most obvious indications of the presence of a warm air heating system are air supply diffusers and extract grilles. It is possible that the air distribution system is just used for ventilation (often

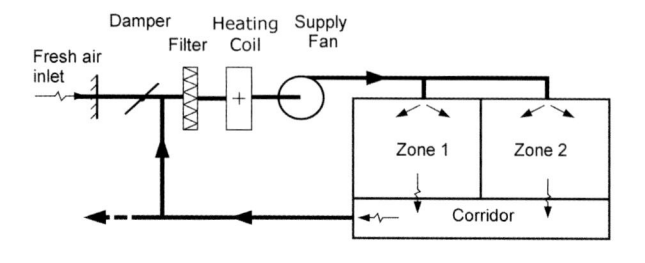

Figure 5.9.1: Schematic of a central heating system using air with one fan

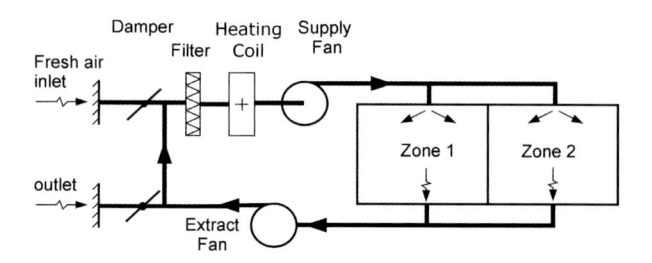

Figure 5.9.2: Schematic of a central heating system using air with two fans and a supply and extract duct system

Figure 5.9: Central heating systems using air

termed 'zone supply/extract' in SBEM documents) or that some form of central air conditioning is present. If there are other heat emitters in the zone (e.g. radiators) then it should be clear that the main heating system is not one using air. If there are no other heat emitters then you must determine if the air distribution system is for ventilation only or air conditioning.

Radiant heating

There are a number of forms of practical heating device that rely on long-wave radiation to emit nearly all of their heat. Unlike hot water radiators that only emit a smaller proportion of their heat by longwave radiation (despite their name) true radiant heating systems operate at high temperature and emit a very high proportion of their heat by radiation. Radiant heating systems found in common buildings are of three types:

- high temperature tube radiant heaters;
- incandescent heaters;
- hot water panels.

High temperature tube radiant heaters (and a number of other heater types) can be classified as 'flued' or 'unflued' depending on whether they discharge their hot gases into the space being heated. Heaters that do not have a flue are, in theory, more efficient as 100% of the heat generated by burning the fuel enters the room. It is consequently important you identify whether such a heater has a flue when carrying out an EPC assessment.

High temperature tube radiant heating

High temperature tube radiant heaters consist of a burner, at which the fuel and air are mixed and burnt. The burner is fixed to the end of a steel tube through which the hot gases pass (see Figure 5.10.1). The outside surface of the blackened tube may have a temperature of a few hundred degrees and so will emit intense longwave radiation. At the end of the tube the hot gases either leave the tube and enter the space being heated (unflued), or are drawn through a fan and discharged out of the building via a simple flue. Radiant heaters of this type are suspended over the occupied space and have their radiant output directed downwards by a mirrored reflector. The tube may be in a 'U' configuration as well as a straight pipe. Such heaters are invariably gas fired.

As radiant tube heaters use a fan at the end of the tube to help expel the hot gas and this uses some electrical energy, some heaters have been developed that use a single fan to draw the hot gases from a number of tubes and use a combined flue (Figure 5.10.2). These are termed 'multi-burner' radiant heaters.

Incandescent radiant heating

Gas heating systems that use incandescent ceramic elements and which have an open flame can be used to heat larger spaces such as workshops and warehouses in non-domestic buildings. This type of device (sometimes called a 'plaque heater') is designed to be mounted at a high position and is tilted towards the occupied space (as in Figure 5.10.3).

Figure 5.10.1: A gas-fired radiant heater. The flue can be seen at the far end

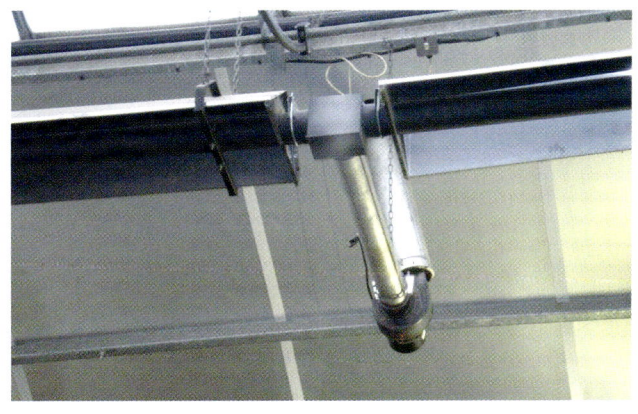

Figure 5.10.2: A multi-burner radiant heater showing the common flue with its fan

Figure 5.10.3: An example of a gas-fired incandescent radiant heater

Figure 5.10: Radiant heating

Figure 5.10.4: A low temperature radiant heating panel mounted in a false ceiling

Hot water radiant heating

Radiant heating systems can also take the form of metal panels that have rows of water passageways incorporated into or attached to the panel. Large panels of this type can be found in tall spaces such as sports halls where they are typically suspended in rows close to the roof. These can be used with LTHW heat sources or higher temperature sources (MTHW and HTHW) where this is available and as long as the panels are mounted where occupants cannot burn themselves on them.

Low temperature radiant heating panels can also be incorporated into suspended ceilings. In this case the panel appears much like a metal ceiling tile and has rows of small diameter pipes attached to the rear and hidden in the ceiling void. Such panels (Figure 5.10.4) operate at water and surface temperatures similar to those of convectional radiators. It is, in fact, suggested (see *iSBEM User Guide*) that low temperature hot water radiant systems are represented as radiator systems rather than 'flued or unflued radiant heating' systems in SBEM calculations. We believe this is a reasonable approach.

Forced convection heating

Natural convection heaters are well suited to distributing heat around the perimeter of buildings to offset heat loss through the fabric. Where it is necessary to deliver more intense heat, forced convection heating that combines a heat exchanger along with a fan may be preferred. Heating equipment that uses the principle of forced convection is found in a wide variety of forms. Such equipment has the advantage of being cost effective and adaptable but suffers the disadvantage of requiring auxiliary electrical energy to drive the fan.

Large tall spaces such as warehouses, sports halls and supermarkets have rather different heating needs than occupied spaces such as offices but can be well suited to equipment that uses forced convection air heating principles. Floor space is valuable in these types of building and so a large variety of equipment can be found that is intended to be mounted near roof level. Equipment for industrial applications such as these is designed to deliver powerful jets of hot air from significant height and over considerable distances.

A 'unit heater' is a common form of forced convection heater that consists of, basically, a propeller fan at the rear that blows air directly through a heat exchanger and out through a nozzle or vanes at the front. This type of heater can be found with a wide variety of heat exchanger and fuel types. Common types are those using hot water heating sources but others can be found that use electrical resistance heating elements (Figure 5.11.1).

In recent decades forced convection heaters have been developed that, rather than being connected to a larger hot water and boiler system, incorporate their own gas or sometimes oil burner. Again, you will usually find that such heaters are directly connected to a flue (sometimes there is second flue pipe for the burner air supply) but may find some that are unflued. You may also find some equipment that uses a condensing system and has a small drain pipe. A variety of convection heating equipment is on the ECA product list.

There are now a variety of gas-fired forced convection heaters that are slightly more sophisticated than 'unit heater' style equipment and may, for example, incorporate multiple nozzles, allow connection of short duct systems or allow introduction and mixing of fresh air from outside the building. Although most equipment is designed to be positioned near the roof or ceiling with a short flue (see Figure 5.11.2), floor standing or external roof mounted equipment can be found. A picture of a floor standing heater is shown in Figure 5.11.3.

If you decide that a zone heating system is a forced convection heating system you must also think about the auxiliary energy used by the fans. In every case you are required to provide this data (or accept a default) in the form of kWh electrical power per kWh of heat. To arrive at an accurate value you would have to consult manufacturers' data.

If a forced convection heater is connected to a hot water heat source it becomes necessary to employ a work-around solution to get a good approximation to the system efficiency and carbon emissions. A work-around method is required for this situation as 'LTHW boiler' is not currently one of the available SBEM heat source options for forced convection heaters or fanned room heaters. This means choosing the appropriate fuel and then adjusting the seasonal efficiency as we discussed earlier in this chapter (see also Appendix C).

Local room heaters

There are situations where it is not feasible or economical to heat a particular room using a centralised system or its needs are very different to the other parts of a building. For example, a factory that has radiant heating but needs another form of heating for a small office or toilet. This can also be the case where a building has been extended or undergone some form of renovation or change of use. In these cases you may find some form of self-contained heater – very often the same sort of equipment as you may find in a domestic property.

Some heating equipment that should be considered as local room heaters include:

- electric fires;
- gas fires;
- gas convector heaters;
- electric storage heaters; and
- solid fuel fires and stoves.

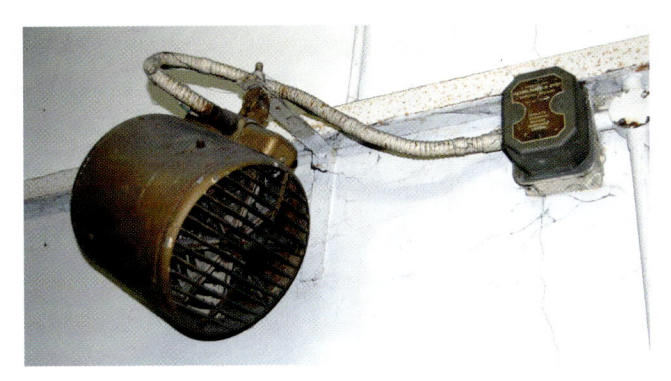

Heating and Hot Water Services Core Knowledge and Recognition

Figure 5.11.1: An example of a forced convection heater with an electric heating element

Figure 5.11.2: An example of a gas-fired forced convection unit heater

Figure 5.11.3: A floor-standing forced convection heater. Air is drawn in at the bottom and discharged through nozzles at the top

Figure 5.11: Forced convection heating

An example of a balanced flue gas convection heater is shown in Figure 5.12.1. Some types of equipment like this may include a fan to enhance heat output (e.g. some types of electric convection heater). The fan in such equipment is associated with auxiliary electrical energy and so fanned local room heaters have a separate category in SBEM calculations. Rooms with larger fanned gas or electric heaters (the limit suggested in the *iSBEM User Guide* is those above 10 kW) should be considered as having forced convection air heating.

Room heat efficiency can vary considerably from one type of room heater to another (this is demonstrated in the data used in domestic building SAP calculations). If you are uncertain about using the default efficiency and don't have documentary evidence about the equipment you should follow the guidance of your Accreditation Scheme Provider. It is reasonable, in our view, to use an efficiency of 100% if the equipment is some form of fanned or unfanned electric resistance heating (Figure 5.12.2).

Fan convection heaters for non-industrial applications such as offices, classrooms and corridors commonly consist of a slim cabinet with a long narrow grille through which warm air is distributed in a wide jet along the wall, ceiling or window surface. Heater units of this basic type can be adapted for wall mounting, ceiling mounting or fixing under a window sill. Heaters like this can also be incorporated into architectural casing below windows (Figure 5.7.3) or concealed within the ceiling void. Examples of this common form of heater are shown in Figures 5.12.3 and 5.12.4.

Fan convection heaters like this are often designed for hot water heating systems and an example is shown in Figure 5.12.5. As 'LTHW boiler' is not one of the available heat source options for room heaters the same sort of work-around has to be employed as we indicated above for forced convection heating.

One particular type of heater that can be classified as a forced convection type is the 'over-door' heater

Commercial Energy Assessor's Handbook

Figure 5.12.1: A wall mounted balanced flue gas room heater

Figure 5.12.2: An example of a panel electric resistance heater

Figure 5.12.3: A common fan convector heater

Figure 5.12.4: A LTHW fan convector heater recessed into a wall

Figure 5.12.5: A fanned convector room heater with LTHW heating source

Figure 5.12.6: An example of an electric over-door heater

Figure 5.12: Local room heaters

(sometimes known as an 'air curtain'). The strong jet produced by this type of heater is directed down from the top of the door and designed to counteract draughts when the door is open. This type of heater can be classed as an 'other room heater' in SBEM calculations. Very large over-door heater systems may be gas fired but most over-door heaters are connected to hot water systems or rely entirely on electrical power (Figure 5.12.6).

Destratification fans

Tall spaces in buildings have the tendency to collect warm air near the ceiling or roof due to natural buoyancy effects and this is detrimental to the efficiency of many forms of heating system. There are a few ways in which this potential stratification can be limited:

(1) incorporation of 'destratification fans';
(2) ensuring a forced convection or nozzle air delivery system produces jets strong enough to generally mix the air in the room; and
(3) open ceiling fans (sweep fans) that are of large diameter and rotate at low speed.

A destratification fan has a short casing that is suspended near the ceiling and draws air in from above and produces a jet of air directed towards the floor (it does not actually provide any heating itself). This basic arrangement is shown in Figure 5.13. Although destratification and other ceiling fans are associated with an electrical energy demand, their use may bring a net energy and carbon saving. You can include the benefits of destratification fans in SBEM calculations.

In each of the points (1), (2) and (3) above the flow of air delivered by the equipment has to be high enough to induce mixing of the room air. If this is the case (and some justification may be required) then the second and third situations can be regarded as equivalent to having destratification fans in SBEM calculations. (The *iSBEM User Guide* provides further information about the minimum required air flow rates.)

HEATING SOURCES

We noted earlier that an important aim in inspecting the building is to be able to associate zones with their heating and cooling systems. Heating systems (and the heating components of air conditioning systems) have, in turn, to be associated with a fuel type. In some cases, such as electrically powered heaters or equipment that incorporates gas or oil burners, once the zone heater type is determined it should become immediately evident what fuel type is applicable.

Many systems use a centralised heat distribution system and you will need to take further steps (perhaps literally going and finding the boiler room) before a zone can be associated with its heat source. You will have to discover the central heating source and determine the type of equipment, after which it is usually not difficult to also identify the fuel type.

The heat sources we shall consider in this section, and which can be explicitly represented in SBEM calculations, are:

- boilers of differing temperature ranges and fuel types;
- combined heat and power systems;
- district heating systems;
- heat pumps – both air and ground source systems.

Boiler heat sources

When we use the term boiler we essentially mean a piece of equipment that generates hot water for heating purposes by a process of combustion. Heating systems that use boilers can be classified according to their operating temperature ranges as follows:

- low temperature hot water (LTHW) systems: water temperatures below 90°C;
- medium temperature hot water (MTHW) systems: water temperatures between 90°C and 120°C;
- high temperature hot water (HTHW) systems: water temperatures above 120°C.

These classes of system correspondingly operate at different pressure ranges and so you may also see the terms low pressure hot water (LPHW), MPHW and HPHW on drawings and in other documents. The large majority of systems you will encounter are LTHW – domestic systems are also in this class. Although it is only necessary to identify the operating temperature range of a boiler and its fuel type for the purposes of SBEM calculations it is worth noting some generic features of different boiler types and system installations.

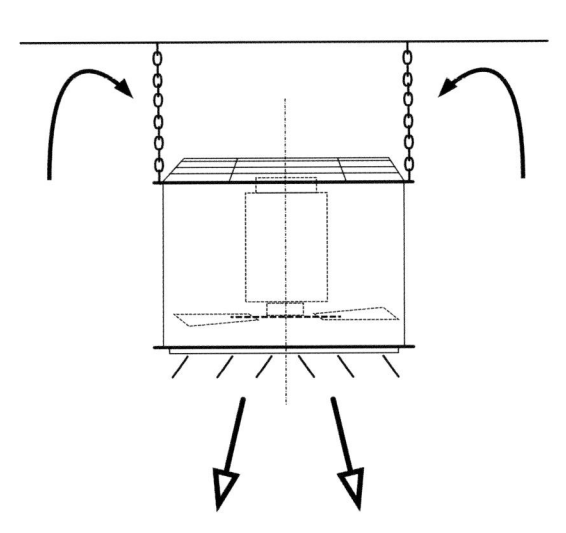

Figure 5.13: An illustration of a destratification fan

Commercial Energy Assessor's Handbook

Boiler flues

You may see boiler installations with a variety of flue arrangements. It is worthwhile inspecting the outside of the building as one of the first exercises in the site visit to identify the presence of heating boiler or hot water generation systems from the presence of flues. There are only a few types of flue type you are likely to see in assessing commercial buildings. These are:

(1) plain natural or forced draught flue;
(2) balanced flue;
(3) fan assisted balanced flue; and
(4) fan dilution systems.

The first three types we list can also be found in domestic boiler installations. It is only small commercial boilers that can be found with balanced flues or fan assisted balanced flue types. Balanced flues can only be used where the boiler can be positioned against a wall. They may only be visible from outside the building and may be suspected from the absence of any other visible flue connected to the boiler. This is illustrated in Figure 5.14.1 and 5.14.2.

Fan assisted balanced flues are similarly only fitted to smaller boilers positioned close to the outside wall (often wall-hung) and with a short horizontal flue pipe from the top of the boiler as shown in Figure 5.14.3.

When we refer to conventional flue stacks we mean plain flues whose height is used to ensure the adequate flow of the flue gases. The height of the flue may also be governed by the need to discharge the flue gases away from any windows or ventilation openings. Flues of this type are used with both small and large boilers. A large flue arrangement for a number of large boilers is shown in Figure 5.14.4.

One type of flue arrangement that can be found in some commercial buildings and that is not found in domestic installations is a flue fan dilution system. In fan dilution systems a fan is used to draw in air from outside the building and is connected to a horizontal duct or flue section to which the boilers are connected in turn. The fan introduces enough air to mix with the flue gases from all the boilers such that the mixture of gases discharged is cooled and the combustion products are diluted to a safe level. This arrangement allows boilers to be installed at basement or ground level in the building without the need for very tall flues. The flue gases can be discharged near ground level without risk to people nearby. The flue gases can be simply discharged through louvres and so are not easily seen from outside the building. The flue dilution system concept is illustrated in Figure 5.14.5 and an example shown in Figure 5.14.6.

Figure 5.14.1: These balanced flue boilers are examples of large domestic boilers used in a non-domestic application

Figure 5.14.2: The balanced flues of the boilers shown in Figure 5.14.1

Figure 5.14.3: Wall mounted condensing boilers with fan assisted balanced flues

Figure 5.14: Boiler flues

Heating and Hot Water Services Core Knowledge and Recognition

Figure 5.14.4: The flue stack of a large boiler installation

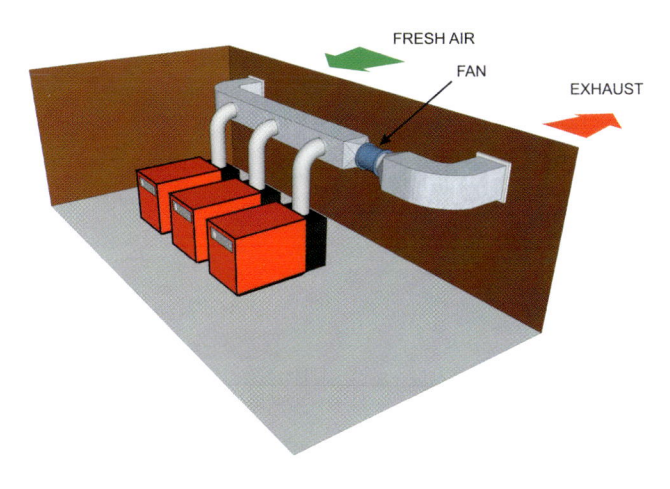

Figure 5.14.5: Diagram of a boiler installation with a fan dilution flue system

Figure 5.14.6: A boiler connected to a fan dilution system (the square duct above the boiler)

Gas and oil fuelled boilers

We have already noted that boilers can be classified according to operating temperature range. Gas (including LPG and biogas) and oil fuelled boilers can also be classified according to their burner type. By burner we mean the device used to control the flow of fuel and air into the boiler and control the combustion flames. The two basic types are atmospheric and forced draught.

In boilers of the atmospheric (or natural draught) type fuel is introduced into the bottom of a combustion chamber through a number of pipes that produce a large number of small flame jets. Air is drawn into the bottom of the boiler below the flames by natural buoyancy forces, so that as more fuel is burnt more air is also naturally drawn in. Boilers with this burner arrangement are invariably of the LTHW class. An atmospheric gas boiler installation is illustrated in Figure 5.15.1.

Forced draught burners are distinguished by incorporating a fan that controls the amount of air in the combustion process to create an intense single flame inside the boiler. The combustion gas pressure inside the boiler is consequently a little higher than atmospheric pressure and this helps the gases to be driven out of the boiler and into the flue. A forced draught burner typically incorporates a fan motor, scroll-shaped fan casing and fuel control valves as shown in Figure 5.15.2. A forced draught boiler arrangement can be concealed inside the casing of the boiler but can be seen attached at the front on larger types. You should be able to see the gas or oil fuel connection to the burner and often identify whether a fuel is gas or oil from the pipe or ID band colour (yellow for gas, brown for fuel oil – see Appendix D).

Boiler installations in commercial buildings of moderate size that have a heating capacity of hundreds of kW often use a modular boiler arrangement. This means a number of boilers of modest size are used side-by-side (Figure 5.15.3) and have their controls linked so that the number of boilers operating is adjusted to match the total heating demand.

A modern forced draught boiler design is shown in Figure 5.15.4. This design, in which the hot gases pass from the main combustion chamber, back towards the front and then through small tubes towards the flue at the rear (a so-called three-pass design) is shown diagrammatically in Figure 5.15.5. This type of boiler is more typical of large installations (probably more than 750 kW).

An example of a large boiler installation using this type of boiler is shown in Figure 5.15.6.

67

Commercial Energy Assessor's Handbook

Figure 5.15.1: A pair of atmospheric LTHW gas boilers

Figure 5.15.2: A close-up picture of a gas fired burner typical of large forced draught boilers (Reproduced with the kind permission of Viessmann Ltd)

Figure 5.15.3: An example of a modular boiler installation

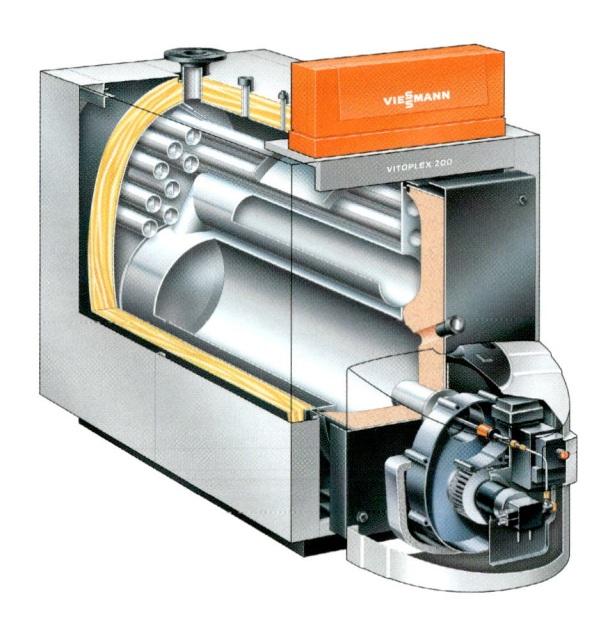

Figure 5.15.4: A diagram of a large steel boiler showing the burner and tube arrangement (Reproduced with the kind permission of Viessmann Ltd)

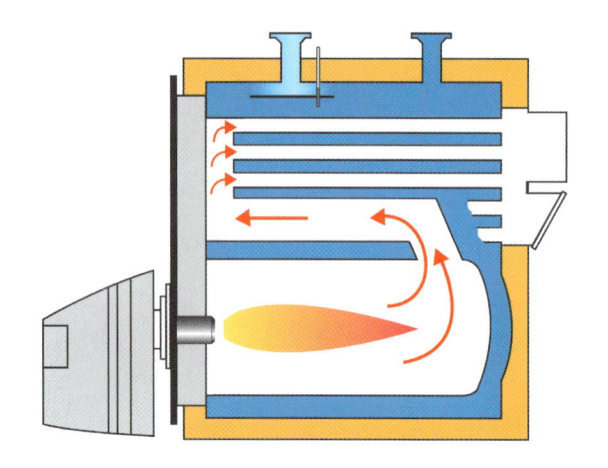

Figure 5.15.5: A diagram of a three-pass boiler design typical of large gas and oil fired boilers (Reproduced with the kind permission of Viessmann Ltd)

Figure 5.15.6: A large LTHW boiler installation

Figure 5.15: Gas and oil fuelled boilers

Mineral fuel and biomass boilers

The vast majority of boilers you will assess will be natural gas or oil fuelled but solid fuel boilers are in use. Biomass boilers have come into use relatively recently in the UK. One key way to identify boilers of these types is by discovering and inspecting any fuel storage facilities. You should also find fuel feed equipment and a means of disposing of ash. Smaller boilers (e.g. Figure 5.16.1) may have a small hopper next to the boiler that is manually topped up from the main fuel store but many boilers that use solid fuel or biomass fuels have an automatic feed arrangement (Figure 5.16.2) that is coupled to the main fuel store (Figure 5.16.3).

Where there is some doubt about the type of mineral fuel it may be worth asking to see fuel delivery notes or invoices. The output of boilers that use solid fuels cannot be varied so easily and rapidly as gas or oil fuelled boilers. Consequently you may find heating water buffer tanks form part of the installation. These may not appear much different to hot water storage tanks but should be distinguishable from the type of pipe connections.

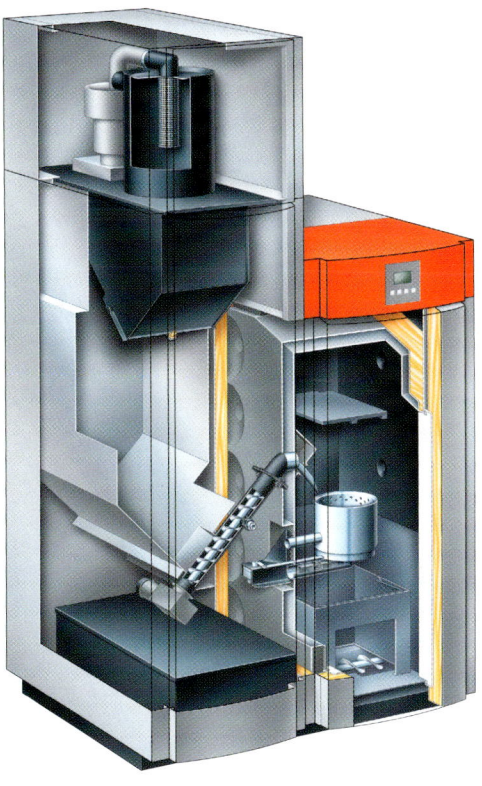

Figure 5.16.1: A diagram of a small biomass boiler. The hopper is shown on the left. A screw feed system transfers the pellets to the combustion chamber on the right. (Reproduced with the kind permission of Viessmann Ltd)

Figure 5.16.2: A medium-size biomass boiler installation. The device at the front feeds the pellets into the boiler from the main storage hopper

Figure 5.16: Biomass boilers

Figure 5.16.3: The storage hopper for the wood pellets used with the biomass boiler in Figure 5.16.2

Medium and high temperature boilers

Heating systems that fall into the categories MTHW and HTHW have efficiency advantages in large scale applications. This type of system makes most sense – both practically and financially – in very large buildings and where more than one building share the same boiler system or generally where heat has to be distributed over a relatively long distance. Situations like this are not common and where you find such systems you are very likely to find they are documented and well known to the staff that operates the building – systems like this are often in boiler rooms that are attended.

Some of the basic features you would associate with MTHW and HTHW boiler installations include:

- large multiple boiler equipment that is of the forced draught type and non-condensing;
- pipe labels and ID bands indicating MTHW, HTHW systems;
- multiple water distribution circuits and sets of circulating pumps.

The boilers themselves may not look that different to large forced draught LTHW boiler installations such as those shown in Figure 5.15.6.

Heating water at the elevated temperatures associated with these types of system cannot be directly circulated through radiator and convector systems. Instead, a small heat exchanger system has to be used to locally reduce the temperature to the LTHW range. The LTHW system receives its heat from the main MTHW or HTWH system but is isolated from it by virtue of the heat exchanger and also has its own circulating pumps. (This is akin to stepping down from high to low voltage at a transformer in an electrical distribution system.) You may well find this type of heat exchanger equipment if you trace a radiator system towards its heating system in a large building that has MTHW or HTHW boilers. Typical heat exchanger equipment is illustrated in Figure 5.17.1 and a typical configuration is shown schematically in Figure 5.17.2.

Combined heat and power (CHP)

Only about 35% of the primary energy delivered to a power station is converted to useful electrical power that can be distributed over the national grid. A large part of the primary energy is dumped to the environment by cooling towers as there is no other use for this heat. Generating the power at the building location using a combined heat and power (CHP) system gives an opportunity for the thermal energy associated with this process to be captured and used in the building for heating and hot water. In most building systems this is achieved using an internal combustion engine (usually a natural gas fuelled spark ignition engine) directly coupled to a generator. It is the heat in the water used to cool the engine that is transferred to the building heating system. Some heat is also transferred from the engine's hot exhaust gases. This arrangement is illustrated in Figure 5.18.1.

Building CHP equipment is often found near the building's boilers but often the equipment is packaged to reduce noise levels so that the engine and generator are enclosed. Like other equipment you should firstly be able to differentiate this equipment by looking at the connections to it. You should find connections to the rest of the heating water system along with an exhaust and a fuel supply line. There must also be a means of ventilation for the air intake. Some CHP equipment is shown in Figure 5.18.2.

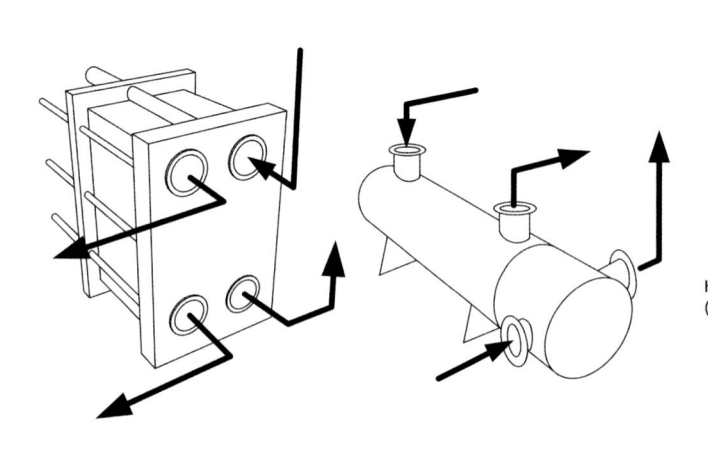

Figure 5.17.1: Heat exchangers of the types illustrated are used in MTHW and HTHW installations to produce water for LTHW heating
A plate heat exchanger is shown on the left and a shell-and-tube type on the right

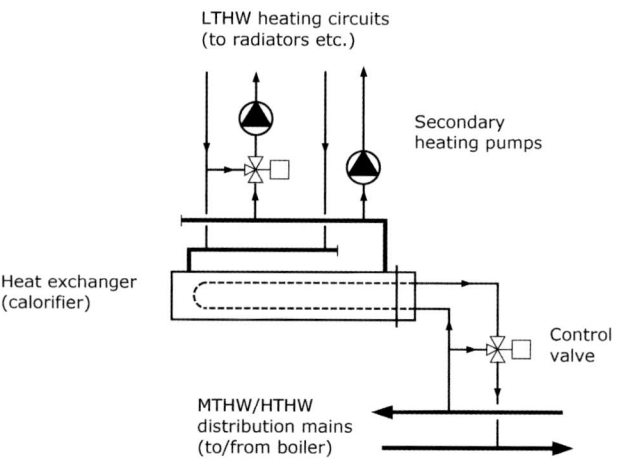

Figure 5.17.2: A schematic showing a heat exchanger and secondary LTHW circuits

Figure 5.17: Heat exchangers

Heating and Hot Water Services Core Knowledge and Recognition

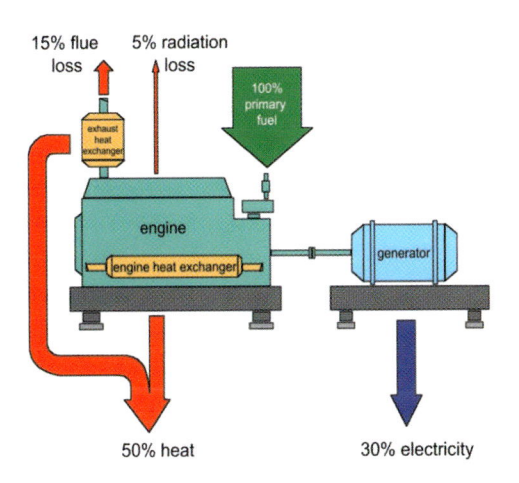

Figure 5.18.1: Diagram showing a building CHP power and heating system (Source: Carbon Trust, GPG388 diagram, p. 6)

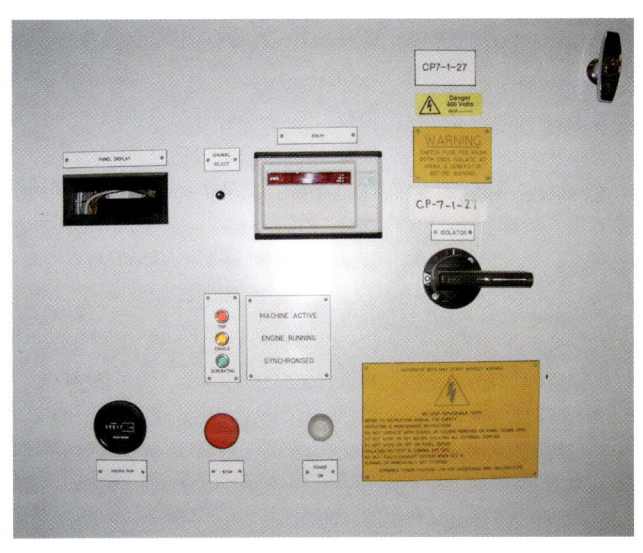

Figure 5.18.3: An example of a control panel found on a CHP unit. This provides information about the engine and gives a clue to the type of equipment

Figure 5.18.2: A building packaged CHP installation

Figure 5.18: Combined heat and power

In order to identify CHP equipment with some certainty it can be helpful to look at the control panel (Figure 5.18.3). You may see information such as engine temperatures, hours run and heat and power kWh that give a clue to the equipment being a CHP system. Often such equipment is provided and maintained by a service company and there is information about the company on equipment or control panel labels. You should record this information in case you need to contact them for further information.

The most important information you need about a CHP system for the purposes of SBEM calculations is as follows:

- the efficiency at which electrical power is generated;
- the efficiency at which heat is generated; and
- the proportion of the building's space heating and hot water energy that is provided by the CHP system. The proportion of heat and power generated by the system is often recorded in terms of respective kWh by the equipment control system and so the building operator or service company may have this information.

A useful source of information about CHP is the Department of Energy and Climate Change CHPQA website (www.chpqa.com).

District heating

There are a number of district heating systems in the UK. These thermal energy systems generate heat and power at a large central plant and distribute the heat to many local properties via underground pipe systems. District heating systems share some of the essential carbon efficiency benefits of CHP systems in that the primary fuel can be used to generate both heat and power without much thermal energy being wasted. A wide variety and mix of fuels may be in use including industrial waste heat, biomass and refuse in addition

71

to oil and gas. Large engine-driven CHP equipment may, in fact, form the central heat generation system.

The mix of fuels used by different district heating systems varies from one system to the next and so does the net carbon efficiency. For this reason an overall CO_2 emission factor has to be found for each system that takes into account the mix of fuels as well as pump energy and pipe heat losses. This calculation has to be fully documented in the EPC report. (There are guidance notes in the *iSBEM User Guide* concerning this.) If the building operator does not have the necessary information you will probably have to contact the district heating energy supplier.

If you assess a building that may use a district heating source you should find the same LTHW heat exchanger arrangement as that described above for MTHW and HTHW systems.

Heat pumps

Heat pumps use refrigeration principles and technology to generate heat for space heating and hot water generation purposes. A heat pump can essentially 'move' heat from one place or system to another and can often work in heating or cooling modes. In heating mode heat is extracted from the outside environment (the air or the ground) and delivered to the building heating system. Cooling is achieved by circulating the refrigerant in the reverse direction. This energy is not 'free' or 'zero carbon' as some electrical power is used by the system compressor to circulate the refrigerant.

Ground source heat pump systems

Ground source heat pump (GSHP) systems use the earth as a heat source and heat store. While ambient air temperatures may vary between -5ºC and +35ºC throughout the UK, ground temperatures vary much less throughout the year. Refrigeration systems that use these moderate ground temperatures can work more efficiently than those that use the air as their heat source.

The ground heat exchanger may be in the form of pipes in horizontal trenches or vertical boreholes through which an antifreeze mixture is pumped. A GSHP is connected to two pipe systems – one being the ground pipe loop and the other being the building heating circuit – and so four pipe connections will be found along with two sets of pumps. The refrigeration equipment is packaged into a single cabinet so that no refrigeration pipework will be visible (Figure 5.19.1).

The heat pump can be placed anywhere in the building – not just on an outside wall. Ground source heat pumps can be used to generate domestic hot water in some applications and so there may be other interconnecting pipes. Large GSHP installations (Figure 5.19.2) can consist of several heat pumps connected in a modular fashion (similar to some boiler installations) and very large heat pump equipment looks very similar to some water cooled chiller equipment.

Ground source heat pumps generate heat at temperatures below that of LTHW systems. They are consequently well suited to underfloor heating and often used this way in the UK.

Figure 5.19.1: A non-domestic ground source heat pump. The cylinder on the right is a buffer tank for the underfloor heating circuit (Reproduced with the kind permission of EarthEnergy)

Figure 5.19.2: A ground source heat pump installation with two heat pumps (Reproduced with the kind permission of EarthEnergy)

Figure 19: Ground source heat pumps

Air source heat pumps

Air source heat pumps use the same refrigeration principles as ground source heat pumps but exchange heat with the outside air rather than the ground. They can be used to generate heating water or to directly heat and cool the building air distribution system. Air source heat pumps may be split into two components – one heat exchanger inside, and an outside unit that includes a fan, heat exchanger and compressor. The outside unit can appear much the same as split-system air conditioners with the two units connected by small diameter insulated refrigerant pipes. We discuss air and ground source heat pumps in the context of cooling in Chapter 6.

HVAC SYSTEM CONTROLS

The way in which a HVAC system is managed and controlled has a significant effect on actual energy use. The object of a good control system is to maintain thermal comfort in every occupied room whilst minimising overheating or overcooling and minimising the amount of time the system has to operate. A number of different control mechanisms – with differing levels of sophistication and effectiveness – may be built into a heating or cooling system to help achieve this. The types of control system to consider, and which we will describe below, are as follows:

- central and local time control;
- local temperature control;
- weather compensation temperature control; and
- optimum start control.

One of the outcomes of the assessment is to provide advice on ways in which energy efficiency might be improved, through the Recommendation Report (RR). Improving a control system can often be a significant cost effective energy saving measure. At present, you can confirm a number of control features in your data entry into SBEM, but these only result in a recommendation being generated in the RR. Presence of these controls does not alter the EPC rating unless you alter the system efficiencies as indicated above. It is nevertheless valuable to identify some common control equipment and have an understanding of the type of advice being suggested by the software.

If you find control equipment that is obviously not being used (broken or not being used) the equipment should still be regarded as part of the buildings assets. In this case it should be taken account of in any assessment as if it is still operational. You should note and confirm this information in your RR. However, some accreditation schemes are likely to have specific rules about this issue, and you should abide by those rules.

Central time control

Unless a HVAC system is required to operate continuously, or is attended, some form of time control is an elementary control mechanism that you would generally expect to find in some form – energy is very simply saved if systems can be turned off when not required. A wide variety of electro-mechanical time switch devices can be found (see Figure 5.20) as well as electronic devices more akin to domestic heating controls that allow a degree of programming.

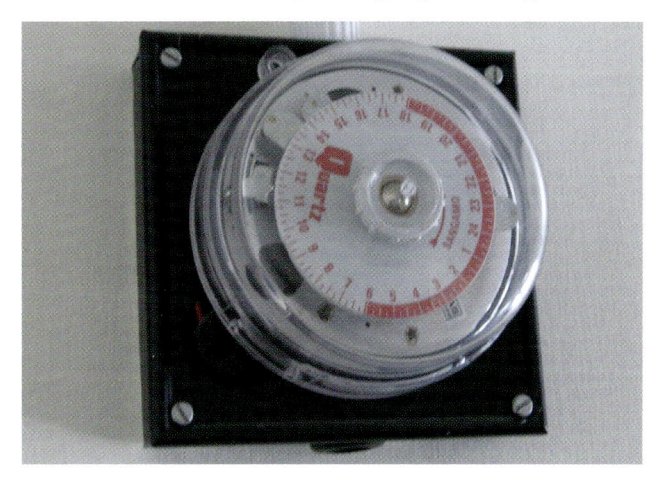

Figure 5.20: A common form of time switch

Local time control

Local time control allows some refinement of system control – further minimising unnecessary operation. This principle may be applied where rooms have self-contained equipment such as a electric room heater (Figure 5.21) or where systems allow local user control – for example a fan coil unit or split system air conditioner. This approach may also be taken to control operation of particular zones or heating system sub-circuits.

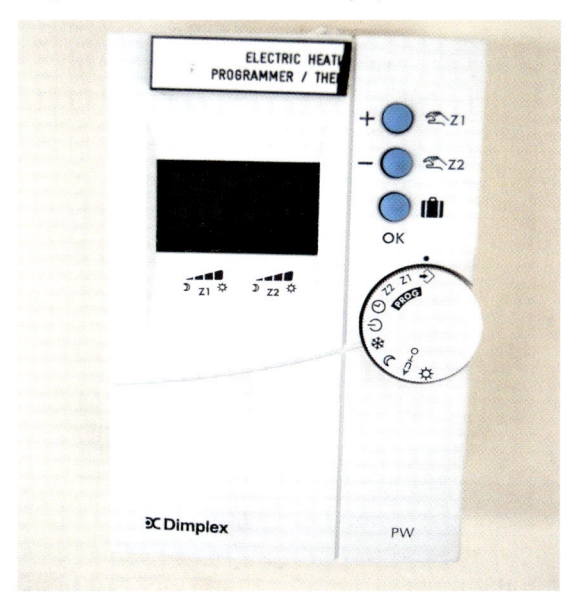

Figure 5.21: A device used to control a local room heater. This provides both local time and temperature control

Local temperature control

Local temperature control can provide a means of controlling overheating and temperatures in unoccupied periods in addition to occupants being able to have some control over conditions in individual rooms. Some forms of heating and cooling system allow some local temperature control. A local temperature sensor may be installed that has its operating temperature set in the building control system rather than a thermostat device that allows room occupants to make adjustments. Typical devices are shown in Figures 5.22.1 and 5.22.2. Both thermostats and temperature sensors can be considered as local temperature controls in the context of SBEM calculations.

Figure 5.22.1: Example of a room thermostat

Figure 5.22.2: Example of a room temperature sensor used for automatic control of local temperature

Figure 5.22: Local temperature controls

Radiators and convectors that have simple thermostatic radiator valves (TRVs) can be considered to provide local temperature control (Figure 5.6.1). Systems providing refrigerant cooling such as split systems, multi-split or VRF, usually have a local controller provided in each room. An example is shown in Figure 6.26.3.

Weather compensation temperature control

Although LTHW boiler systems are often designed to provide hot water at a constant temperature maximum temperature may not be required at all times. In many radiator or convector heating systems the water temperatures can be lowered to reduce the heat output according to how cold it is outside. This weather compensation temperature control mechanism uses an outside air temperature sensor and adjusts the temperature of the water flowing around the system in proportion to the difference between the outside temperature and room temperature. This is illustrated in Figure 5.23.

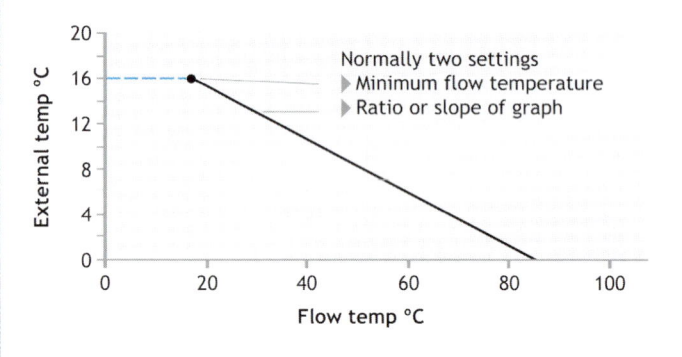

Figure 5.23: The weather compensation control (Reproduced with the kind permission of Carbon Trust)

Optimum start/stop control

In buildings that are intermittently heated it is necessary to operate the heating system some time before the occupied period in order to raise the temperature of the building in good time. This period can be smaller when the outside temperature has been warmer than is necessary in the most severe conditions. Optimum start control can be used to automatically adjust the preheating period according to the recent outside conditions and according to the responsiveness of the building and its heating system – consequently minimising the time the system needs to operate and saving energy (Figure 5.24.1).

This type of control can be found in older sophisticated electro-mechanical controllers such as shown in Figure 5.24.2. This control facility may well be found in modern integrated boiler controllers and certainly BMS systems. You may need to gather further information about an electronic programmer from the

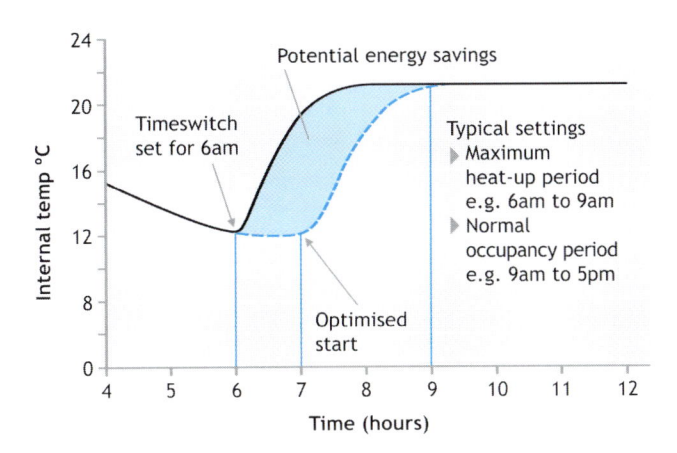

Figure 5.24.1: The optimum start control principle (Reproduced with the kind permission of Carbon Trust)

manufacturer literature before being sure what control facilities are in use or are available.

Figure 5.24.2: An example of an electro-mechanical optimum start controller

HOT WATER GENERATION

Hot water for domestic purposes is required in virtually all commercial buildings and can, in some cases, constitute as large a part of a building's carbon emissions as space heating. Establishing a building's hot water generation characteristics and establishing the correct demands are an essential and important part of the energy assessment process. Your essential task is to gather the information you need to associate each zone with a hot water generation system, and to associate hot water generation systems with their fuels.

Hot water demands

Hot water energy demands and the corresponding annual carbon emissions depend, in the first place, on the volume of hot water that is used in the building over a year. A fundamental concept in SBEM calculations of hot water energy demand is that the demand for hot water depends essentially on an occupant density and the activity of the occupants. Occupant density and the amount of hot water associated with a particular activity are data contained in the NCM databases.

The volume of hot water that is used in the hot water energy assessment depends on establishing the building zones and deciding what is the most appropriate activity to associate with each zone.

It is important to note that demand is *not* related to the point at which hot water is delivered, the number of taps, or the area of the toilet or kitchen facilities. For example, the consumption of hot water in a tea room depends on the floor area of the adjacent office rather than that of the tea room. This is because the size of the office gives the best estimate of the number of occupants expected to use the facilities.

Hot water energy losses

The energy required to provide a building with domestic hot water depends on how much energy is lost between the location at which it was heated and the point at which it is drawn off as well as the energy required to heat the water from cold. There are a number of hot water heat losses that should be considered.

Distribution losses

Heat is inevitably lost from all hot pipes passing through a building and can represent a noticeable loss of energy in domestic hot water systems. Losses are exacerbated by the fact that hot water systems are operated intermittently and so water remaining static in the pipe becomes unusable once cooled too much by any heat losses. Some quantity of water typically has to be flushed through the pipe before water arrives at the point of use at an acceptable temperature.

Hot water that has to be flushed out and replenished by other hot water represents an energy loss that has to be offset by additional hot water generation and fuel usage. You should take this energy loss into account in the assessment exercise by estimating the length of any dead leg in the rooms where the water is delivered (toilets, kitchens, etc.) and entering this data into the software on a zone-by-zone basis.

It is difficult to find a universally applicable rule for determining dead leg lengths in existing buildings. Where you know there is a recirculation system it is reasonable to measure the dead leg assuming the supply to the edge of the zone does not include any dead leg, i.e. the main supply and return pipes are close by. Your accreditation scheme may recommend you adopt this practice in all cases and you should become familiar with their advice.

If the building does not have any recirculation system – as we find is the case in many larger converted domestic properties – then it is reasonable, in our view, to include some of the pipes outside the zone in your estimate of the dead leg.

In many buildings you can measure the exact dead leg length because the pipes are visible. Where this is not possible, e.g. the pipes are behind the wall finish, or in the floor or ceiling, you should estimate the length based on what seems reasonable based on where the outlets are and the position of the hot water generator.

There are three common ways, in addition to good insulation, that the effects of heat loss in a domestic hot water system can be mitigated:

(1) local instantaneous electric heating;
(2) secondary circulation of the hot water (storage systems only);
(3) supplementary electric trace heating.

You are likely to find these measures taken in some buildings (although trace heating is a more recent approach) and, as these can be taken into account in SBEM calculations, it is necessary to look for their presence in any building.

Local instantaneous electric heating (under the sink heaters, etc.) may be found in some locations but, as well as being not very carbon efficient, this is not practical on a large scale.

Secondary circulation is the most common approach in medium and large buildings. In these systems a pair of hot water pipes is used to distribute the hot water to every point of use. One pipe is to deliver hot water on demand, and the second smaller pipe is used to circulate a small proportion of the water and return this to the hot water generator and its storage tank. This ensures the heat loss is always compensated for, but, as the water has to be circulated by a pump, this system has some auxiliary energy associated with it.

The energy demand associated with a secondary circulation system can be estimated by SBEM and this is carried out on the basis of the rate of heat loss along the pipes, the length of the pipe loop and the pump power. These factors can be specified by the user but are calculated by SBEM if not. In the iSBEM software you will be able to see the value of these parameters when the EPC calculation has finished.

You may be able to decide if a hot water system has secondary circulation by either identifying a circulation pump and a return pipe near the hot water generator or tank, or by inspecting the pipes at the point of use. If you can see secondary pipes are extended to the point of use you can justify entering the dead leg as '0' in that particular zone. Figure 5.25 illustrates how dead legs may vary.

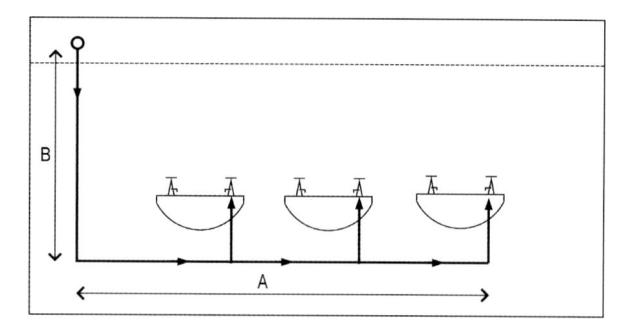

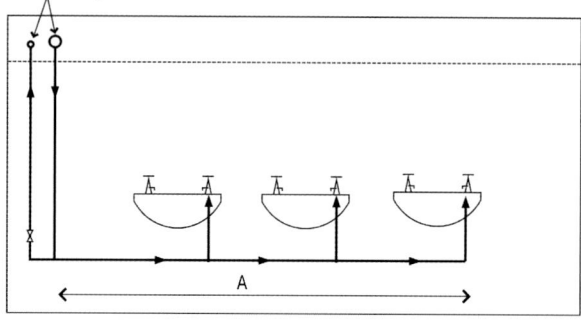

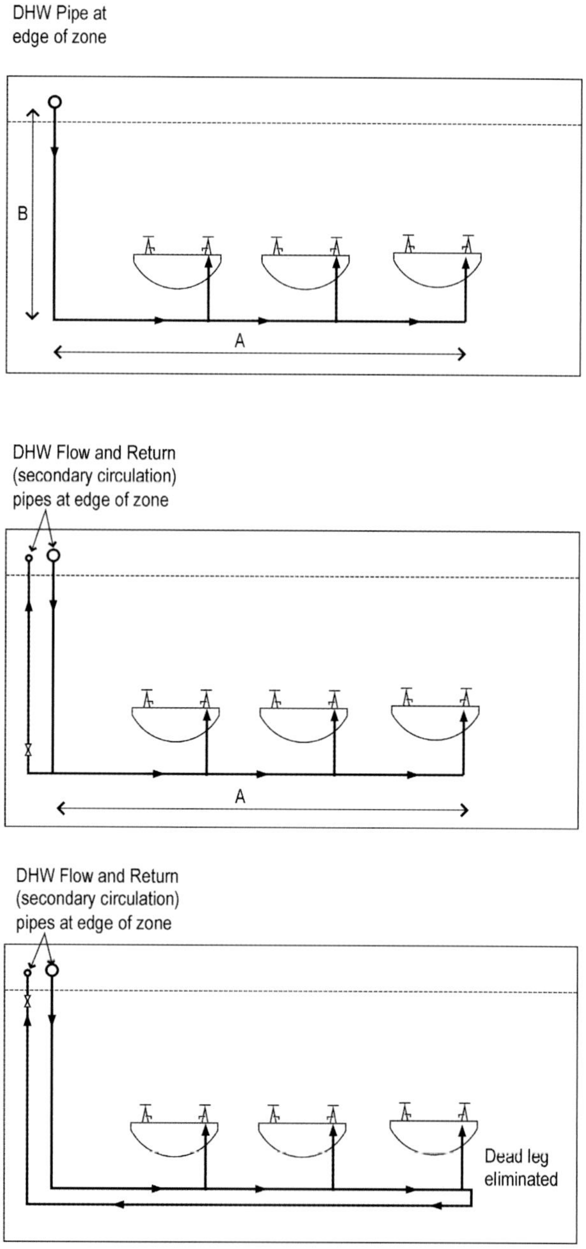

Figure 5.25: Variations in hot water supply and secondary circulation pipe connections to wash hand basins. The dead leg length is indicated by dimensions A and B. In the bottom figure the dead leg length is effectively zero

An alternative method of offsetting the heat losses from the hot water pipes and eliminating the effect of dead legs is to use a system known as 'trace heating'. In this system there are no secondary circulation pipes or pumps, rather, the supply pipes themselves are kept warm by electric heating along their whole length. This very modest amount of heat is delivered through a special heating cable along the outside of the pipe and underneath the insulation. This type of installation is illustrated in Figure 5.26. This system can be represented in an approximate way in SBEM calculations (details are given in the *iSBEM User*

Guide). The electrical demand associated with the system can be accounted for by specifying a circulating pump system but setting the dead leg for every zone served by the system to zero.

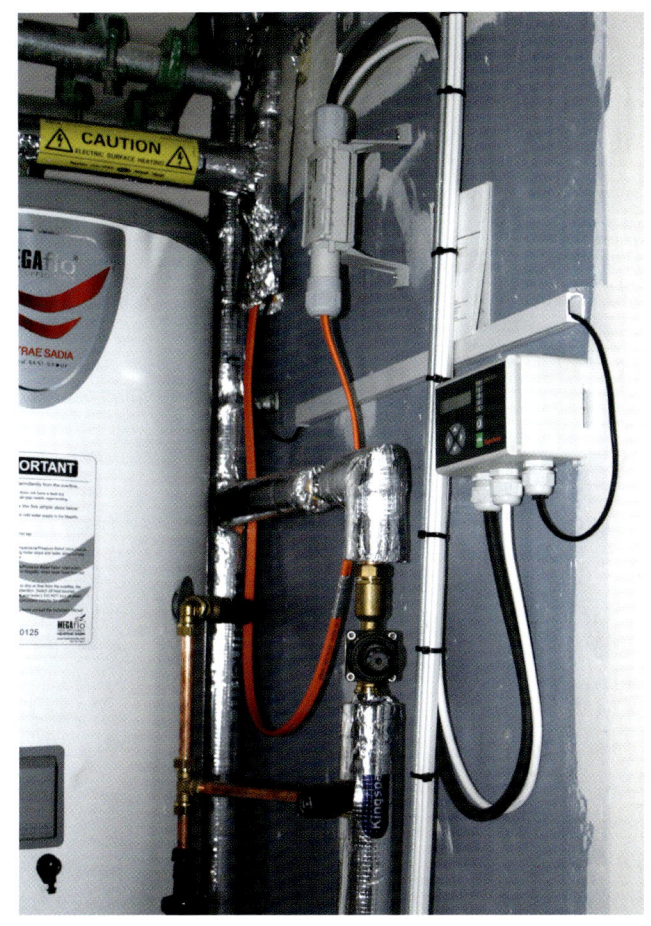

Figure 5.26: An electric trace heating installation for hot water distribution pipes. The orange heating cable is installed under the pipe insulation

Storage losses

Heat is lost through the surface of domestic hot water storage tanks through the insulation and due to the intermittent operation of the system. SBEM software can be relied upon to make default calculations of these losses but may allow you to input detailed information if you have it. The designers of the system are more likely to have information such as the heat loss per metre of pipe than assessors of existing buildings, but you may well be able to find the size of the storage tank and enter that data.

Hot water storage tanks found in commercial buildings have the same common features as domestic cylinders but are of large scale and always indirectly heated. They are commonly vertical cylinder form and appear with their insulation covered with protective cladding (Figure 5.27.1). Other water tanks such as heating system buffer tanks can be found in commercial building plant rooms and so it is worth noting the types of pipe connections you should find to help identify a tank as one used for storing domestic hot water:

- heating source supply and return pipes;
- cold water feed pipe (cold feed);
- hot water outlet pipe.

These pipe connections are often identifiable from pipe labels or identification bands. Other pipes you may see, that are usually uninsulated, are for drainage and safety valve discharge. Some hot water systems use a second source of heat and so have a second heat exchanger in the storage tank along with corresponding insulated flow and return pipe connections. Such systems include those with solar hot water generators and those that use waste heat sources or heat recovery systems. Some systems also have electric heating elements but this is less common than in domestic systems.

Your main task in the inspection process is to identify the presence of a hot water storage system, and then establish the storage capacity (for data input to the storage heat loss calculation). This can be estimated by taking the main measurements of the cylinder (with some allowance for insulation thickness) but can be best done by inspecting any manufacturer's label (Figure 5.27.2), talking to the manufacturer's technical department or looking at the information on their website.

Figure 5.27.1: Domestic hot water storage tanks (calorifiers) in a large commercial building

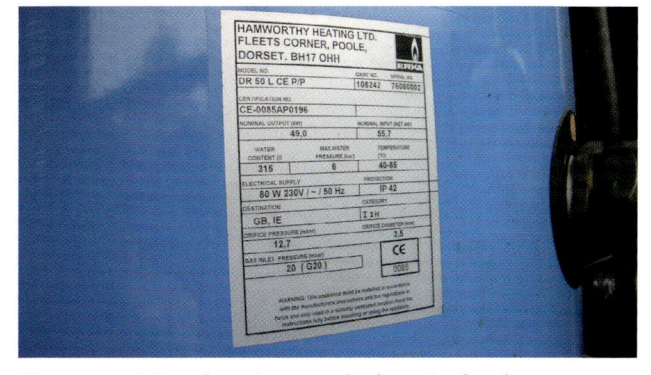

Figure 5.27.2: It is good practice to record information from hot water storage tanks

Figure 5.27: Hot water storage tanks

If you are using iSBEM and rely on the default data in your assessment, the software will estimate the storage tank size in proportion to the monthly hot water use and the heat loss is calculated assuming a modest level of insulation. You should see the results of these default calculations once the final EPC rating calculation has been completed.

Hot water generators

The carbon emission rate associated with a particular level of total zone hot water demand and heat loss depends on the efficiency of the hot water generator and the fuel type. As we noted earlier, an essential task is for you to associate each zone with the correct hot water system and associate each of these with a fuel type. It is important that you do not assume there is only one hot water source (even in a small building) as; for example, instant electric hot water heaters are often added in isolated locations after construction.

It has been conventional to generate domestic hot water using the same heat source as that used for space heating. The most common arrangement is that the boiler that provides heating water for a radiator system also provides heat to a hot water tank – the same concept as that in many domestic properties. This arrangement is denoted 'Same as HVAC' in iSBEM. This arrangement can include heating systems with heat pumps, CHP and district heating sources – the hot water storage equipment will be much the same.

The other forms of hot water generating system that can be represented in SBEM calculations are:

- dedicated hot water boiler systems;
- stand-alone water heaters;
- instantaneous water heaters;
- instantaneous combi systems;
- heat pumps.

Dedicated hot water boiler systems

One disadvantage of using the building's main boilers as a heat source for generating hot water is that the boiler and circulation system have to be operating all the year round. Otherwise the boiler system may well be shut down outside the heating season. This can be one reason that a separate (possibly quite smaller) boiler is used that is dedicated to hot water generation. The boiler is only used to heat the hot water tank in this situation. Another reason this arrangement may be used is that space heating is provided by something other than a wet heating system such as a gas fired radiant system.

Hot water generation systems that use a dedicated boiler have similar characteristics to other boiler and hot water storage systems. The same approach to identification and assessment of the boiler and its fuel type can be taken as when assessing a boiler used for space heating.

Stand-alone water heaters

Water heating equipment that operates independently of any other HVAC systems and is self contained can be advantageous in terms of reliability, maintenance, performance as well as space saving compared to conventional boiler heated indirect storage systems. Self contained stand-alone water heaters combine a gas or oil fired heat source with a hot water heat exchanger and some storage capacity in a single piece of equipment. They can be found in both domestic and commercial buildings.

The most common stand-alone water heater equipment has a natural gas fired burner at the bottom, a vertical insulated storage cylinder that contains a hot water heat exchanger, and a central natural draught flue at the top (as in Figure 5.28.1). Such equipment is intended to form part of a pressurised hot water supply system that is directly fed from a cold water supply. The burner and flue are usually easily visible and so both the equipment and fuel source are usually not difficult to identify. You may find condensing varieties of this type of equipment, and with a range of flue arrangements.

Hot water generators that combine a modest storage capacity and electrical heating elements also fall in the category of 'stand alone water heater' in SBEM calculations. As no flue is connected to this type of equipment it can be sited in a wide variety of locations – sometimes well concealed, as in Figure 5.28.2.

Instantaneous water heaters

Hot water generation equipment that is able to provide hot water to one or a few outlets on demand can be found in situations where:

- hot water use is small and very intermittent;
- the location is remote from other hot water and HVAC systems;
- the only energy supply is electricity;
- refurbishment or extension makes this economic and convenient.

Small electric instantaneous water heaters can be found installed over or under wash hand basins and sinks in a wide variety of building types (Figures 5.29.1 and 5.29.2).

Often this equipment is deliberately well concealed and so it is worthwhile examining locations such as cupboards below sinks to see if there is an electric water heater (electrical isolation switches may be visible nearby) even if there is another type of hot water system for most of the building (Figure 5.29.3). You may find very similar gas fuelled appliances where there is an outside wall for the flue.

The *iSBEM User Guide* description of this type of system is 'a water heater with no (*or limited*) storage capability' (our emphasis). We suggest the word 'limited' includes storage up to around 15 litres. If the storage in the system is not 'limited', then it will be one of the other systems, with a greater storage capacity.

Heating and Hot Water Services Core Knowledge and Recognition

Figure 5.28.1: A typical small gas fired hot water generator

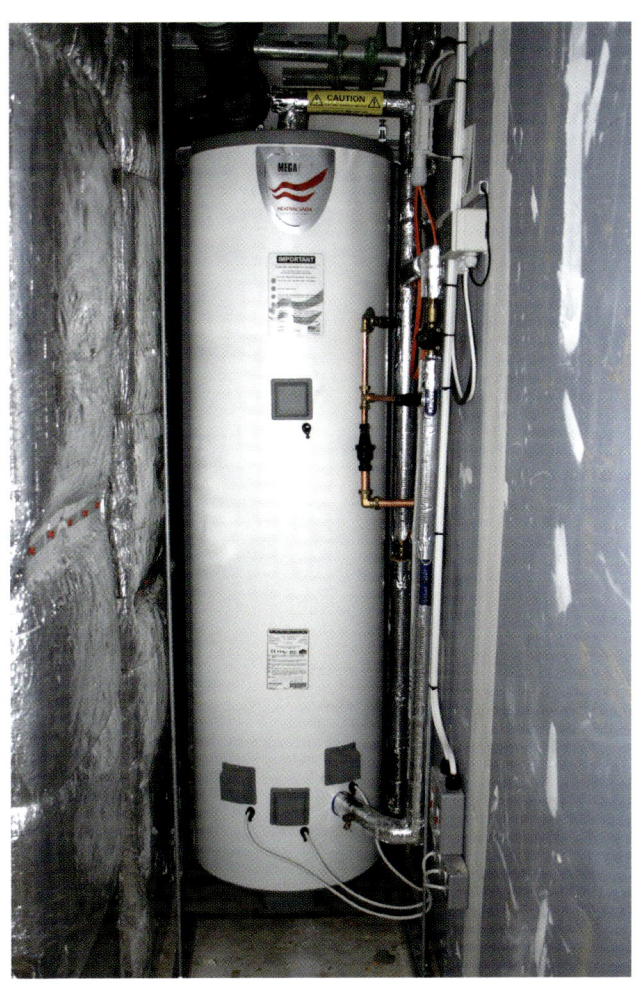

Figure 5.28.2: Example of an electric stand alone water heater

Figure 5.28: Stand-alone water heaters

Figure 5.29.2: An electric instantaneous water heater in a wash room

Figure 5.29.1: An example of a simple instantaneous water heater for hand washing

Figure 5.29: Instantaneous water heaters

79

Figure 5.29.3: Example of an electric instantaneous water heater hidden under a sink

Instantaneous combi systems

Heating equipment that provides both a heating water source and an instantaneous supply of hot water – so called 'combi' boilers – have become prevalent in recently built or refurbished domestic properties. They can be found in a variety of smaller commercial buildings where heat losses are of the same magnitude to domestic properties or are very small. This can be the case where only a small portion of a large building has a need for both hot water and radiator/convector heating. One example would be where a factory has a comparatively small office and kitchen/toilet facility. A good deal of information about such systems can be found in the *Domestic Energy Assessor's Handbook*.

Heat pump hot water generators

Heat pumps can be used as a source of energy for generating hot water. This can be achieved with equipment that only provides a source of hot water, or which combines both hot water and space heating/cooling functions. The alternative configurations can be summarised as:

(1) air-to-water (air source) heat pumps that generate heating water and are combined with a storage cylinder;
(2) water-to-water ground source heat pumps with a heating and hot water generation facility;
(3) water-to-air ground source heat pumps with a built-in heating and hot water generation facility.

It has been feasible to use heat pumps to provide domestic hot water for some time but it is only recently that equipment of this type has become commercially available in the UK. Air-to-water heat pump equipment that can be used to provide water for heating can be used, as boiler heating equipment is, to generate hot water by an indirect heat exchanger in a storage tank (system type (1) above). Domestic scale systems like this are sometimes referred to as 'heat pump boilers'.

Ground source heat pumps that provide a source of heat for hot water as well as heating have become available quite recently. In the case of water-to-water ground source heat pump systems with a built-in hot water generator you will find cold water feed and hot water supply pipe connections in addition to the four building and ground loop circuit pipe connections. Ground source heat pumps of the water-to-water type are also used to generate hot water by connection to an indirect hot water cylinder but most only provide heating. Heat pump hot water systems are not common and are often found with backup electric water heating elements in the storage tank. Systems in the third category may come into use in the near future.

You will see that there is a wide variety of equipment that is used for generating and storing hot water. It is often useful to note any manufacturers' information that you can find on the equipment – by carefully recording the information in your site notes and taking a photograph of the label or name plate if possible. This information may be useful later for finding out the age of equipment (pre- or post-1989) whether it is on the ECA list, or its storage capacity.

SUMMARY

In this chapter we have tried to give you enough information and practical guidance for you to:

- understand the basic operating principles of commonly occurring heating and hot water systems;
- identify key system components and distinguish between systems;
- collect the HVAC data you need to carry out SBEM calculations.

The objective in inspecting the heating and hot water systems is to identify:

- room heat emitters;
- heating sources;
- which fuels are being used by each heating source.

Although the range of heating systems in use is very large, we hope you will have seen that there are enough key distinguishing features for you to identify everything you need to.

In the next chapter we discuss:
- ventilation;
- air conditioning equipment;
- heating and cooling system types;
- cooling sources;
- renewable energy systems.

If you are working as a CEA at Level 3 you need to be familiar with the material relating to ventilation, small cooling systems and renewable energy sources that appear in this next chapter.

6 Ventilation and Air Conditioning Core Knowledge and Recognition

INTRODUCTION

In this chapter we aim to provide the core knowledge required to undertake assessment of building ventilation and air conditioning systems and associated cooling sources. This chapter also deals with renewable energy systems. In this chapter we try to give you enough information and practical guidance for you to:

- understand the basic operating principles of commonly occurring air conditioning systems;
- identify key system components and distinguish between system types;
- identify types of cooling sources;
- assess building renewable energy systems; and
- collect the data you need to carry out SBEM calculations.

Chapter 5 dealt specifically with heating and hot water systems and we build on this material here. The major topics dealt with later in this chapter are:

- ventilation;
- air conditioning equipment;
- air conditioning system types;
- heat recovery systems;
- cooling sources; and
- renewable energy systems.

VENTILATION

Ventilation is the introduction of fresh air from outside and corresponding removal of stale or contaminated air. Every litre of air that is drawn into the building has to be heated to room temperature in winter, or cooled down by any air conditioning system in summer. This requires energy and has an impact on carbon emissions. The quantities of fresh air that are assumed to flow through a particular zone for ventilation purposes are determined according to the NCM databases depending on what activity type you assign. For the purposes of SBEM calculations the energy associated with heating and cooling fresh air is always included – whether you specify natural ventilation, mechanical ventilation or some form of air conditioning.

The amount of energy consumed by any fans can be significant and depends on the form of ventilation being used and whether there are additional local extract fans. The aim of the software, and your aim in collecting data, is to ensure that the right quantity of fresh air is being accounted for and the correct number of fans and their respective efficiency are taken into account.

There are four means of providing ventilation that you need to consider in making your assessment:

- natural ventilation;
- central air condition system ventilation;
- mechanical ventilation (zone supply/extract); and
- local extract.

When we say 'mechanical ventilation' this means a system with both supply and extract air distribution but without cooling.

The overall efficiency of ventilation systems using fans depends on a number of things:

- fan power;
- whether there are high pressure drops;
- the presence of heat recovery systems;
- application of demand controlled ventilation.

We will explain each of these in some detail a little later. Heat recovery systems can be applied to all systems that have both supply and extract ducts and so you can specify such features where there are conditioning systems or zone mechanical ventilation systems – we will consider the different techniques used to achieve this later.

Ventilation openings

To achieve ventilation by mechanical means clearly requires that openings in the building fabric are provided. This might mean a cowl, open duct, door grille, or louvres – these all provide a certain amount of open area and weather protection. These features are one of the important features that you should be looking for when you inspect the outside of the building – acknowledging that some such features may only be visible at roof level. Finding such features should cause you to make a note to look for a corresponding mechanical system inside.

Louvres come in many forms and can be made from a number of materials. Louvres built into or above the doors of ground level plant rooms are quite common. Larger mechanical ventilation systems are often associated with large areas of metal louvres built into the walls of the building – or indeed forming a wall – or in a 'penthouse' form on top of a roof (Figures 6.1.1–6.1.6). You should note that louvre openings can be also used for smoke ventilation.

Louvres above doors at ground level may or may not indicate the presence of a mechanical ventilation or local extract system. Doors of electrical equipment rooms and boiler rooms are often provided with louvres as a simple means of securely providing natural ventilation. The louvres above doors bear closer inspection as these can be used for the same purpose, blanked off behind, or used by smaller mechanical ventilation and local extract systems. An example of the later arrangement is shown in Figure 6.1.4.

Natural ventilation

Many buildings you assess are likely, at least in some zones, to rely on natural ventilation. It is not necessary to say much about identifying opening windows but it is worth remembering that rooms with opening windows can also have mechanical ventilation systems and so you should still check for inlet and extract grilles as well as local cooling systems. For example, a room may have opening windows for summer ventilation but there may be air distribution grilles because the main heating system is warm air central heating.

Figure 6.1.1: Example of a penthouse louvre arrangement (Reproduced with the kind permission of Gilberts (Blackpool) Ltd)

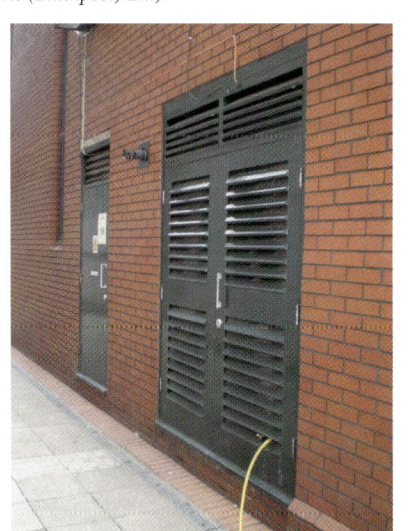

Figure 6.1.3: Common plant room louvre door

Figure 6.1.4: Duct connection above a plant room louvre door

Figure 6.1: Louvres

Figure 6.1.2: A large vertical louvre forming part of the facade (Reproduced with the kind permission of Gilberts (Blackpool) Ltd)

Central air conditioning system ventilation

Central air conditioning systems such as those that can be specified in SBEM software are designed (and assumed) to distribute adequate fresh air through the duct systems. If a zone uses a central cooling system of any type there is no need to do anything else in the software to define ventilation in that zone – the energy required to heat or cool the fresh air is taken into account automatically. You will need to account for any local extract fans as we explain below.

Mechanical ventilation

In buildings that do not need cooling, but do need heating, natural ventilation is appealing but is not always practical (in noisy or polluted environments for example). You will find that mechanical ventilation (zone supply and extract) is relied on in certain heated buildings and this has the advantages that air can be filtered and some heat can be potentially reclaimed. You may also find zones in naturally ventilated buildings that are far from the perimeter that have mechanical ventilation. Supply and extract mechanical ventilation systems (as we define them) are usually only designed to preheat the incoming fresh air during winter and not provide any room heating.

Any heating system that can be specified in SBEM can be combined with a mechanical supply and extract system. This cannot be specified as a property of the system but must be specified on a zone-by-zone basis. The only heating system that is treated any differently is warm air central heating. In this case the heating system is assumed to also provide all the required fresh air.

Refrigeration cooling systems such as single-room air conditioners and mini-split systems do not usually, in themselves, provide any fresh air ventilation and so may be found alongside mechanical ventilation systems. Again this is specified on a zone-by-zone basis.

An important thing to note is that unflued gas and oil heating devices – as they discharge the products of combustion into the room – need to be carefully ventilated. There might be natural means for doing this but in any case you should specify a mechanical ventilation system in the zone.

The combination of evidence you should consider to decide if a zone has simple supply and extract mechanical ventilation are:

- ventilation openings in the building fabric with duct connections;
- room supply and extract grilles (as in Figure 6.1.5 for example);
- room heat emitters;
- absence of building cooling sources.

Systems in corridors and larger toilet facilities are possible applications of zone mechanical ventilation. If there are no room heat emitters you should consider if the system is a central heating system using air. If the building has central cooling source you would need to investigate the central equipment in more detail to see if the room grilles were part of an air conditioning system.

Figure 6.1.5: Common supply (right) and extract (left) diffusers

Local extract

Local extract systems are generally used in environments where air becomes directly contaminated by a particular activity or processes. By asserting a negative air pressure in the room, airflow is guaranteed to be drawn from either the outside, or surrounding rooms, and to flow directly out.

Typical environments for these systems are as follows:

- *Kitchens*. Extract systems in large kitchens often consist of a hood located over the range which has an associated fan and duct discharging directly outside the building.
- *Toilets and bathrooms*. The extract system can consist of a small wall or window mounted fan (e.g. Figure 6.2.1) but may be one serving multiple rooms via extract ducts and a series of grilles. In the later case a central fan system will be provided (Figure 6.2.2).
- *Industrial buildings*. In these buildings the purpose of extractor ventilation is to remove warm air and contaminants such as moisture, dust and fumes. Roof-top extractor fans are common (Figure 6.2.3). Air drawn through local extract systems such as dust or fume extraction systems (Figure 6.2.4) does have an energy impact but whether you include this in the assessment would depend on whether you view this as a permanent part of the ventilation system

Not every room sharing a particular HVAC system may have a local extract system and so this is specified on a zone-by-zone basis. You should check for local extract devices or systems as you inspect every room – not least because small extract fans are often retrofit after construction. You may be able to find product information to determine an appropriate flow rate but you can also find some good guidance on the appropriate flow rate by consulting the *CIBSE Guide F Energy Efficiency in Buildings* (CIBSE, 2004) and the *Approved Document for Part F of the Building Regulations* (CLG, 2006).

Commercial Energy Assessor's Handbook

Figure 6.2.1: An example of a wall mounted extract fan. These can be found in a very wide range of applications

Figure 6.2.2: An extract fan unit used in a system servering several rooms

Figure 6.2.3: A common form of roof-top extract fan

Figure 6.2: Local extract units

Figure 6.2.4: A fume extract outlet associated with a workshop

Mixed-mode ventilation

Research shows that occupants are usually able to accept higher temperatures where they have some control of their conditions (i.e. opening windows and blinds) and where they can adapt clothing levels appropriately. One approach to low energy design that takes account of this is to make use of both natural ventilation operation for most of the year and air conditioning in peak heat gain conditions. This is what is meant by 'mixed-mode' ventilation.

In simple terms this means a building designed for mixed-mode ventilation would have opening windows but also air conditioning. This usually means the air conditioning does not operate until the room temperature rises towards the upper limit of acceptability – something like 27 or 28°C. This clearly results in lower carbon emissions compared to an equivalent building with sealed windows and air conditioning. We should say that just because a building has a cooling system and opening windows does not mean it should necessarily be classed as a mixed mode building. True mixed mode buildings have been purpose built to operate this way and often have features such as automatically controlled windows advanced BMS controls and highly efficient HVAC systems.

VENTILATION AND AIR CONDITIONING EQUIPMENT

In order to develop some skills in assessing air conditioning systems it is valuable to gain some core knowledge of the common equipment and system components you may find. Being able to identify key components and how they are combined helps distinguish one type of system from another.

This section discusses the primary components that are commonly used and combined together in air conditioning systems. The components of most interest are:

- fans;
- filters;
- dampers;
- heating and cooling heat exchangers; and
- heat recovery devices.

It is particularly worthwhile being able to identify the main heat exchangers in the system as this can help identify the heating or cooling source. In some cases this will enable you to determine the fuel source directly. In other cases this will help you decide what heating or cooling sources you need to look for elsewhere.

You do not need to collect specific information about components such as filters or dampers for SBEM calculations. However, as you will often find these components, it is worthwhile having sufficient knowledge to be able to identify them and to distinguish them from the components you are most interested in.

Fans

All mechanical ventilation and air conditioning systems rely on fans to circulate air through the heating and cooling components and through the duct distribution system. Two types of fan are in common use: centrifugal and axial. Axial fans (Figure 6.3.1 and 6.3.4) have propeller like blades and draw air along the axis of the fan. These are easily incorporated into duct systems and are common in local exhaust applications.

Centrifugal fans (Figure 6.3.2) have sets of blades mounted in a scroll shaped casing that draw air through the sides, into the centre, and discharge it tangentially

Another form of centrifugal fan is known as a 'plenum fan' and this uses blades similar to those of conventional centrifugal fans but not mounted inside a scroll shaped casing. Instead, air is drawn through a plate or bulkhead on one side – as shown in Figure 6.3.3.

Figure 6.3.1: An example of an axial fan mounted inline with a duct system (the fan is to the left of the label)

Figure 6.3.2: A centrifugal fan with a belt drive mounted inside an air handling unit

Figure 6.3.3: An example of a plenum fan arrangement. Air is drawn in through the connection on the right. The fan blades are clearly exposed. The black box on top of the motor is a variable speed drive

Figure 6.3.4: A small axial fan used in a toilet extract system

Figure 6.3: Fans

Filters

Filtration of particles of various sizes is an important function of mechanical ventilation and central air conditioning systems. Filtration is used to control room air quality but also to prevent dust building up on the system components.

One of the most common filter arrangements is to have an initial set of filters of the panel type (these are much like domestic filters) and a second set of finer filters of the bag type. Bag filters have an extended area by virtue of the filter material being formed into a series of bags or pockets. Examples of these types are shown in Figures 6.4.1 and 6.4.2. Very fine filters (some are referred to as HEPA filters) may be provided for controlling very small particles such as pollen and contaminants detrimental to health or where such particles are detrimental to a manufacturing process. These are associated with high pressure drops that you may need to consider when entering specific fan power.

Dampers

Dampers are devices fixed in the air stream that can open and close to restrict the flow of air. They consist of rows of blades that can be rotated to be parallel to the flow in the 'open' position, or rotated 90° to be perpendicular to the flow, with their edges touching, to close off the flow. Rotating the blades to angles between open and closed allows the flow to be varied.

Dampers can be used for a number of purposes in ventilation systems. In central ventilation equipment they are used with the blades connected to motorised actuators to automatically adjust the flow according to the required conditions (Figure 6.5). Two or more dampers can be arranged to operate together and divert the airflow from one path to another. This is typically done in air handling units to control recirculation and the introduction of fresh air or to shut off air flow when the system is off.

Figure 6.4.1: Example of a panel filter installation in an air handling unit

Figure 6.4.2: Example of a bag filter installation in an air handling unit. The flow direction is right to left

Figure 6.4: Filters

Figure 6.5: A motorised damper installed in an air handling unit

Air heat exchangers

Heating and cooling of the air that is distributed through a building is achieved indirectly by exchanging heat with another medium such as water, gas or refrigerant. This requires using some form of 'heat exchanger' component. The commonest form of heat exchanger in air conditioning systems is commonly called a 'coil'. These are connected to the chilled water supply system and used to cool the air, or are connected to the heating water system to heat the air. Where a coil is used for heating it is sometimes termed a 'heater battery'.

In such heat exchangers, rows of tubes, through which the heating or chilled water flow, are arranged perpendicular to the direction of air flow. The rows of tubes have closely spaced fins fixed to them between which the air passes. There are usually several rows of tubes that are arranged in a serpentine form (coil) and are connected to a larger header pipe that is, in turn, connected to the main heating or chilled water pipe system. A typical coil component is illustrated in Figure 6.6.1 and an installation is shown in Figure 6.6.2.

In the case of a heating coil the water passing through the tubes heats the fins which, in turn, heat the air as it passes through. (Car radiators can be classified as water-to-air heat exchangers and work on the same principle.) In the case of cooling coils, chilled water is circulated through the tubes to cool the fins. If the coil is cold enough the air may be cooled to the point at which moisture is caused to condense on the fin surfaces so that the air downstream of the coil becomes dehumidified. Condensate formed in this process must be drained away from the bottom of the cooling coil.

It is often possible to distinguish whether a heat exchanger is one used for heating or cooling by looking for ID bands on the connected pipes (e.g. LTHW or chilled water). One other way to distinguish whether a coil is one used for cooling is to see if there is a condensate drain near the bottom. A typical drain arrangement is shown in Figure 6.6.3. Similar, but smaller, heating and cooling coils are used in terminal equipment such as fan coil units as we will see later.

Refrigerant-to-air heat exchangers are also used in air conditioning equipment. These also consist of rows of tubes with closely spaced fins but in this case refrigerant passes through the tubes rather than water. The air is either heated or cooled depending on whether the refrigerant is made to condense or evaporate as it passes through the tubes respectively. These 'direct expansion' or 'DX' heat exchangers have efficiency benefits as no intermediate water circuit and associated pumps are necessary. Direct expansion heat exchangers are used in small terminal equipment such as split-systems but also in air source heat pumps and some air handling units (Figure 6.6.4).

Air can be heated using a gas fired heat exchanger. In this type of heat exchanger a forced draught burner is used to control combustion and drive the hot gases through a series of tubes or passages that cross the air stream. This type of heat exchanger is found in forced convection air heaters and also central heating and cooling systems. A gas supply pipe and the flue are normally visible (Figures 6.6.5 and 6.6.6) although the burner may be hidden.

Air heat exchangers can also use electrical power as their source of energy. The resistance heating elements in electric air heat exchangers are normally 'U' shaped elements mounted in rows perpendicular to the air stream. The elements may have a plain tubular shape, coils of wire or have small disc shaped fins. Electric resistance heating elements can be employed in air handling units, small terminal equipment or at other points in a duct system.

If you examine a piece of air conditioning equipment one of the first things to determine is whether it is able to both heat and cool the air, or just heat it. Some heat exchangers, such as those with a gas burner, enable the fuel source to be readily identified. Heat exchangers connected to a heating or chilled water circuit imply that the energy source is elsewhere and that you will have to look for this equipment in another part of the building before you can determine the fuel type. It is possible to give a few simple guidelines to identify air heat exchanger equipment and we do this in Table 6.1.

Commercial Energy Assessor's Handbook

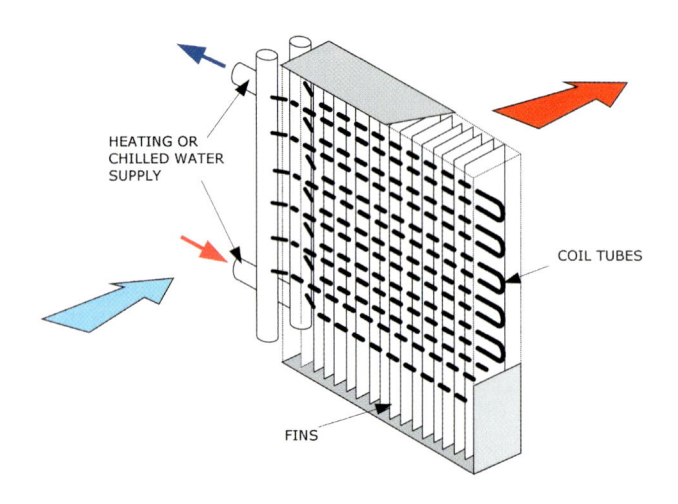

Figure 6.6.1: An illustration of typical heating or cooling coil construction

Figure 6.6.2: A heater coil installed in an air handling unit

Figure 6.6.3: An example of a cooling coil condensate drain

Figure 6.6.4: A direct expansion (DX) coil used to provide cooling in an air handling unit. The small refrigerant pipes typically have black foam insulation. There is a control valve but no other valves

Figure 6.6.5: An example of an air handling unit with a gas fired heat exchanger. The gas pipe connection is quite visible

Figure 6.6.6: An example of a flue associated with a gas fired heater. This balanced flue is found at the rear of the gas fired heater in figure 6.6.5

Figure 6.6: Air heat exchangers

90

Table 6.1: Tips to identifying air heat exchangers

Heat exchanger type	Identifying features
Hot water	• Flow and return pipe connections to the top and bottom – usually with isolating valves nearby and drain cock at the bottom. An automatic control valve is also usually nearby • The type of heating system (LTHW, MTHW, etc.) may be identified from pipe colour codes (see Appendix D)
Gas fired	• External gas pipe – uninsulated and colour coded yellow • Air inlet – combustion requires air and so you may see an inlet, such as a small louvre or grille in the side of the heat exchanger section. Fan assisted flues can be used on smaller units • Flue – combustion products are seldom mixed directly with the air supply and so a flue may be visible at the rear or top of the heat exchanger section
Chilled water	• Flow and return pipe connections to the top and bottom – usually with isolating valves nearby and drain cock at the bottom. An automatic control valve is also usually nearby • Chilled water pipes can be differentiated from heating pipes by their colour coding (see Appendix D) • Condensate drain pipe at the bottom of the cooling coil. A plastic or copper pipe with a U bend and sloping towards a drain should be visible
Refrigerant (DX)	• Relatively small pipes with foam insulation – usually black and without any colour ID bands • You may find a condensate drain – just as with a chilled water cooling coil • Refrigerant pipes do not have isolating valves as water coils do • Control valves may be visible but these look quite different to those in water systems. A very small capillary tube connecting the refrigerant control valve and the AHU may be visible. The control valves may be hidden inside the AHU

Air handling units

Complete air conditioning systems require a number of the components discussed above to be coupled together so that the air is 'conditioned' as it passes from one component to the next. There are three approaches to installing the equipment, each of which achieves the same air conditioning function. These can be summarised as:

- using inter-connecting sheet metal duct sections between each component;
- installing components within purpose made metal, brick or concrete chambers formed inside the building;
- using factory made modular air handling units (AHUs). This is the most common situation.

Air handling units (sometimes termed air handlers) are modular in construction so that different combinations of components can be arranged to suit particular applications. AHUs have a framed construction that is divided along its length into sections with fixed or opening panels enclosing each component. AHUs are installed in plant rooms within a building, or have a weather proof design that allows them to be installed outside on the roof or next to the building. AHUs in plant rooms need to be connected to fresh air intakes and exhausts elsewhere in the plant room or elsewhere in the building by intervening ductwork.

At this point we should note that although we think of AHUs being part of systems that provide both heating and cooling, the same type of modular equipment can be used in large systems that purely provide heating (Figure 6.7.1). This type of heating supply can be treated as an air central heating system in SBEM calculations.

AHUs are designed in a modular fashion and so it is difficult to identify 'standard' configurations. However, we have identified three types of AHU configuration that have commonly occurring features and illustrated these in diagrammatic form below. The symbols used in the diagrams are similar to those found in schematic drawings of HVAC systems (see Appendix E). In these diagrams the green arrows indicate flow of fresh air from outside. The blue arrows indicate the supply of conditioned air to the system. The grey arrows show flow of air extracted from the building or exhausted from the system depending on whether the arrow points inwards or outwards respectively.

Supply systems with heating cooling and filtration

The arrangement shown in Figure 6.7.2 does not include any extract fan or recirculation facility and is designed to only condition fresh air. This configuration would be used in a system that had a extract fan and duct system elsewhere in the building or plant room.

Supply and extract systems with heating, cooling, filtration and recirculation.

This AHU shown in Figure 6.7.3 is an in-line configuration with two fans and connections to the zone supply and extract ductwork at each end. The sections of the AHU are arranged end-to-end and recirculation is controlled in the sections with dampers (the mixing box) near the centre.

Figure 6.7.4 shows a 'double deck' configuration where the extract system components are arranged above those of the supply. In this configuration recirculation is controlled by three dampers in a mixing

Commercial Energy Assessor's Handbook

Figure 6.7.1: Example of an AHU installation inside a plantroom (some of the ducts are connected to louvres elsewhere in the building). This unit has only a heating coil. The three-port control valve is visible on the wall to the right

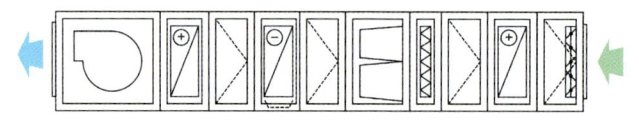

Figure 6.7.2: A diagram of an AHU used for conditioning a fresh air supply.

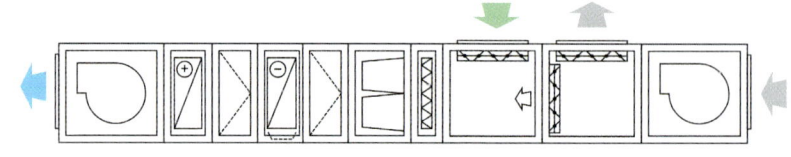

Figure 6.7.3: A diagram of an inline AHU

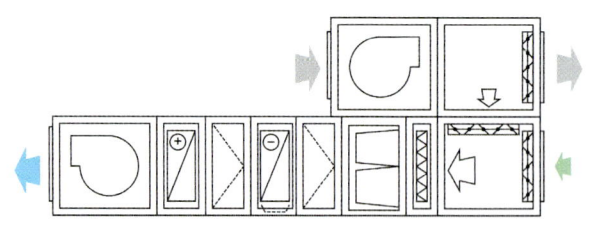

Figure 6.7.4: A diagram of a 'double deck' configuration AHU

Figure 6.7: Air handling units

box arrangement near the inlet and exhaust connections. An equivalent arrangement is with the exhaust components alongside those of the supply rather than on top.

There is no functional difference among the later two types of AHU configuration. The only difference is the physical arrangement of the supply and exhaust sections. The choice of configuration is mainly driven by the space available for HVAC equipment.

As we noted earlier it is difficult to identify 'standard' configurations of AHU. This is partly because the same function can be achieved with certain components placed in a different order in some cases. For example, the diagrams show the supply fan as the last component in the supply sections and so 'draw through' the air. An alternative is to locate the fan before the heat exchanger components in a 'blow through' arrangement. The heating and cooling coil positions can vary and some units have more than one heating coil. It is also quite possible to see the second set of filters located near the end of the supply section. We briefly mentioned humidification equipment earlier (although this can not be explicitly represented in SBEM calculations). You may find this type of equipment in AHUs in which case a humidification section will be included somewhere downstream of the heating and cooling coils.

HEAT RECOVERY

In all mechanical supply and extract ventilation systems air extracted from any zone has a temperature similar to the average room temperature. On a cold day heat can potentially be captured from this warm air and transferred to the fresh air being drawn into that system. This achieves some preheating of winter fresh air and a reduction in heating demand. Similarly in summer conditions, when it is hotter outside than the room temperature inside, incoming fresh air can be precooled by the exhaust air. The heat recovery systems that can be represented in SBEM calculations and their respective efficiencies are shown Table 6.2.

Table 6.2: Summary of the heat recovery systems available in SBEM software

Heat recovery method	Typical efficiency ratio (SBEM default values)	Specific fan power increase (W per l/s)
Plate heat exchanger	0.65	0.3
Thermal wheel	0.65	0.15
Run-around coils	0.5	0.3
Heat pipes	0.6	0.3

Plate heat exchangers (recuporator)

These devices consist of a sandwich of closely spaced flat plates between which air can flow. The heat exchanger is connected to the inlet and exhaust air streams such that the passages between the plates carry, alternately, inlet and exhaust air. Heat is conducted through the plates from the hot air stream to the cold. Plate heat exchangers can be incorporated into central air handling equipment or packaged separately for connection to the supply and extract duct systems. Where the plate heat exchanger is incorporated into an AHU it will be of the 'double deck' configuration and take up a good deal of space near the centre. A diagram of an AHU that includes a plate heat exchanger is shown in Figure 6.8.

Thermal wheels

A thermal wheel or rotary heat exchanger consists of a rotating wheel divided into two sealed sections. The exhaust air stream passes through one section and the inlet air stream passes through the other. Air is able to pass through small passageways in the wheel structure. As the wheel rotates, each sector of the thermal wheel is exposed to the exhaust air flow and then, when that sector has rotated further, it becomes exposed to the inlet air flow. The thermal wheel material changes its temperature and stores some heat as it rotates. Some heat is consequently collected from the warm air stream and carried to the cooler air stream. The heat exchange effectiveness of these devices can be high but additional energy is required to drive the small motor that slowly rotates the wheel. An example is shown in Figure 6.9.1 and an AHU diagram shown in Figure 6.9.2.

One of the ways that the presence of a thermal wheel can be checked is to see if there are controls for the motor on the AHU control panel – as in Figure 6.9.3.

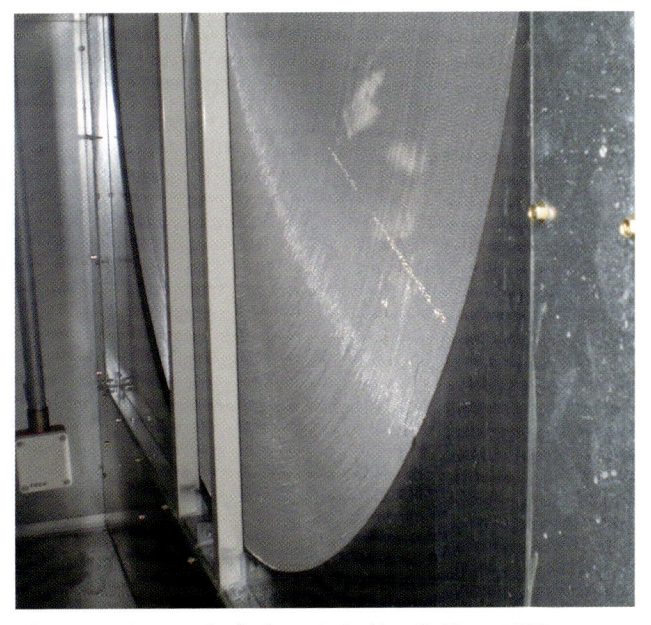

Figure 6.9.1: An example of a thermal wheel installed in an AHU

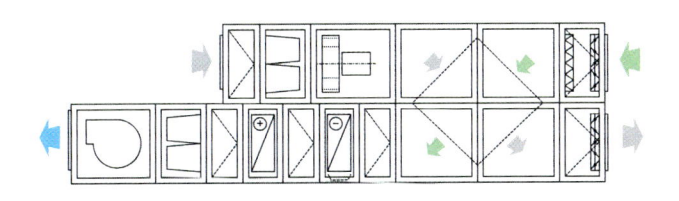

Figure 6.8: A diagram of a plate heat exchanger heat recovery device incorporated into an AHU

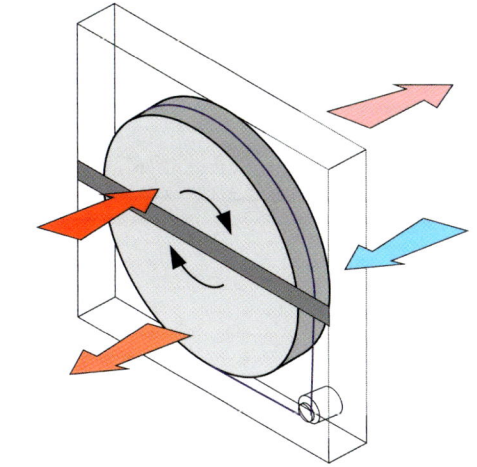

Figure 6.9.2: A diagram of a thermal wheel heat recovery device

Figure 6.9.3: This control panel for an AHU gives a clue to the existence of a thermal wheel

Run-around coils

A typical run-around coil system comprises a pair of finned tube coils in the supply and exhaust airstreams connected by pipes in a closed loop. A small pump simply circulates a water or antifreeze mixture around the pipe circuit. The warm air flowing through one coil heats the circulating fluid as it passes through the tubes of the coil. This heat is circulated to the other coil where it heats the cooler air. Efficiency is limited by their being two heat exchange processes and auxiliary energy being consumed by the pump.

In an AHU that provides heating and cooling and has a run-around heat recovery system, you should find the heat recovery coil adjacent to the fresh air inlet (Figure 6.10). The heat exchangers may be some distance away in separate pieces of equipment.

Figure 6.10: Example of a run-around coil arrangement. Note that this example has pipes that are not labelled and does not have valves like other coils

Heat pipe devices

A heat pipe heat recovery device consists of a pair of finned air heat exchangers that are connected together by heat pipes containing a special heat transfer fluid. The fluid inside the vertical heat pipes is evaporated by the warm air and circulates naturally to the neighbouring heat exchanger where heat is released to the cooler air stream. Alternatively horizontal pipes are used and fluid moves along a wick inside. No pump or compressor is necessary. The two heat exchangers have to be fixed close together in these devices and may be connected into the inlet and exhaust duct system rather than incorporated into an AHU. Such devices are not common.

Other heat recovery systems

You may find heating and cooling systems with forms of heat recovery other than the four types you can explicitly represent in SBEM calculations. One other system that is used relies on a reversible heat pump to transfer heat between air streams. The coils in each air stream are connected by refrigerant pipes. In these cases you should use a 'work-around' by selecting the most similar type of heat recovery system and entering more appropriate values of efficiency and specific fan power.

Fan speed control

Fans rarely have efficiencies above 65% and account for a significant part of the energy demands of air conditioning systems. Varying fan speed (and hence flow rate) in response to cooling or ventilation demand can save considerable energy. Controlling fan flow rate has three common applications:

(1) variable air volume (VAV) air conditioning systems;
(2) adjustment of flow rate in 'constant volume' systems; and
(3) demand controlled ventilation.

Speed control devices are known as 'variable frequency drive' (VFD), 'variable speed drive' (VSD) or sometimes 'inverter drives'. These electronic devices are coupled inline with the fans power supply. They tend to be rather bulky packaged devices with their own control interface as illustrated in Figure 6.11.

A VSD may be fixed in a number of places such as:

- within a motor control panel;
- mounted adjacent to a motor control panel;
- adjacent or on the outside of the AHU;
- inside the AHU.

It is good practice to adjust the speed of the system fans during commissioning to ensure no more energy is being used than strictly necessary. This is a lot easier if there is a VSD and fans with motors above 1.1kW are required, as of 2006, to include one (an example of a VSD used for this purpose is shown in Figure 6.3.3). You may find this provision has been retrofit to older systems.

Figure 6.11: Variable speed drive device installed in a plantroom

Demand controlled ventilation

Demand controlled ventilation systems seek to reduce energy consumption by automatically controlling the flow of fresh air according to indoor air quality conditions. In practice this means that the flow is automatically controlled according to CO_2 levels. This feature can be highlighted in SBEM calculations and some energy efficiency improvements claimed. In this type of system you would expect a fan speed controller and some form of carbon dioxide sensor – either in each room, in the extract ducts or in the AHU (Figure 6.12).

Figure 6.12: Example of a CO_2 sensor installed in a plant room used for demand controlled ventilation

Specific fan power

The amount of electrical power a fan requires is dependent on the flow rate but also the pressure losses in the system. It makes most sense to make comparisons and set standards on the basis of how much power the fans consume per unit of flow. This is what is meant by specific fan power (SFP) and this is defined with units Watts per litre per second (W/l/s). The electrical power in this context is that consumed in the electrical circuits as well as all the fan motors.

Specific fan power is one piece of data that is used to work out the efficiency of a heating and cooling system in SBEM calculations. A range of 'typical' values of SFP are adopted by SBEM software according to system type specified by the user. However, these values do not necessarily reflect the values limited by the 2006 Building Regulations.

The other situation in which you need to consider whether larger values of SFP need to be used in the calculations, is where you know particular components are installed that result in large pressure drops. You are required to increase SFP values where heat recovery equipment is present for example (Table 6.2).

AIR CONDITIONING SYSTEMS

Air conditioning systems can be subdivided according to whether the heating and cooling energy is distributed by air alone, or a combination of water and air. These categories are commonly called 'all-air' and 'air-water'. We can add another category of systems that use a combination of refrigerant and air. The heating and cooling systems that can be represented in SBEM calculations are listed below and are categorised in this way as follows:

All-air systems:

- single-duct VAV;
- dual-duct VAV;
- indoor packaged cabinet (VAV);
- constant volume system (variable fresh air rate);
- terminal reheat (constant volume);
- constant volume system (fixed fresh air rate);
- multi-zone (hot deck/cold deck);
- dual duct (constant volume).

Air-water systems:

- fan coil systems;
- induction systems;
- chilled ceilings or passive chilled beams and displacement ventilation;
- active chilled beams;
- water loop heat pumps.

Refrigerant systems:

- split and multi-split systems;
- air-to-air heat pumps;
- variable refrigerant flow (VRF) – also known as variable refrigerant volume (VRV).

It is also helpful to realise that these systems differ according to:

- the type of air distribution system:
    - single supply ducts,
    - dual hot and cold supply ducts,
    - direct room cooling (no ducting);
- the form of capacity control:
    - constant flow, variable temperature,
    - variable flow, constant temperature,
    - variable flow, local mixing;
- the ability to provide local occupant control;
- the type of terminal equipment local to each room. This could include:
    - control dampers,
    - heat exchangers,
    - secondary fans.

You will see from the descriptions that follow that the heating and cooling systems that can be represented in SBEM calculations can be distinguished in these ways. The features that distinguish the various types of system are summarised later in Table 6.3.

It is worth pointing out – before we get into the details of each type of system – some of the key things that may help you decide which type of system is being used in a particular zone. In most cases being able to examine any room terminal would enable you to readily identify the type of central air conditioning system. However, this is not always practical as terminal units are normally concealed above false ceilings or behind window sill panels. You generally have to rely on non-intrusive inspection of each zone and also plantroom or roof-top air handling equipment. We try to provide some suggestions for identification tips but you should bear in mind that these are not definitive indicators, access is sometimes poor and you should aim to collect a number of items of evidence before reaching a conclusion about the air conditioning system type.

- **AHU configuration.** You will see from the descriptions of each system that follow that certain types have more than one supply duct or divide the air stream into separate heating and cooling supplies near the outlet to the AHU. Consequently it is worth inspecting the supply outlet end of the AHU for such features.
- **Fan control.** Whether a system uses constant speed fans or has some means of varying the flow is a key distinguishing feature. There are three types of variable flow system that can be represented in SBEM as we describe below.
- **Outlet position.** You will see that certain terminals are able to be installed below windows and deliver their air supply upwards. Other systems are not suited to this arrangement.
- **Air diffuser type.** Inspecting a room's air supply diffusers can help in the identification of certain system types – or at least eliminate some possibilities. With a little experience you will get used to identifying diffusers and associating them with certain types of system. Systems such as displacement ventilation have very distinctive diffusers.
- **Heating system.** Many central air conditioning systems are, naturally enough, able to provide all the heating demanded and so no other form of heating system will be present: but this is not universally true. Systems such as single duct VAV are sometimes installed alongside conventional heating systems, for example. Buildings designed for hybrid operation are also found with heating systems in addition to some form of cooling.
- **Application.** You will see that certain types of system are best suited to only certain applications. It is not possible (or wise) to classify systems based only on application but, nevertheless, you should begin to appreciate the design intent of each system and the most likely applications. This knowledge may help focus your inspection.

We offer a number of tips and key points that will help you identify HVAC systems. We emphasise that these are not universally applicable or allow definitive identification. We also want to emphasise that we do not suggest opening equipment such as AHUs (not least because of health and safety hazards) or stopping/starting systems during assessment. This should not be necessary for the purposes of identification.

Variable air volume (VAV) systems

There are three varieties of all-air VAV system that can be defined in SBEM calculations:

- single duct VAV;
- dual duct VAV;
- packaged cabinet VAV.

Their basic operating principle is that the amount of heating or cooling that is delivered is varied by keeping the supply air temperature constant, but varying the air flow rate.

There are some common characteristics of variable volume air conditioning systems that help in distinguishing them from other systems:

- the presence of variable speed fans;
- the presence of local room temperature sensors; and
- the presence of room slot or variable geometry diffuser outlets.

Finding just one of these features is not enough, in itself, to show a zone has a VAV system but finding all three is a good indicator.

One difficulty with operating with reduced air flow rates in some rooms is that the low air speeds at the outlet of the terminal can lead to 'dumping' of the cold supply leading to draught problems. The preferred form of ceiling diffuser for variable flows is the slot type. Consequently this is quite common to see where single or dual duct VAV systems are being used. Having said this we should say that constant flow systems can also use slot diffusers. This is one reason we said earlier that one of the indicators is not enough, in itself, to identify a system as VAV.

Single duct VAV systems

In this type of system the fan in the AHU pressurises the main duct distribution system with air at a constant low temperature. In each room, a terminal (VAV box) controls how much air is allowed to flow from the main distribution system into the room. The box is coupled to a local temperature sensor and allows more air into the room as the cooling demand increases. The overall effect of the various terminals adjusting the flow is that the total system flow requirement varies and the main fan speed can be adjusted to match this. A schematic of a single duct VAV system is shown in Figure 6.13.1.

A single duct VAV system is primarily designed for situations where there is a predominant cooling demand (the AHU will still have coil for preheating the air). There needs to be additional means of heating the building. It is also sometimes necessary to provide heat to avoid overcooling certain rooms when heat gains are low. This heating capability can be provided in the form of a small coil built into the terminal box (a reheat coil) or by room heat emitters such as radiators or convectors.

VAV terminals are very simple and consist of a flow measuring device, a damper, an actuator that is connected to a controller to automatically adjust the flow. These features can be packaged into a 'VAV box' that is hidden in the ceiling void. A number of proprietary variable geometry diffusers achieve the same function. Some products combine the function of the flow controller and a slot diffuser.

The most common applications of this type of system are purpose built office buildings. An illustration of an office perimeter zone application is shown in Figure 6.13.2.

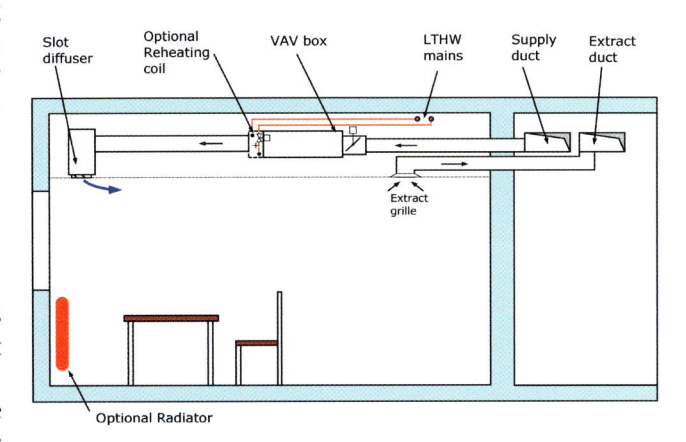

Figure 6.13.2: Illustration of an office single duct VAV system installation

Dual duct VAV systems

Dual duct VAV systems use separate supplies of hot air and cold air distributed to terminals in each room. Each terminal is able to control and mix the supplies of hot and cold air so that the air delivered is exactly at the right temperature according to the demand for heating or cooling. Figure 6.14 is a schematic of a dual duct VAV system.

The most obvious difference from other systems is that there are two independent supply systems (that may or may not be from the same AHU package) with variable flow fans. There are three fans in this system and three duct systems – two for supply and one for extract – and you should be able to verify this during inspection of the AHU installation. This form of VAV system can provide space heating through the air distribution system and so you would not expect to see any other form of heating in rooms with this system. This form of VAV system is more costly than others and is not common.

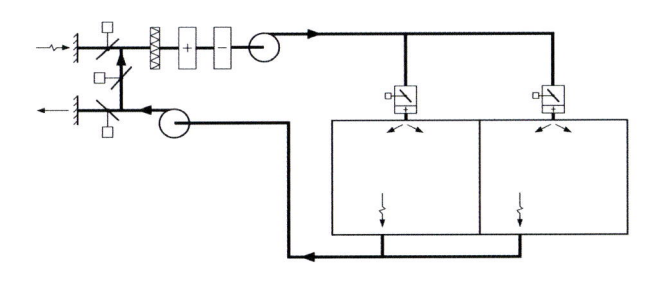

Figure 6.13.1: Schematic of a single duct VAV system. This example shows room terminals that have reheating coils

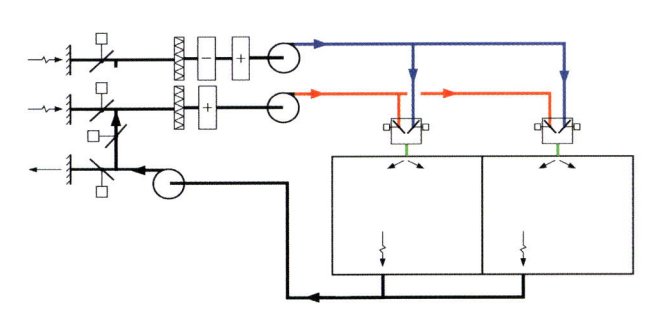

Figure 6.14: A typical dual duct VAV system – a number of variations can be found

Indoor packaged cabinet VAV systems

There is a variety of air conditioning equipment that could be called 'packaged cabinet' equipment. This type of equipment is designed to primarily provide cooling and not heating or fresh air. The air is recirculated using a single fan and simply drawn into the bottom of the equipment and discharged either through a large diffuser or possibly a small duct or underfloor air distribution system. A diagram showing cabinet equipment is shown in Figure 6.15.

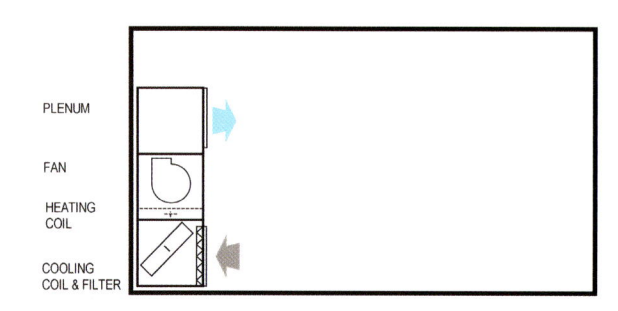

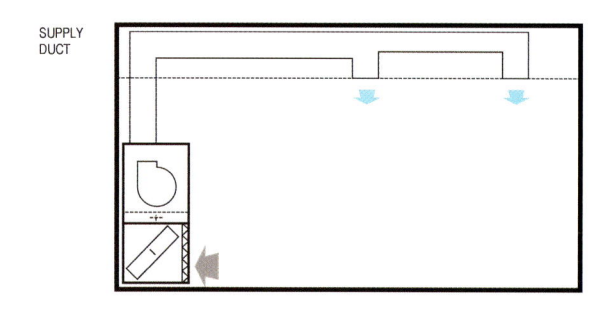

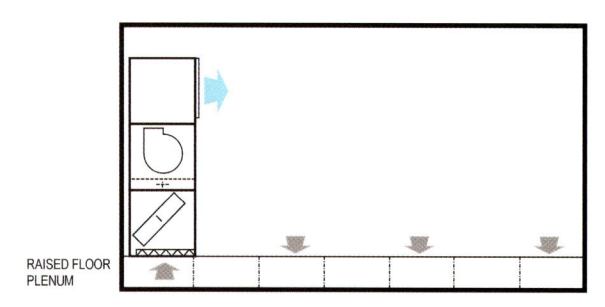

Figure 6.15: Possible cabinet system arrangements

Cooling may be achieved through a refrigeration heat exchanger which is coupled to an outdoor unit (something bigger than common split systems) or can use a chilled water heat exchanger. Where there is some requirement for heating the refrigeration equipment can be operated like a heat pump or some heat provided by a supplementary electric air heat exchanger.

One of the difficulties with identifying this type of system is that there is a variety of equipment of similar appearance that is designed for constant volume operation rather than VAV. This is probably a situation where you have to rely on collecting some manufacturers' information and doing further research. Cabinet equipment that has constant volume flow could be represented by another system in SBEM with a similar heating and cooling source and suitable choice of SEER.

Constant air volume systems

The state of a room's air is controlled by means of varying the temperature at the air inlet while maintaining a constant fan flow rate in 'constant volume' or 'constant air volume' (CAV) systems. As the fan runs at full speed even when the overall cooling demand is low, the fan energy can become a large part of the systems energy consumption. You can represent two simple types of CAV system in SBEM. These are described as:

- variable fresh air CAV; and
- fixed fresh air CAV.

In these common forms of air conditioning system the conditioned air is simply distributed to every room at the same temperature and a predetermined flow rate. This does not allow local control of temperature when multiple spaces are serviced. Such systems are more appropriate for single spaces. This includes large spaces such as cinemas, theatres and supermarkets. Constant volume systems have the advantage of being simple, relatively low cost, and not requiring any room terminal boxes. The difference between the two variants is related to how fresh air is controlled and recirculated in these systems. Other forms of system that can be grouped together as CAV systems are:

- terminal reheat CAV;
- dual duct CAV; and
- multi-zone (hot deck/cold deck).

These types of constant air flow system are more complex but allow local control of individual room temperatures.

Variable fresh air CAV

In many CAV systems it is possible to recirculate some of the air rather than discharge all the air extract from the rooms directly to the outside. Being able to vary the fresh air quantities (variable recirculation) saves energy for three reasons:

(1) minimising fresh air quantities during normal operation limits the associated energy demand;
(2) completely recirculating the air is efficient when the building is being preheated and no fresh air is required; and
(3) avoiding recirculation (using 100% fresh air) when cooling is required but it is cool outside reduces cooling demand (so called free cooling).

Recirculation is controlled by a set of three dampers between the supply and return duct systems – usually at the inlet to the AHU in a section called the mixing box ('economiser') as we noted earlier. This variable fresh air form of CAV system is shown schematically in Figure 6.16. An AHU configured with a mixing box was shown in Figure 6.7.4. Dual duct and multi-zone systems can also use recirculation and include a mixing box in the AHU.

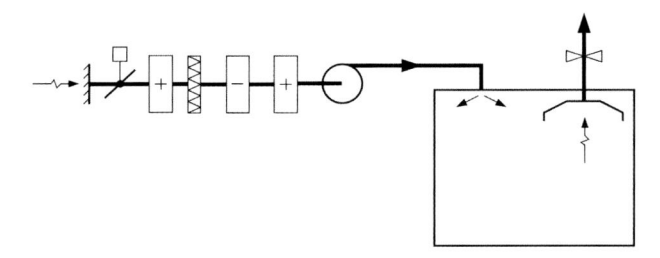

Figure 6.17.1: Schematic of a constant volume system designed to condition a fresh air supply

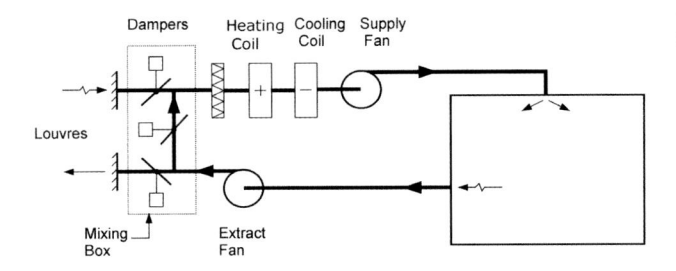

Figure 6.16: Schematic of a constant volume system that has dampers to allow variable fresh air flow

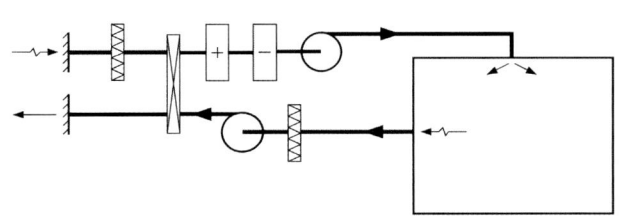

Figure 6.17.2: Schematic of a fresh air system incorporating a thermal wheel heat recovery

Figure 6.17: Fixed fresh air CAV

Fixed fresh air CAV

The are a range of systems that can be placed in this category and we will indicate the most common.

There are a number of situations where it is not possible to allow any of the air extracted from the rooms to be recirculated. Examples of applications where cross contamination of the air must be avoided include certain healthcare zones, kitchens and certain laboratory or manufacturing facilities. Where there is generally a high demand for a constant supply of fresh air, for example in densely occupied spaces, there is no advantage in having a recirculation capability. A fixed (or full) fresh air system like this is shown schematically in Figure 6.17.1.

In these types of system the extract fan and ducts are separate from the AHU. The extract fan may be near the zone served or be somewhere else in the same plantroom. A full fresh air AHU often has an additional heating coil near the inlet that is used in the very coldest conditions. If a large quantity of fresh air has to be supplied then heat recovery can be very worthwhile. A schematic of a fresh air system that uses a thermal wheel is shown in Figure 6.17.2.

You may find a variety of other CAV systems that incorporate recirculation but with a fixed proportion of fixed fresh air. Schematically they may be arranged like Figure 6.16 except that any dampers are not motorised to allow automatic control. This is less likely with AHU equipment but is quite possible with simple systems and a variety of packaged equipment. Systems that have a recirculation facility but use only one fan are generally fixed fresh air systems. If you think that the air conditioning system is a constant volume system but can't determine if there is a variable fresh air capability it would be conservative to assume that it is of the fixed fresh air type. Your Accreditation Scheme may recommend this approach.

Terminal reheat CAV

In applications where a CAV system supplies conditioned air to a number of rooms (e.g. partitioned offices), using a common supply temperature can become a problem. Different rooms may have varying cooling requirements but, as all rooms have a common air supply temperature in simple CAV systems, some may be overcooled and some may be too hot. There can be no local control of the temperature in simple CAV systems.

The only way to overcome this problem in CAV systems is to operate the system at the lowest required temperature and then 'reheat' the air at each room or zone where cooling demand is not so great. This means having a small heating coil fixed in the supply duct near the point when it enters each room. The reheat configuration is shown schematically in Figure 6.18. The reheat coils may be electric but are usually connected to a LTHW distribution system.

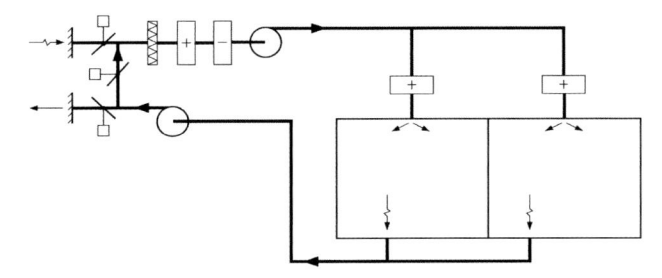

Figure 6.18: A schematic showing a constant volume system with terminal reheat

The default heating and cooling system for SBEM calculations is a constant volume system with terminal reheat and fixed fresh air. This is a conservative assumption as this type of system is not as energy efficient as most others.

Dual duct CAV systems

A dual duct CAV system conditions the air and distributes it to each room using two systems of supply ducts – commonly at high pressure. One duct carries cold air and the other carries hot air. In each conditioned space or zone, a terminal box mixes the warm and cold air in proper proportion to satisfy the room heating or cooling demand (Figure 6.19). A single constant speed fan is used in the AHU that has the flow divided and between parallel heating and cooling coils near the outlet.

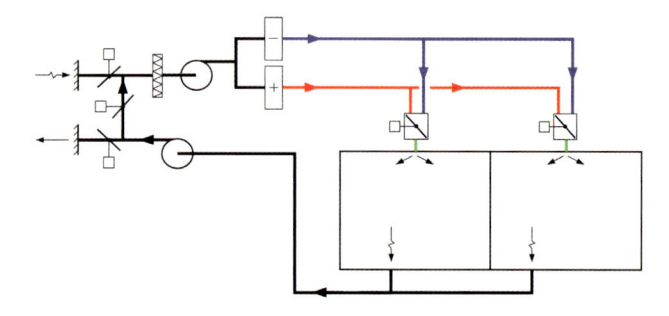

Figure 6.19: Schematic of a dual duct constant volume system

Dual duct CAV systems suffer the same cost, space and noise disadvantages as dual duct VAV systems and are a less common type.

Multi-zone systems

A multi-zone air conditioning system can be used to distribute conditioned air to a number of zones at a constant flow rate. The temperature of the air distributed to each part of the building is modulated by controlling the mixture of hot and cold air at the point where it leaves the AHU.

The air supply in the AHU is separated into two streams. One stream passes through a cooling coil to produce air at a steady low temperature, and the other stream passes through a heating coil (Figure 6.20). The two outlet sections of the AHU form what are termed a 'hot deck' and 'cold deck'. Motorised mixing dampers at the air handler outlet control the proportions of the hot and cold air leaving the AHU and subsequently supplied to each zone. The mixed, conditioned air is distributed to each zone of the building by a single supply duct.

An AHU providing conditioned air for a multi-zone system should be distinguishable by the multiple supply duct connection and the division of the air into parallel 'hot deck' and 'cold deck' paths. Multi-zone systems are only likely to be found in large buildings. Unlike dual duct CAV systems that may supply a large number of zones, multi-zone systems can be used to service only a limited number of zones.

Air-water air conditioning systems

Air-water systems distribute heating water and/or chilled water as well as some conditioned air from a central system to individual rooms. In every form of air-water system the terminal units in each room are connected to the chilled and heating water system and deliver the required heating and cooling energy. Each terminal unit can provide local temperature control.

Heating and cooling energy is distributed primarily by the heating and chilled water pipe systems in this class of air conditioning. Every room requires ventilation and so air-water air conditioning systems also comprise a central air distribution system and corresponding AHU – usually a modest CAV system – to distribute preheated or precooled fresh air. The types of air-water system that can be explicitly represented in SBEM calculations are:

- fan coil units;
- induction units;
- chilled beams;
- chilled ceilings;
- water loop heat pumps.

Chilled beams and chilled ceilings only provide cooling and not heating.

Fan coil unit systems

The basic elements of fan coil units are a pair of heating and chilled water coils, filter, and fan. The fan recirculates air at a constant flow rate by drawing it from the room, blowing it through the heating and cooling coils, and delivering it to the room. Where the fan coil unit is hidden in a false ceiling, short ducts and simple diffusers are connected to the outlet.

This type of terminal unit is available in various configurations to fit under window sills, above false ceilings, in ceiling bulkheads or simply to be fixed to or recessed in walls. As these terminals include a cooling coil some dehumidification can occur and so a condensate drain is also required. Heating and cooling functions suggest two heat exchangers and four pipe connections to each fan coil unit. This is usually the

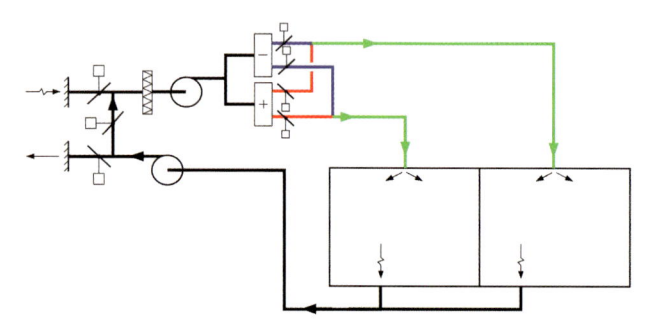

Figure 6.20: Schematic of a multi-zone system

Ventilation and Air Conditioning Core Knowledge and Recognition

case but some systems (termed two-pipe systems) use only one heat exchanger connected to a single water distribution system. In this case the pipe system is used to distribute either heated or cooled – a change over from one to the other being necessary at times. Fan coil unit systems are relatively cost effective and can be found in many purpose-built office buildings in the UK.

Induction units

Induction units differ from fan coils in that they operate without fans in the room terminal. The primary air supply enters a small chamber in the room terminal and then is forced out at high velocity through small nozzles. This creates a low pressure near the inlet to the terminal and an induction effect that draws air in from the room. This room air flows across the water coil and is mixed with primary air before being discharged back into the room in the form of a wide jet of air (Figure 6.21.1). As the air in the terminal is forced out of the jets at high speed they tend to be rather noisy.

Induction unit terminals are mostly suited to application under windows. Horizontal discharge of the air supply is possible but only when mounted behind a ceiling bulkhead or behind a wall. Induction terminals discharge air into the room through relatively long thin slots. Consequently they are usually found with linear grilles such as those shown in Figure 6.21.2.

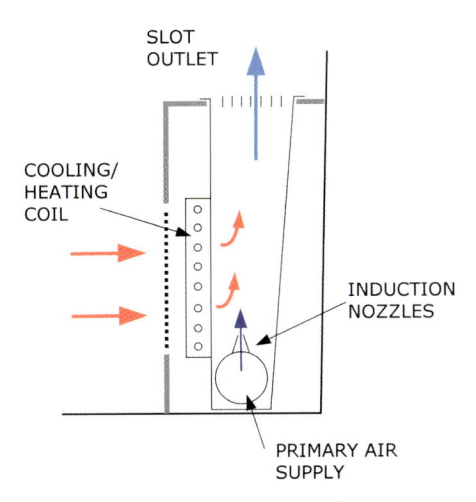

Figure 6.21.1: Diagram of air flow through an induction unit

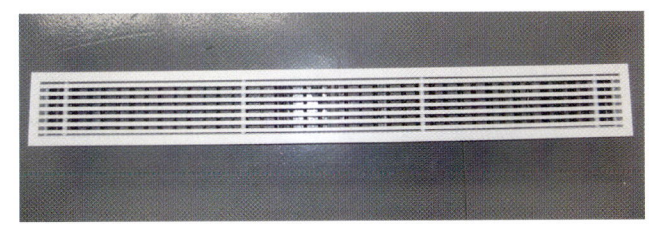

Figure 6.21.2: Linear grilles like this are often used with terminals under windows such as induction units or fan coil units (Reproduced with the kind permission of Gilberts (Blackpool) Ltd)

Figure 6.21 Induction units

Chilled ceilings

A chilled ceiling achieves a cooling effect by absorbing long-wave radiation from the room and cooling the air near the ceiling. The system is composed of cooling panels installed into a ceiling that results in a combination of radiation and convective cooling of the room.

Chilled ceilings comprise metal panels with small chilled water pipes incorporated into them. The panels are either suspended below the ceiling or constructed as part of the false ceiling. The panels are usually designed to look the same as any other ceiling tiles.

Chilled beams

Passive chilled beams are a water-air cooling system that uses convection cooling elements mounted at ceiling level. In this type of system nearly all the cooling effect is via convective heat transfer. The cooling elements consist of simple pipes with finned elements (Figure 6.22). Passing cooled water through the elements induces buoyancy forces that cause a flow of cooled air down towards the occupied space. This is the same principle as convection heating except that the fins are cooled and the flow of air is downwards.

Chilled beams are usually arranged in rows in open plan spaces. They are often designed to fit in well with the false ceiling and lighting design but can also be suspended in rooms without false ceilings. As with chilled ceilings it is often the case that chilled beams are combined with a displacement ventilation primary air system.

Chilled beams can have their cooling output enhanced by incorporation of a direct connection to the primary air supply. This form of chilled beam is called an active chilled beam. This type of system has a separate category in SBEM calculations.

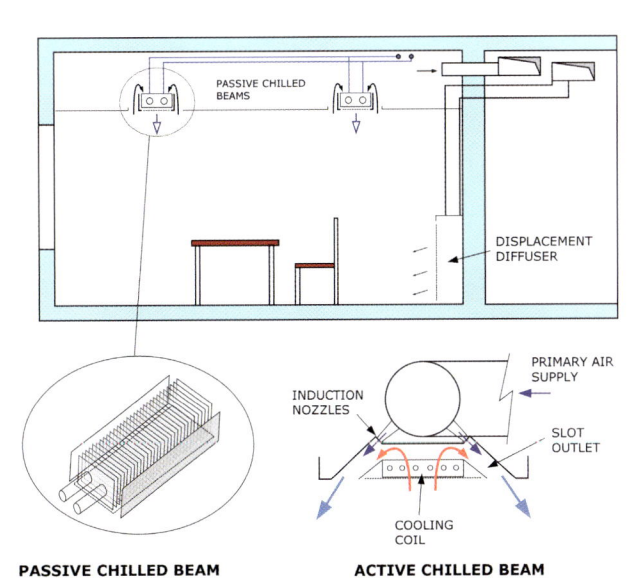

Figure 6.22: Chilled beams

Displacement ventilation

Common mechanical ventilation systems deliver air near the ceiling and use diffuser outlets that are designed to produce jets of air moving across the ceiling. The jets of air are strong enough that the air in the room is stirred up and becomes well mixed. This makes the air about the same temperature everywhere in the room. Displacement ventilation is a form of air distribution system that uses a very different principle. In this form of ventilation air is introduced at very low speed near the floor. The cool air is not mixed but is drawn upwards wherever there is a source of heat – e.g. computers and occupants. This happens naturally as the plume of warm air generated by heat sources draws air in and moves it towards the ceiling. A layer of warm air forms near the ceiling and leaves the room via a conventional grille.

Displacement ventilation diffusers tend to be devices with a large vertical surface area and positioned at the edge of the room (Figure 6.23) or possibly next to a column.

It is assumed, by default, in SBEM calculations that displacement ventilation is the type of ventilation used with chilled ceiling and passive chilled beam systems. If the chilled ceiling or passive chilled beam installation does not include a mechanical ventilation system (i.e. no primary air supply from a duct system) you should set the system specific fan power to zero. If a mixing ventilation system is being used (i.e. conventional diffusers) then this can be represented by doubling the specific fan power according to the *iSBEM User Guide*.

Water loop heat pumps

Small self-contained heat pumps of the water-to-air type can be used as room terminal units in a centralised air conditioning system. In large buildings small heat pumps of this type can be connected to a water loop that circulates water through all the heat pumps and is also connected to central heating and cooling sources. Heat pumps do not need high temperature heating or low temperature chilled water like other room terminals. Instead, water is circulated around the loop and maintained within a moderate temperature range of between 15 and 35°C. In this temperature range a heat pump can efficiently cool or heat the room by exchanging heat with the water in the pipe loop. The water in the loop is heated by the heat pump when it is cooling the room and vice versa.

A system like this (sometimes called 'versa-temp' systems after the brand name) has the advantage that when some parts of the building require heating, and other parts require cooling, the heat transferred to and from the water loop by the heat pumps partly balances out. Heat can be added to the water loop by a heat source such as a boiler when there is a larger heating demand. Similarly, when there is a predominant cooling demand, the water loop can be cooled by a central chiller or heat rejection system such as dry cooling. A larger heat pump can also be used as the central heating and cooling source. This type of system can be found in buildings such as large offices and hotels. A diagram of this system is shown in Figure 6.24.

Figure 6.23: Example of a displacement ventilation diffuser (Reproduced with the kind permission of Gilberts (Blackpool) Ltd)

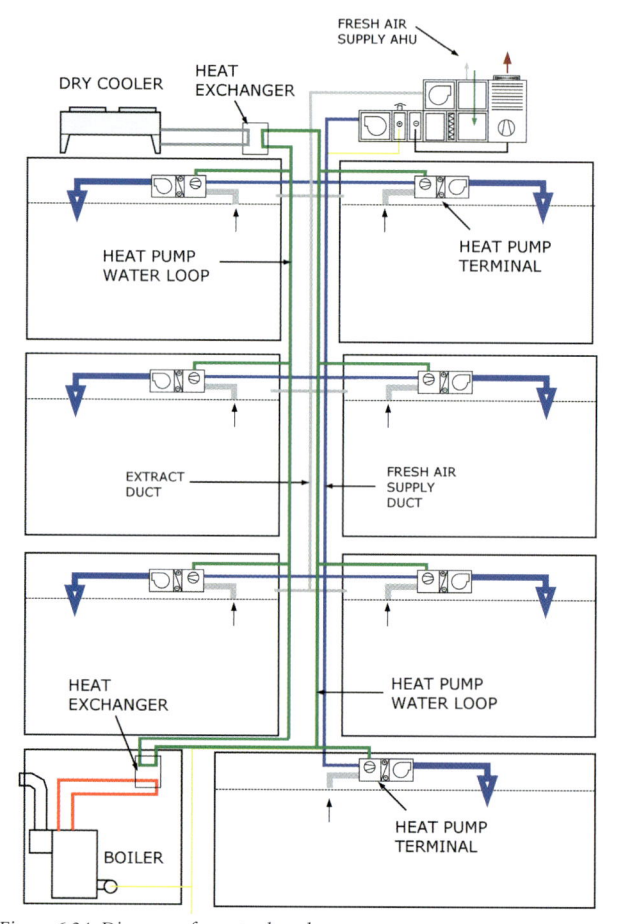

Figure 6.24: Diagram of a water loop heat pump system

Table 6.3. A summary of air conditioning system features

KEY FEATURES

	Fan speed/flow control	Supply ducts from AHU	Local temperature control	Diffuser types	Room terminal	Room terminal heating coils	Room terminal cooling coils	Room terminal fan	May be found with secondary heating
VAV systems									
Single duct VAV	Yes	1	Yes	Slot or variable geometry	Yes	Possible	No	No	Yes
Dual duct VAV	Yes	2	Yes	Slot or variable geometry	Yes	No	No	No	No
Indoor packaged cabinet VAV	Yes	None or 1	Yes	Slot, variable geometry or none	No	No	No	No	Yes
CAV systems									
Constant volume, fixed fresh air	No	1	No	All types	No	Possible	No	No	No
Constant volume, variable fresh air	No	1	No	All types	No	Possible	No	No	No
Constant volume, dual duct	No	2	Yes	All types	Yes	No	No	No	No
Constant volume with reheat	No	1	Yes	All types	No	Yes	No	No	No
Multi-zone, hot deck/cold deck	No	2 or More	Yes	All types	No	No	No	No	No
Air-water systems									
Induction system	No	1	Yes	Linear grilles	Yes	Yes	Yes	No	No
Fan coil units	No	1	Yes	All types	Yes	Yes	Yes	Yes	No
Chilled ceilings or passive chilled beams with displacement ventilation	No	1	Yes	Low level displacement types only	No	No	No	No	Yes
Active chilled beams	No	1	Yes	Beams have built-in slots	No	No	Yes	Yes	Yes
Water loop heat pump	No	1	Yes	All types	Yes	DX type	DX type	Yes	No

The heat pump room terminal can be installed in a similar way to fan coil units, i.e. under a window sill, above a false ceiling, on a wall, or in a ceiling bulkhead.

Refrigerant air conditioning systems

There is a wide variety of mass produced air-conditioning equipment intended for use in individual rooms and small premises. This equipment uses vapour compression refrigeration to directly cool the room air and discharge heat to the air outside. Such equipment incorporates a fan and high velocity outlet used to recirculate the air in the zone. Heating and ventilation is generally achieved by some other means.

Such systems have the advantage that they are suitable for retrofit applications. Small zones or individual rooms in larger buildings without central air conditioning can be conveniently cooled with these systems. Applications also include retail and restaurant premises and computer server rooms.

Single room cooling systems

Single room cooling systems, in their simplest form, comprise self-contained factory-made air conditioners that are installed into an opening in an outer wall or window. A self-contained refrigerant system produces the desired cooling. The indoor room side of the unit is separated from the outside. Air is drawn by a fan from the room, through a filter, over the evaporator coil and then is returned, chilled, back to the room. At the same time the outside air is blown over the

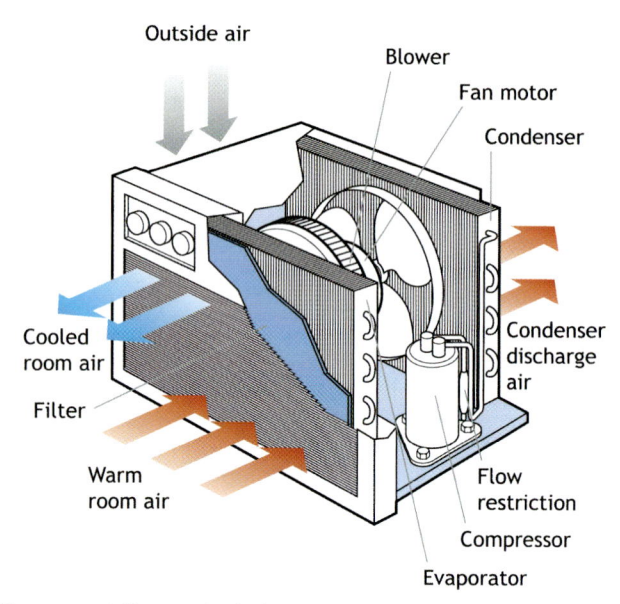

Figure 6.25: A diagram of a single room air conditioner. (Reproduced with the kind permission of Carbon Trust)

condenser to carry away the 'rejected' heat to the outside (Figure 6.25). There are no external connections to the unit other than an electrical supply. These systems can be retrofit quite easily, and are used in large numbers for residential air conditioning in some parts of the world but are not as popular as mini-split systems in UK commercial buildings.

Split and multi-split systems

Split system air conditioners are so called because the system is separated into indoor and outdoor units. The indoor unit that provides cooled air simply includes a fan and an air-refrigerant heat exchanger (the evaporator) along with its control system. The outdoor unit includes the refrigerant compressor, air-refrigerant heat exchanger (condenser) and a propeller fan. The refrigerant pipes connecting the units are small in diameter, often covered in black foam insulation and fixed alongside the system's electrical cables.

Although some split systems may have a capacity of several kilowatts the most common are known as 'mini-split' systems and have units with capacities below 3kW. The room units of split systems come in a number of forms. Some are intended for wall mounting (Figure 6.26.1) and others for incorporation into a ceiling – so-called 'cassette' units (Figure 6.26.2). You may also find floor standing systems.

Multi-split type air conditioners are based on the same principle as single split air conditioners except that a number of indoor units (up to about eight) can be coupled to a single outdoor unit.

These types of cooling system allow a good deal of occupant control. You will often find rooms with split, multi-split or VRF systems have multi-function controllers that display a variety of information and allow times and temperatures to be adjusted. An example of such a controller is shown in Figure 6.26.3. Some systems have infrared remote control.

Figure 6.26.1: A wall mounted split system room air conditioning unit

Figure 6.26.2: A ceiling mounted (cassette) split system room air conditioner

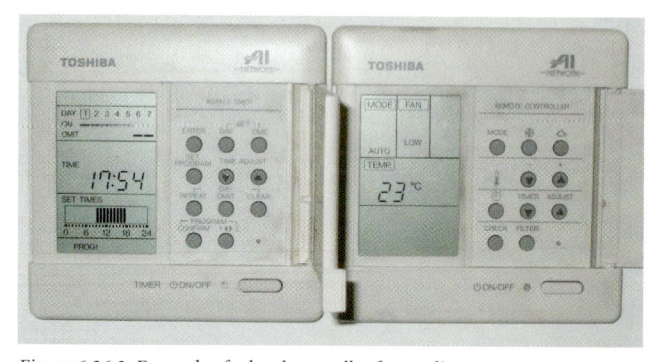

Figure 6.26.3: Example of a local controller for a split system

Figure 6.26.4: An example of a split system outdoor unit (condenser)

Figure 6.26: Split systems

Condenser units may be found on an outside wall nearby or further away on the roof of the building. It is good practice to check, at the end of the survey, that you have associated all the outdoor units you have found with equipment found inside the building. Split and multi-split systems are mass produced but by relatively few companies so that you may soon recognise common brands. It can be useful to note brands and product labels as you carry out the survey as this may help you match up indoor and outdoor equipment if there are several systems.

Split and multi-split systems are primarily intended for cooling and other forms of heating, such as radiators, are often found in the same room. However, some split system equipment can also provide heating by operating as an air-to-air heat pump. This may be the case if no other form of heating is provided (remembering some internal rooms and equipment rooms will not necessarily have heating systems). It will be difficult to tell if a system provides heating in this way without taking a careful look at the controls and product labels. If heating is provided by the split system, this can be accounted for in SBEM calculations by specifying the heat source as an 'air-to-air heat pump'.

Variable refrigerant flow systems

A variable refrigerant flow (VRF) system, also known as variable refrigerant volume (VRV) system, is essentially a sophisticated multi-split system. The difference between true VRF and multi-split systems is the ability to provide heating and cooling simultaneously in different parts of the building.

Where several indoor units are connected to the same system there is a good chance that not all the units require maximum refrigerant flow at the same time. In some systems this is taken advantage of and the overall flow of refrigerant is reduced in these conditions by slowing down the compressor. The compressor speed control system is often referred to as an 'inverter drive'.

Figure 6.27: VRF rooftop equipment

The variable flow principle can be applied in larger multi-split systems as well as true VRF systems. In the case of a multi-split system the whole system must work in either cooling or in heating mode. In the case of a VRF system different rooms can be heated and cooled at the same time. These systems use three pipes in the refrigerant pipe network rather than two. An example of some VRF rooftop equipment is shown in Figure 6.27. The indoor units of VRF systems are not easily distinguishable from those of other systems and it is probably only by looking carefully at outdoor equipment and taking details from labels that you will be sure a particular system is a VRF type.

For the purposes of SBEM calculations VRF systems are (at present) placed in the same category as split/multi-split systems. The efficiency of these systems is arguably better than common split and multi-split systems and so it is worthwhile making some effort to find if a refrigerant cooling system is of the VRF type, checking some product details, and deciding if you should use a higher seasonal efficiency in the calculations.

Air source heat pumps

Air source heat pumps (also called air-to-air heat pumps) take a number of forms and vary in scale. Some split or multi-split systems are able to operate as heat pumps. This means that the refrigeration system can be reversed so that the indoor unit can provide heating (these systems are often termed reversible split systems). This feature may not be obvious when inspecting the room unit and may require you to check the manufacturers' information. The controller can also give some clues about any heating function. If you find a perimeter room that appears to have split or multi-split equipment this might be something to look into if you find the room does not have any other means of heating.

Larger packaged air source heat pump systems can be found. This equipment can be in the form of self-contained rooftop equipment that is designed to be connected to supply and extract duct systems below. Such equipment is usually in the 10–50kW range. Air-to-air heat pump systems can also be incorporated into larger rooftop AHU equipment.

The key point, in terms of SBEM calculations, is that air source heat pumps are not listed as a specific type of 'heating and cooling' HVAC system. Rather, 'heat pump (electric): air source' is defined as a type of heat source or cooling source. This means that in order to represent an air source heat pump you must first choose one of the available HVAC system types and then assign 'heat pump (electric):air source' as the heating system and cooling system. For example, to represent a reversible split system our suggestion is to assign 'split or multi-split' as the HVAC system type and then select 'heat pump (electric): air source' as the heating system.

Similarly, in the case of larger rooftop packaged equipment you could select the most similar air conditioning type (based on the fan control and air distribution system) and then select 'heat pump (electric): air source' as both the cooling system and heating system. For example, if the unit uses a CAV principle and introduces a fixed quantity of fresh air our suggestion would be to assign 'constant volume system (fixed fresh air rate)' as the main HVAC system type and assign the air source heating and cooling types. There are similar oil/gas fuelled air source heat pump options.

Cooling sources

The chilled water used in many central air conditioning systems is distributed from a central cooling source in the form of a chiller of some type. A chiller is a refrigeration system that extracts heat from the chilled water circulation system and discharges it to the environment. Just as heating systems have to be associated with heat sources such as boilers in the assessment process, so air conditioning systems have to be associated with cooling sources.

Chillers vary in two principal ways. First, in the refrigerant compressor type used, and secondly the way in which heat is rejected to the environment. Heat extracted from the chilled water by the refrigeration system ultimately has to be rejected to the air outside and this can be achieved in more than one way. Heat can be rejected from the system:

- directly to the air outside via a built in refrigerant-air heat exchanger when the chiller is an air cooled chiller;
- to a heat rejecter outside via a secondary water circuit when the chiller is a water cooled chiller;
- to a condenser system outside via refrigerant pipes when the chiller is one with a remote condenser.

Heat rejection equipment is located on the roof of the building, or next to the building, and you may notice from the points above that the pipes connecting the equipment to inside differ according to the type of chiller. Understanding this can help in recognition of the type of cooling source. The different arrangements are summarised in Table 6.4.

Table 6.4: A summary of chiller and heat rejecter types

Chiller type	Pipe circuit outside	Heat rejecter type
Air cooled	Chilled water	Built-in refrigerant-air heat exchanger
Water Cooled	Condenser water	Dry air cooler or cooling tower
Remote condenser	Refrigerant	Dry or evaporative remote condenser

We noted earlier that chillers vary according to the heat rejection method and also compressor type. Compressor type does vary, in broad terms, according to the size of the equipment. You do not need to be concerned with identifying compressor types but you do need to estimate the capacity of the equipment within four ranges for SBEM calculations. These ranges are: (i) less than 100kW; (ii) 101–500kW; (iii) 501–750kW; (iv) 750kW–3.5MW. This categorisation is based on the size of each piece of equipment in a multiple chiller installation rather than the total. You may need to collect manufacturers' information to determine the relevant capacity range. The fuel type associated with all chillers in SBEM calculations is grid supplied electricity.

Although you are not required to collect information about the heat rejection equipment used in any chiller installation for SBEM calculations, having some understanding of air cooled condensers, dry air coolers and cooling towers can help in identifying the chiller system.

Air cooled chillers

Air cooled chillers incorporate all the refrigeration and heat rejection components necessary to produce chilled water in a single piece of equipment. This type of chiller is usually what is meant by the term 'packaged chiller'. All the refrigeration and heat rejection heat equipment is factory assembled so that the only pipe connections made during installation are those with the chilled water circulation system. This arrangement is shown schematically in Figure 6.28.1.

Air cooled chillers have a few common features that make them relatively easy to identify. In every case the refrigerant-air heat exchanger, with its closely spaced fins, will be quite visible at the sides of the chiller. A number of fans on the very top of the chiller are used to draw air through this heat exchanger. In small air cooled chillers such as that shown in Figure 6.28.2 other components may be hidden behind the heat exchanger. In much equipment the heat exchangers are fixed in a 'V' formation with the fan above. The refrigeration equipment such as the compressor with its motor and the chilled water heat exchanger is often visible below (Figure 6.28.3 and 6.28.4).

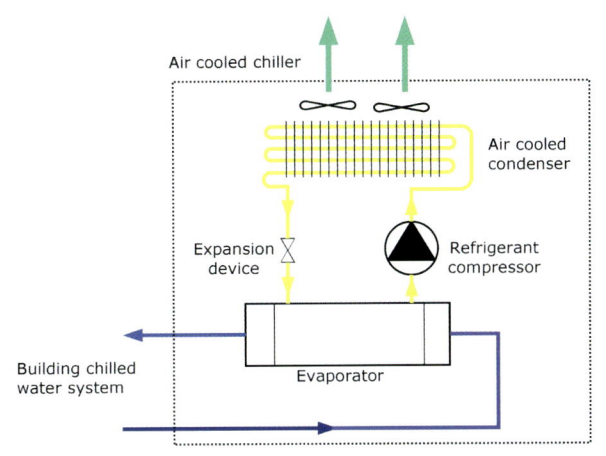

Ventilation and Air Conditioning Core Knowledge and Recognition

Figure 6.28.1: Schematic of a chilled water system with an air cooled chiller

Figure 6.28.2: An example of a small air cooled chiller. (Reproduced with the kind permission of Coolmation Ltd)

Figure 6.28.3: An example of an air cooled chiller. The chilled water pipe connections to the refrigeration equipment are quite visible

Figure 6.28.4: An air cooled chiller. The 'V' form of the heat exchanger coils is very noticeable

Figure 6.28: Air cooled chillers

Water cooled chillers

A water cooled chiller uses a condenser in its refrigeration system that is a refrigerant-water heat exchanger. This is packaged with the compressor and chilled water heat exchanger in a single piece of equipment designed to be located inside a plant room. The complete cooling plant consists of the water cooled chiller connected by a second pipe circuit to heat rejection equipment outside.

A water cooled chiller is connected to two pipe systems. One system is the building's chilled water system and the second is a separate circuit connecting the chiller to the heat rejection equipment outside (Figure 6.29.1). This circulation system is known as the condenser water system and may have the pipes identified using a single white band.

Small water cooled chillers (<100kW) can be enclosed but very often the three main parts of the chiller will be quite visible (Figure 6.29.2). You can expect to see two tubular heat exchangers – one the condenser and one the evaporator – and one or more compressors.

Chillers with remote condensers

Chillers with remote condensers have the main components of the refrigeration system packaged

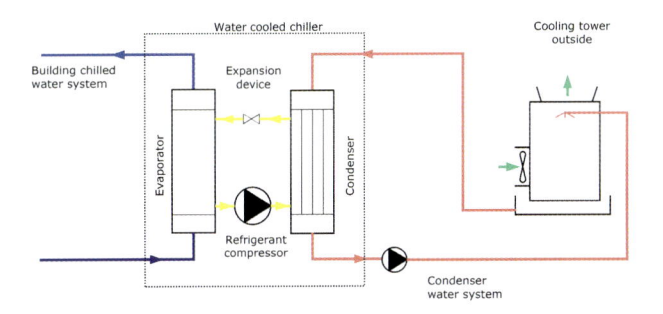

Figure 6.29.1: Schematic of a water cooled chiller system and cooling tower

Figure 6.29.2: Water cooled chiller plant room installation arrangement (Reproduced with the kind permission of Coolmation Ltd)

Figure 6.29: Water cooled chillers

107

differently to air and water cooled types. The chiller equipment consists of the main chilled water heat exchanger (evaporator) and the compressor system. The condenser part of the system is a separate piece of equipment located outside the building (Figure 6.30.1). In this type of chiller, refrigerant pipes connect the chiller inside to the condenser outside.

The condenser equipment used with remote condenser chillers is often quite simple and consists of just refrigerant-air heat exchangers in a frame with a series of horizontal fans above. Air is drawn through the heat exchangers from below. This type of heat exchanger may appear quite similar to a dry air cooler except that the pipes connecting it to the systems inside the building are refrigerant pipes rather than water pipes. Pipe connections are often flanged and bolted in water systems but never in refrigerant systems. An example of a condenser is shown in Figure 6.30.2.

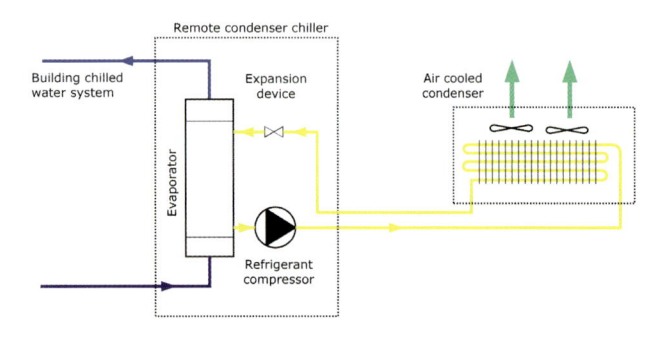

Figure 6.30.1. Schematic of a chilled water system with a remote condenser

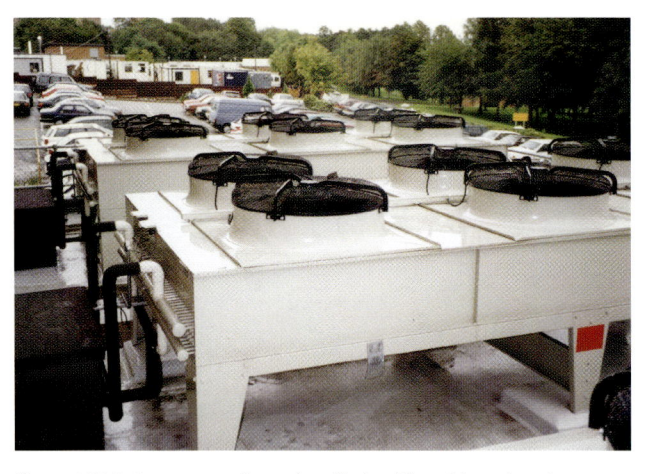

Figure 6.30.2: A remote condenser installation. The refrigeration pipe connections are exposed on the left. (Reproduced with the kind permission of H. Guntner (UK) Ltd)

Dry air coolers

Water cooled chillers require some sort of heat rejection equipment to cool the condenser water circuit. This can take the form of a cooling tower that uses evaporative cooling principles – as we discuss below – or what can be termed a dry air cooler (sometimes called an 'air blast cooler' or just 'dry cooler'). The operating principle is simply that heat is rejected by passing the warm condenser water through water-air heat exchanger coils and drawing large amounts of air across these using a number of fans. The heat exchanger coils can be horizontal (Figure 6.31.1) or arranged in a 'V' in larger systems (Figure 6.31.2).

Figure 6.31.1: A flat dry air cooler installation. The heavily insulated condenser water pipes can be seen in the foreground. (Reproduced with the kind permission of H. Guntner (UK) Ltd)

Figure 6.31.2: A large dry air cooler installation. (Reproduced with the kind permission of H. Guntner (UK) Ltd)

Figure 6.31: Dry air coolers

Cooling towers

Evaporation of water from a surface, or evaporation of water droplets in the air, creates a cooling effect. This principle is employed in cooling towers and these can be used in large chilled water systems to cool the condenser water. In all cooling towers water is sprayed downwards at the top and falls through the centre as air flows upwards and out of the top. The water cools as some portion of it evaporates and the rest is collected in a tray at the bottom of the tower. Cooling towers only tend to be used in very large chiller installations in the UK (Figure 6.32).

Cooling towers have been recognised as the source of outbreaks of legionella infections and now fall under the *Notification of Cooling Towers and Evaporative Condensers Regulations* 1992. This can be helpful in assessment as, if you believe the building uses a cooling tower you should be able to confirm this by consulting the authority. This can be little trouble as many local authorities list the properties online (e.g. www.ealing.gov.uk/services/environment/cooling

_towers). You will usually see that much of this equipment is registered to companies using them in industrial processes rather than HVAC cooling.

Figure 6.32: A large cooling tower installation (Reproduced with the kind permission of Coolmation Ltd)

RENEWABLE ENERGY SYSTEMS

The types of renewable energy source that can be explicitly represented in SBEM calculations, and which you may have to assess, are:

- solar energy systems (for hot water);
- wind turbines; and
- photovoltaic systems.

It is worth noting these renewable technologies are suitable for retrofit. Biomass boilers and ground source heat pumps were discussed in Chapter 5 in the context of heating sources.

In assessing these features of the building you will need to collect geometric information but also make reference to manufacturers' data. Manufacturers' information is likely to be available from the owners or directly from the manufacturers or their websites.

Solar energy systems

Solar hot water collectors absorb solar irradiation and transfer heat to a circulating water/antifreeze fluid that, in turn, transfers heat to a storage cylinder. Such systems are able to provide some proportion of the total hot water demand and are always found alongside a primary hot water generation system. You should consequently see some evidence of the solar hot water system when inspecting the main hot water generation system. You may see a second hot water storage cylinder or connections to a second heating coil in the main storage cylinder. The solar hot water system will also have its own pumps.

Two forms of solar heating panel are in common use:

- flat panel solar collectors; and
- evacuated tube solar collectors

Although it is not necessary to distinguish between these two types for data entry in to your SBEM software you do need to be aware of their different appearance in order to identify them.

Solar heating panels are found fixed directly to the roof structure in a frame or, where there is a flat roof, they may be fixed into a separate frame that tilts the panel to an optimum angle. Flat panel types are sometimes fully integrated with the roof structure.

Flat panel solar collectors in their simplest form consist of rows of metal tubes bonded to a metal sheet and this may be finished with a special coating to encourage absorption of solar irradiation. The flat collector plate is enclosed to form an insulated panel with a transparent cover (Figure 6.33.1).

Evacuated tube solar collectors consist of rows of glass tubes that contain fins bonded to a heat pipe. Each of the heat pipes is fixed into a header pipe fixed across the top of the panel. An example is shown in Figure 6.33.2.

Figure 6.33.1: A flat panel solar hot water panel installation (Reproduced with the kind permission of Viessmann Ltd)

Figure 6.33.2: An evacuated tube solar collector panel (Reproduced with the kind permission of Viessmann Ltd)

Figure 6.33: Solar energy systems

The data you need to collect to account for the energy and carbon saved by the solar hot water system is:

- panel area;
- orientation (SW, S, SE, etc.); and
- inclination angle – zero means horizontal.

You do not need exact measurements of the angle and orientation as inclination is entered in 15 degree increments and orientation in 45 degree increments of the compass. You also need to associate the system with another conventional hot water system in the building (it is not possible to specify a solar system as the only hot water generator).

Photovoltaic systems

Photovoltaic cells are made from semi-conductor materials that are able to directly convert solar irradiation to low voltage electrical energy. A photovoltaic panel (module) consists of rows of cells connected together to provide a direct current power source. Several of these panels or modules are connected together to provide a power source at a suitable voltage. Panels are often mounted, much like hot water panels, in frames fixed in the plane of the roof or in a separate tilted frame on flat roofs. It is also possible to closely integrate certain types of PV panels into the fabric of a building so that they can be found built into vertical wall surfaces or even glazing systems.

Photovoltaic cells have developed as lower costs and higher efficiencies have been sought so that you may come across a few different types of panel. These types and the efficiencies SBEM assigns to them are shown in Table 6.5.

Table 6.5: Photovoltaic panel types and their efficiency

PV cell type	Assumed efficiency
Mono-crystalline silicon	15%
Poly-crystalline silicon	12%
Amorphous silicon	6%
Other thin films	8%

The cells used in mono-crystalline PV panels consist of rows of these circular cells (approximately 120mm diameter) and this makes them quite recognisable (Figure 6.34.1).

Poly-crystalline cells are made using a casting process so that the finished cell material consists of a large number of crystals. These cells are less efficient but, as they are cheaper and can be cast in rectangular shapes, they are useful in building applications. This is probably the most common type of panel at present. If you look closely at the panel you can usually see the crystal structure of the cell surfaces (Figure 6.34.2).

Amorphous silicon is not crystalline but consists of silicon atoms deposited on a substrate material. This technology is often called 'thin film' PV and this is potentially cheaper although less efficient than other types. As the material is not crystalline it has a rather plain colouring. You made need to consult some of the building documents or manufacturers data before classifying a thin film panel as 'amorphous silicon' or 'other thin films'.

All photovoltaic systems generate direct current power and this has to be converted to alternating current at a suitable voltage to allow connection to the building's electrical system. The electrical equipment that does this is commonly called an 'inverter' and you may find this type of device mounted in a plant room or next to the building main meter. You may also find a meter used to record the PV system output (Figure 6.34.3).

Figure 6.34.1: Mono-crystalline PV panel (Reproduced with the kind permission of Viessmann Ltd)

Figure 6.34.2: Polycrystalline PV panel showing the distinctive colouring panel (Reproduced with the kind permission of Viessmann Ltd)

Figure 6.34.3: Example of electricity meter arrangement that includes an additional meter for the photovoltaic power generation system

Figure 6.34: Photovoltaic systems

Besides identifying the PV cell type; the other data you need to collect is the same as for a solar hot water panel – area, orientation and inclination.

Wind turbines

There are two basic classes of wind turbine – vertical axis and the more common horizontal axis. Horizontal axis wind turbines, as their name suggests, rotate about a horizontal axis and so the blades rotate in a vertical plane. In addition to the main rotor blades you should see some mechanism to rotate the whole turbine so that it always faces into the wind (Figure 6.36.1).

Figure 6.36.1: Horizontal axis wind turbine mounted on a building

Vertical axis devices are found on some buildings and consist of a vertical shaft at their centre about which the blades rotate. An example of a vertical axis wind turbine is shown in Figure 6.36.2.

Figure 6.36.2: Modern building integrated vertical axis wind turbine

Wind energy can be calculated from knowing wind speeds, the swept area of the turbine blades and the power conversion efficiency. The swept area of the blades is calculated differently depending on the class of turbine. This can be calculated from the rotor diameter of a horizontal axis wind turbine, or from the height and width over the blades in the case of a vertical axis wind turbine. This is the most important information that needs to be collected.

The wind speed that is really needed to estimate the annual energy is that at the centre of the blades. Wind speed varies very noticeably near the ground and so this is calculated according to the height – the hub height in the case of a horizontal device or height at the centre of a vertical device. Local wind speed can vary according to how the surrounding terrain disrupts the airflow near the building. SBEM accordingly adjusts the effective wind speed according to the type of terrain you select. Local wind speeds can be affected significantly even by relatively unobtrusive obstacles, so we suggest you are fairly conservative in your choice.

You need to specify the capacity (kW) of the turbine in order to make your SBEM calculation. You may be able to get some idea of the capacity from the size but more likely building documentation or manufacturers' data (there are very few manufacturers so finding data should not be too difficult). Wind turbines tend to be maintained by third party companies so you may find information if the client has invoices or other information about maintenance visits.

Power generated by on-site wind turbines has to be connected into the building's main power system in a carefully controlled way – much as PV systems. You may be able to find evidence of power control and metering equipment inside the building probably near the main meter. Again power from the wind turbine system is likely to be separately metered.

SUMMARY

Your aim, in carrying out site inspection and examining documentary evidence, is to build up a picture of relationships between zones and their HVAC systems. Although the range of air conditioning systems in use is very large, we hope you will have seen that there are enough key distinguishing features for you to make clear identifications.

7 Lighting Core Knowledge and Recognition

INTRODUCTION

In this chapter we consider the building service installation that currently makes the largest contribution to CO_2 emissions: lighting systems. We have divided the chapter into four distinct sections:

- a very brief review of lighting theory and practice, including glossary;
- lighting systems and practical lamp recognition;
- lighting in SBEM;
- lighting controls and recognition; and
- some practical aspects of lighting.

Satisfactory identification of lamp types, control gear and controls means 'getting up close' and really interrogating the systems. In practice, this can be quite difficult.

LIGHTING THEORY AND PRACTICE

The nature of light

The *CIBSE Lighting Guide – The Industrial Environment* (p. 1) states:

> 'Lighting, as used in industry, has three objectives: to facilitate quick and accurate work, contribute to the safety of those doing the work and to create a good visual environment.'

You should remember those requirements when you consider recommendations generated by SBEM.

Light in the everyday environment is a form of electromagnetic radiation in a relatively narrow band of wavelengths. Light visible to the human eye ranges from a wavelength of around 380nm (nanometres), i.e. violet, through indigo, blue, green, yellow and orange (in that order) up to around 760nm, i.e. the colour red. What we perceive as white light comprises a combination of all of those different colours and their various hues. Natural light comprises light visible to the human eye, infrared and ultraviolet light.

Ultraviolet light sits in the electromagnetic spectrum between 100 and 400nm, i.e. below violet, and infrared has a wavelength of around 800nm, i.e. above red. Although artificial lighting can emit some ultraviolet and infrared radiation, the human eye cannot detect these. The electromagnetic spectrum is illustrated in Figure 7.1 and shows where light 'sits' amongst the other types of radiation.

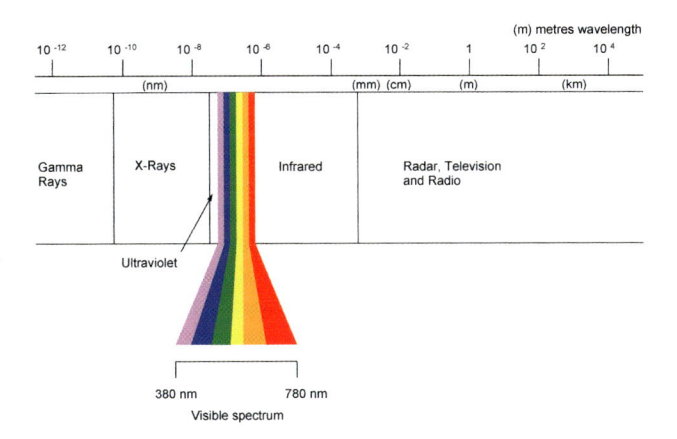

Figure 7.1: Electromagnetic spectrum

In lighting practice, a light 'bulb' (the source of the light and so named because of the shape) is known as a 'lamp' – bulbs are for gardeners and as an energy professional you should never refer to the electric lamps in your home or business as 'bulbs'. Lighting professionals also refer to what most people would call a 'light fitting' as a luminaire. We explain a number of lighting related terms in a glossary later in this chapter.

It is helpful to consider all light, artificial or natural, in the following simple way:

- you measure the power of a lamp or natural light (from the sun, our local star) in 'luminous intensity', using candela (cd) as a method of measurement;
- as the light travels from the source to the object it will illuminate, it is known as 'luminous flux', measured in lumens (lm); and
- when the light arrives at the object and illuminates the surface, the amount of illumination is measured in 'lux'.

As light (or as you now know it 'luminous flux') travels, it tends to spread outwards. The technical name for this

Commercial Energy Assessor's Handbook

effect is the 'inverse square law' whereby illuminance from a light source reduces in inverse proportion to the square of the distance from that source. In simple terms, the further you are from a lamp, the less light falls on you. Methods of concentrating light in beams are hence much sought after.

Colour temperature

Figure 7.2 shows the colour temperature scale of light in K ('Kelvin'), with some typical lamps shown for reference. The lamp temperatures are generalised and are subject to change as lamp characteristics alter and, generally, improve.

Counter-intuitively, higher colour temperatures (5300K or more) are 'cool' colours (green–blue) and the lower colour temperatures (2700–3300K) are 'warm' (yellow–red).

You can consider different lamps in terms of the practical application of light to various tasks:

- 'cool' light, 5300K and higher, is considered better for visual tasks – i.e. fluorescent lamps for offices;
- colour temperatures in the 2700–3600K range are recommended for most general indoor and task lighting – e.g. tungsten and tungsten halogen lamps for retail space;
- 'warm' light is preferred for living spaces that are more flattering to skin tones and clothing.

Colour temperature of different lamps

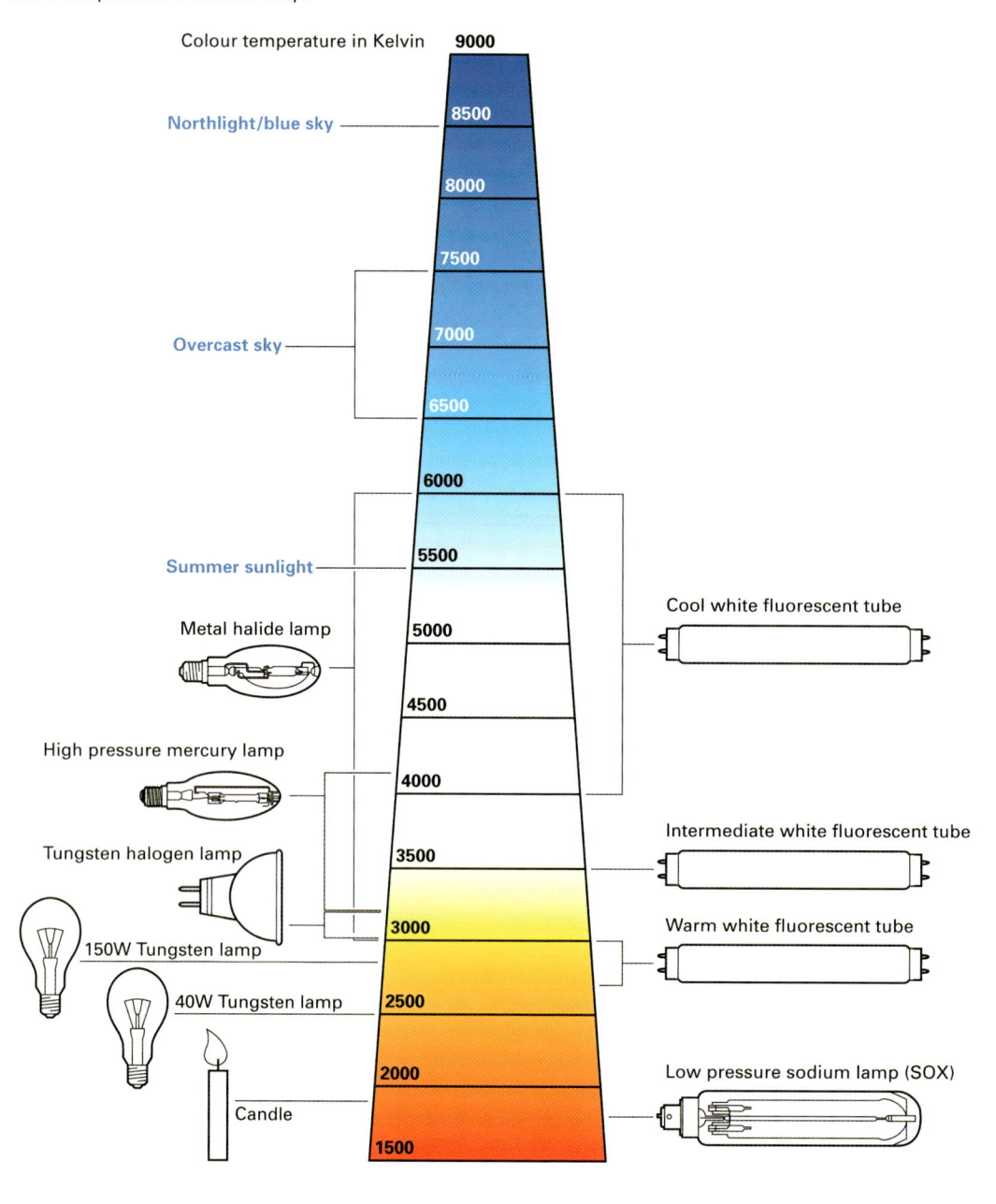

Colour temperature in Kelvin	Reference
9000	
8500	Northlight/blue sky
8000	
7500	
7000	Overcast sky
6500	
6000	
5500	Summer sunlight
5000	Metal halide lamp / Cool white fluorescent tube
4500	
4000	High pressure mercury lamp
3500	Tungsten halogen lamp / Intermediate white fluorescent tube
3000	150W Tungsten lamp / Warm white fluorescent tube
2500	40W Tungsten lamp
2000	Candle / Low pressure sodium lamp (SOX)
1500	

Figure 7.2: 'Colour temperature' of light, p. 29, Appendix A, 'Display lighting, Technology Guide CTG010', Carbon Trust, 2008

Colour rendering

This concept is a quantitative measure of the ability of a light source to reproduce the colours of various objects 'faithfully' in comparison with an ideal or natural light source. The international 'colour rendering index' (CRI or R_a) has a scale of 1 to 100 with higher numbers indicating better colour rendering.

- 100 is equivalent to daylight;
- 90–100 is required when people need to accurately identify colour in circumstances such as in a hospital or museum, or fine technical work;
- 80 and above is good and considered appropriate for most conditions when people are working;
- less than 80 is appropriate for areas where people are not always present, e.g. a storage room, or car park.

Examples of lamp colour rendition are:

- tungsten 'incandescent', lamp, i.e. a typical household 'bulb' – R_a = 100;
- fluorescent lamps – R_a between 50 and 90;
- low pressure sodium 'discharge' lamp, i.e. the normal amber street light – R_a ~5 (thus it is difficult to differentiate between car colours at night).

The colour rendering of fluorescent lamps varies considerably depending on the coating inside the tube. Some tubes give quite poor colour rendering but modern coatings are able to provide good colour rendition.

Identification of lamps

Many lamps do not have adequate identification, or manufacturers' identification numbers or letters are not universally recognised. This makes lamp identification difficult in many instances.

The international colour code is often, but not always, used to indicate the temperature of a lamp's light on the lamp. This code is a three digit number. The first digit refers to the colour rendering index; thus, if the number is 8, then the R_a (CRI) is between 80 and 90. The next two numbers are the colour temperature, to the nearest hundred, in hundreds.

As an example, consider a fluorescent lamp with the following information stamped on the tube – '18W/8 27':

- '18' refers to the lamp power in watts – this may not always be the total wattage for the fitting as you might also need to consider 'ballast' power. A fluorescent lamp is a discharge lamp, so will have a ballast that will contribute to the total 'circuit watts' – see later;
- '8' confirms the CRI or R_a of 80+; and
- '27' is an indicator of the colour temperature – in this case 2700K.

Older lamps tend not to have the R_a indicator, but will only have the colour temperature, e.g. 18W/35 – 18 watts power input and colour temperature of 3500K.

Levels of lighting

The lighting industry recommends certain minimum standards of artificial lighting (illuminance) for different activity types. This is referred to in SBEM data as the 'design illuminance'. Examples include 100 lux for a restaurant, 150 for a public house, 300 for a classroom and 500 for an office and shop or supermarket.

You can review the levels of artificial light SBEM assumes in the activity database.

Building Regulations

Approved Document L2A indicates that 'reasonable provision' for compliance with that document would be where average 'initial efficacy' is 45 luminaire-lumens/circuit watt for office, industrial and storage areas in all types of building. The calculation 'luminaire-lumens' is arrived at by multiplying the lumens for the lamp, or the lamp plus control gear (see later) by the 'LOR' – the 'light output ratio' (see glossary). The description 'office' includes zones with activities that involve sitting at a desk, i.e. classrooms, consulting or meeting rooms and similar.

For other spaces except for areas with display lighting the requirement is met with an efficacy of not less than 50 lumens/circuit watt, including an allowance for the ballast. For display lighting, the equivalent figure is 15 lumens/circuit watt.

Lighting glossary

You will find many different descriptions of lighting installations in manufacturers' literature. Here is a glossary of some of the terms you may encounter. We have indicated the most important definitions for SBEM in **bold**.

Table 7.1: Lighting glossary

Item	Description and/or explanation
Ampere (amp)	Unit of electrical current, in lighting terms a measurement of the amount of current passing through a cable or fitting. A = W ÷ V
Ballast or choke	Component of discharge lamp control gear
Batten	A simple fluorescent lamp fitting for surface mounting and usually a single bare tube
Candela	Measurement of 'luminous intensity' – 1cd is approximately equivalent to the power/energy of one candle
CIE (Commission Internationale de l'Eclairage)	International lighting standards commission. The commission coordinates the system that describes all colours and has defined Colour Rendering Index
Colour	Visual effect on the brain when the eye is stimulated by light in the various wavelengths
Control gear or system	Combination of electrical or electronic components including ballast, power factor correction capacitor and starter. High-frequency (HF) control systems sometimes include other components in order to allow dimming of the output
Diffuser	Translucent screen that is part of the luminaire, fitted on the user side to protect the lamp and also soften the output of light – sometimes a major problem for energy assessors since it can prevent you identifying the lamp type
Discharge lamp	Type of lamp that produces illumination by 'exciting' phosphors through electric discharge via two electrodes through a gas, a metal vapour or a mixture of gases and vapours that become ionised and act as an electrical conductor. The ultraviolet radiation caused by the 'excitation collisions' is converted to visible light by a fluorescent phosphor coating applied to the inside of a glass outer casing or tube. This type of lamp requires control gear to start the lamp and maintain and regulate the voltage level
Efficacy (luminous efficacy)	Ratio of light (luminous flux) emitted by a lamp to the power consumed by it, i.e. the energy input. Calculated in lumens per Watt, i.e. **luminous efficacy = luminous flux output ÷ electrical energy or power input**. It is by implication therefore a measure of efficiency, i.e. how efficiently the lamp can convert the electric power it consumes (in 'Watts') compared with the light it produces (measured in 'lumens'). When the control gear or ballast losses are included, it is sometimes referred to as 'lamp circuit luminous efficacy' or lumens per circuit Watt (lm/W). Lamps with high efficacies are more efficient
Filament lamp	See 'incandescent lamp'
Hertz (Hz)	Unit of measurement of frequency, measured in cycles/second
HID	High intensity discharge lamp
Illuminance	Amount or density of light, or **luminous flux**, falling on a surface, measured in lux. Lighting levels are specified in terms of illuminance on a working plane, e.g. 500 lux at desktop height in an office
Incandescent lamp	Lamp which produces light by heating a filament (length of metal) to incandescence ('white heat') by passing an electric current through the filament, e.g. a tungsten lamp
Initial efficacy	Efficacy of a lamp before any deterioration arising from use or age
Kelvin (K)	Absolute temperature. This is the unit of measurement of colour temperature
Lamp-lumens	Amount of luminous flux emitted by a lamp
Light output ratio (LOR)	Ratio of the total light output from a lamp together with the luminaire in which it is fitted, compared with the light output from the lamp only
Low voltage lamps	Refers to 12 volt tungsten halogen lamps
Lumen (lm)	Unit of luminous flux, used to describe the amount of light produced by a lamp or falling onto a surface
Luminaire	Light fitting and lamp together with all of the associated components for holding, containing and protecting the lamp(s) along with the control gear
Luminous flux	Defines the total amount of light emitted from a lamp, usually measured in 'lumens' (lm) at a standard temperature of 25°C
Lux	Unit of measurement of the intensity of light falling on a surface and has the units lux. A convenient quantity to specify 'lighting levels'. 1 lux is equivalent to illuminating a surface 1m distant from one candle, or one lumen/m²
Photometer	Light meter or device that measures luminous intensity of a light source
PIR	Passive infrared, usually presence/motion detectors
R_a	Colour performance of a lamp defined by the colour rendering index ('CRI' or 'R_a') which describes the ability of the lamp to show colours accurately or faithfully
Rated average lamp life	Period when 50% of lamps in a system fail under test conditions

Table 7.1: Lighting glossary (cont.)

Item	Description and/or explanation
Re-strike	Time a lamp needs to illuminate after it is switched off and on again
Soft start	A lamp control feature that helps to prevent the first 'shock' after switch-on. This can result in darkening of the lamp surface and reduction in lamp life
Start-up	Time the lamp requires for illumination, after being switched on from cold – a significant consideration for some lamps, e.g. metal halide and high pressure mercury
Task lighting	Lighting designed to provide high levels of illuminance where this is only required locally. This lighting is in addition to the main ambient lighting system. Examples include desk mounted lights and those built into work benches
Utilisation factor	Proportion of luminous flux emitted by a lamp that reaches the working plane
Volt (V)	Unit of electromotive force. $V = W \div A$
Watt (W)	Measure of energy. Watts $= A \times V$. One kilowatt (kW) is 1000W

ARTIFICIAL LIGHTING SYSTEMS AND THEIR RECOGNITION

Types of lamp

Humankind first used artificial lighting in the form of fire, and then progressed to candles, gas and oil lamps. In 1841 electric arc lamps lit parts of Paris. The incandescent lamp was developed by Edison and Swan in 1878. In the UK there are currently three main sources of artificial lighting systems:

- incandescent lamps – do not require control gear;
- gas discharge (HID) lamps – require control gear; and
- light emitting diodes (LEDs) – 'solid state' technology.

Incandescent, or filament, lamps are the simplest and oldest design of lamp. They are also generally the most wasteful in energy terms. A typical gas discharge lamp is a fluorescent tube, or sodium street light – usually more efficient than incandescent lamps. LEDs develop light by exciting a semi-conductor crystal – see later.

Recognition of lamp type

We now describe some of the characteristics of different types of lamp and how to recognise those differences, together with some pictures of them. Look at figures in this chapter and the other publications we refer to for help with identification You will find the information about efficacies, colour temperature, lamp life, etc. varies depending on the publication you consult. This is due to different manufacturers' data and the fact that efficacies and life of lamps are continuously improving due to technological advances.

Tungsten filament (GLS) lamps – incandescent

These are the 'general lighting service' lamps we know from before the days when use of 'low energy' lamps became important. They usually have a standard shape but it is easy to confuse modern versions with other types of lamp. Start-up and re-start ('re-strike') of tungsten lamps is immediate, i.e. they achieve instant maximum output of illumination.

The lamp has a filament made from tungsten, a metal that melts at 3400°C; all contained within a glass (soda-lime cilicate) container (the 'bulb' shape) filled with inert gases such as nitrogen and argon to help oxidisation, or burning, of the filament. The lamp produces light by using electrical energy to heat the filament to incandescence. During use, small pieces of tungsten evaporate, or flake off from the filament and lodge on the inside of the lamp (commonly visible as darkening of the glass). This process eventually causes the filament to break and the lamp therefore to fail.

Emitted light has a 'warm', usually 'off-white' appearance.

Only 5–20% of the energy a tungsten lamp consumes converts to light. The balance of the energy becomes heat. Such a lamp can be prone to glare; but this affect can be reduced by adding a frosting or opal effect onto the glass (one manufacturer describes the effect as 'soothing'), containing the lamp within a luminaire, or adding a diffuser (cover) between the lamp and the object it is illuminating.

Figure 7.3: Tungsten lamps

The EU is working towards almost total banning of production of GLS lamps by 2012 (and low efficiency halogen lamps by 2016), beginning in September 2009. They will be replaced by either compact fluorescents or a new high pressure halogen lamp currently under development.

Tungsten halogen (TH) lamps – incandescent

A tungsten lamp filled with halogen added to inert gases such as krypton, argon or xenon. When the filament warms the halogen heats up and glows – this provides most of the visible light. During use the tungsten that evaporates from the filament does not adhere to the inside surface of the lamp due to presence of the gas, but instead re-attaches to the filament, albeit at the hotter parts which causes thinning and eventual failure of the filament but nevertheless extends the life of the lamp when compared with 'ordinary' tungsten lamps and means the bulb can be smaller than a GLS lamp. The xenon gas-filled and infrared internally coated lamps are recent improvements that increase efficacy of TH lamps.

The shape of this lamp type is more varied than standard tungsten lamps. Start-up is immediate. They produce a brilliant white light to enhance texture and colour in contrast with the 'warmer' light of GLS lamps. Since tungsten is not deposited on the inside of the bulb during use, TH lamps do not darken with age – light quality remains constant throughout their lives.

Because the glass bulb is so close to the filament, tungsten halogen lamps can be much smaller than conventional lamps. However, they operate at high temperature and an electrician must take care when fitting them that they are not fitted close to combustible materials. Some versions are prone to premature failure

Typical applications are in display areas and for 'down-lighters'.

Fluorescent 'strip' discharge lamps – (MCF)

This type of lamp works by forcing electrons from cathodes at each end of the lamp to collide with mercury atoms in an atmosphere of low pressure mercury, argon or argon-krypton mixture contained in a glass tube with a fluorescent coating on its inside surface. This type of lamp requires additional electrical equipment to control starting and regulate the current and we will discuss this in more detail later in this chapter.

Older lamps incorporate a 'calcium halo-phosphate' coating on the inside of the tube. Modern lamps incorporate 'tri-phosphor' coatings developed in the

Figure 7.4.1: 12v tungsten halogen lamp

Figure 7.4.2: 500w tungsten halogen outdoor lamp

Figure 7.4.3: Halogen master capsule

Figure 7.4.4: 2.38w 2.8v tungsten halogen lamp

Figure 7.4: Tungsten halogen lamps

Lighting Core Knowledge and Recognition

1980s that give better colour rendering, longer lamp life and efficacy. The latest NO (nitrous-oxide) tri-phosphor lamps are even better.

The tri-phosphor coating used on most new fluorescent lamps provides a more faithful rendering (R_a >80) for the whole life of the tube. In addition, they provide more light for longer. Older halo-phosphate lamps deteriorate in their quality of light quicker than modern fluorescent lamps.

On-site, you can begin your identification of fluorescent lamps based on the diameter of the tube, or lamp, and the length. Younger readers need to know that 1 inch, or 1", is equivalent to 25.4mm. You can divide 1" into 'eighths', i.e. ⅛", or approximately 3mm. In lighting terms, the letter 'T' denotes 'Tubular'. The following descriptions of fluorescent lamps therefore refer to old ⅛" units of imperial measurement:

- T12 = 1½" = 38mm diameter tubular lamp;
- T8 = 1" = 26mm;
- T5 = ⅝" = 16mm;
- T4 = ½" = 12.5mm; and
- T2 = ¼" = 7mm.

Once you have identified the generic type of fluorescent lamp, you will need to narrow down the lamp type, i.e. whether it is 'halo-phosphate' or 'tri-phosphor'. There is no method you can use to visually identify the type of light emitted by these different lamp types. You will need to identify the type of lamp by recording the details (if any) printed on the lamp, or the boxes containing spares and/or consulting the manufacturer's information. T5 lamps are modern, but you can still buy T5 halo-phosphate lamps.

Most fluorescent lamps work best, i.e. have high efficacy, in a temperature of 20–30°C. However, the optimum temperature for T5s is around 35°C. In general, longer lamps have higher efficacies as the significant electric power losses occur at the ends of the tube.

Lamps with a colour rendering index lower than 80 should not be used where people work or stay for long periods. Therefore, where fluorescent lamps are used in normal commercial interiors, designers should specify tri-phosphor lamps. A typical application for a fluorescent lamp is in an office, commercial or educational building.

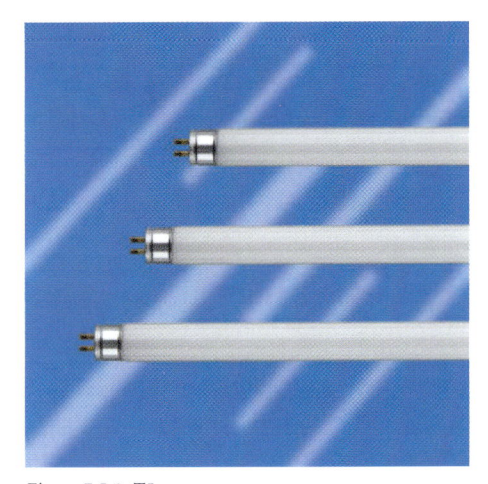

Figure 7.5.1: T5

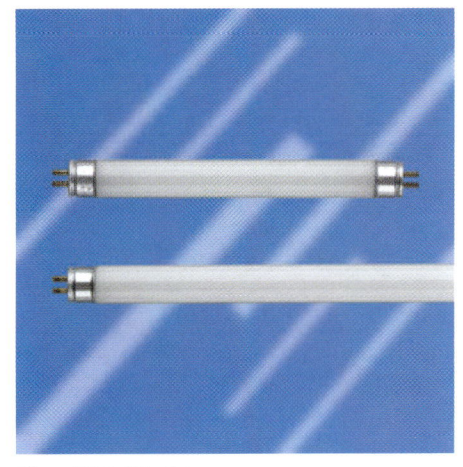

Figure 7.5.2: T5 mini

Figure 7.5.3: T8 tri-phosphor

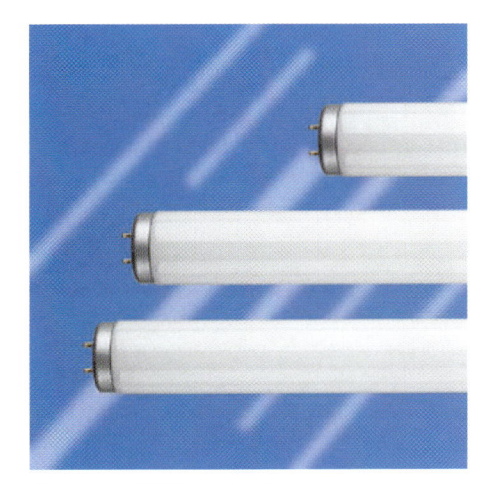

Figure 7.5.4: T12 halo-phosphate

Figure 7.5: Fluorescent tubular lamps

Compact fluorescent discharge lamps – (CFLs)

The CFL developed from tubular fluorescent lamps. Increases in efficacies in tubular lamps meant tube diameters reduced, until the tubes could be folded and compacted. They meet the need for lamps that will fit into 'conventional' GLS lamp fittings.

CFLs operate in the same way as tubular discharge lighting and emit a gentle, diffused, white light. Most lamps are a significantly different shape though. All CFL tubes are coated with tri-phosphors.

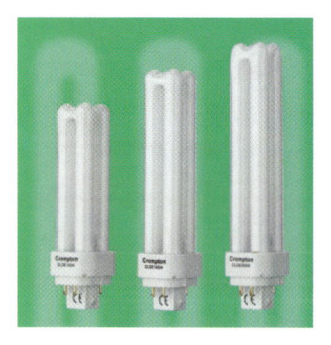

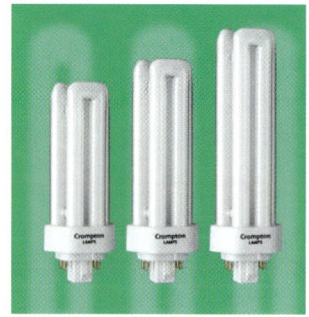

Figure 7.6: Variety of compact fluorescent lamps

Low pressure sodium discharge lamps – (SOX)

This is a 'discharge' lamp that works by discharging electricity into an atmosphere of low pressure sodium in an arc tube within a glass outer casing. Low pressure results in low energy density and the tube must be longer than a high pressure sodium lamp of equivalent output.

This lamp type serves as most of the 'traditional' street lights. Lamps emit a strong orange/yellow light. They can require an 8–9 minute warm up and a 1 minute re-strike time when warm.

Figure 7.7: Low pressure sodium lamp

High pressure sodium discharge lamps – (SON)

This lamp has an atmosphere of high pressure sodium within a translucent alumina oxide ceramic arc tube (to withstand the high pressure) in a glass outer casing. ST denotes 'tubular', SE 'elliptical'. DL is a deluxe version, with an improved CRI.

These lamps produce what has been called a golden white, or white, light. SON lamps with a golden-white light have better efficacy than those that give off a white light. This type of lamp is also used in street lighting. You will often find SON lamps in warehousing or cold storage areas as they operate well in cold temperatures, but CRI is not crucial in those buildings.

These lamps require a 4–7 minute warm up from cold and a 1 minute re-strike time.

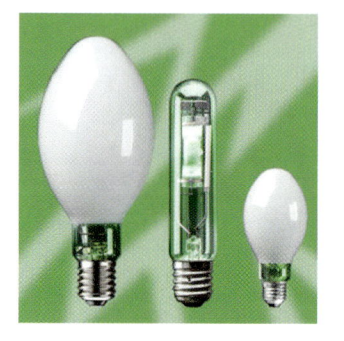

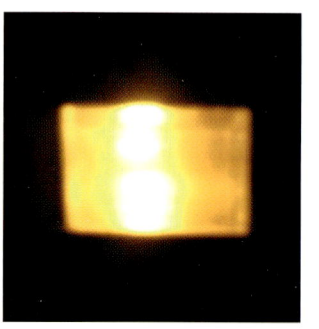

Figure 7.8: High pressure sodium lamps

Metal halide discharge lamps – (MBI or HPI or MBI/MH or CDM)

Operates in same fashion as, and developed from, a high pressure mercury lamp, i.e. discharging electric current into high pressure mercury atmosphere, but metal halide additives are contained in the arc tube. They provide a very bright, crisp white light with good colour rendering. A CDM (ceramic discharge) version provides a warm colour and reduces discolouration of the bulb.

In general, metal halide lamps require a glass cover to protect people and property in the event of the lamp exploding. These lamps usually require a 1–5 minute warm up and a 7–10 minute re-strike time. However, modern switchgear enables immediate switching. You are likely to find them in an increasing number of applications, including industrial and entertainment buildings.

Figure 7.9: Examples of metal halide lamps

High pressure mercury discharge lamps – (MBF)

A lamp developed in the 1930s with an atmosphere of high pressure mercury within an arc tube in a fluorescent-coated glass outer casing. They emit a 'cool' white light, generally with poor colour rendering. They require a 5–7 minute warm up and a 2–7 minute re-strike. You are likely to find these lamps types in warehouses and other areas where their poor efficacy is not important. In time they are likely to become totally obsolete.

Figure 7.10: Examples of high pressure mercury lamps

Induction discharge lamps – (QL)

Developed from fluorescent lamps, an induction coil in a hollow centre in the lamp connected to a high frequency supply generated by control gear mounted in the lamp cap produces a magnetic field which creates an electric current by reacting with gas in the lamp. The current excites mercury atoms and UV radiation is converted to visible radiation by the powder coating, as in a fluorescent tube lamp; but lack of cathodes and electrodes means a very long lamp life – induction lamps are sometimes called 'electrode-less' lamps.

You will usually find them in locations where access is difficult and costs of regular replacement would therefore be high.

Figure 7.11: Induction lamps

Light emitting diodes (LEDs)

In LEDs the production of light occurs in a semi-conductor crystal. The crystal is excited electrically to produce the light, or luminance – the process is known as 'electroluminescence'. They give a cold, neutral light.

LEDs are currently available in colours orange, red, green, blue and yellow. LEDs produce white light by mixing all the wavelengths. LEDs are a relatively new process so far as lighting is concerned, having developed within the last 50 years. The first systems were inefficient, and the existing units are not as efficient as the best lighting systems. However, science is continuously pushing the boundaries and there is an expectation that use of LEDs will progress to a very significant extent as they become even more effective.

LEDs have many advantages including:

- light emitted by LEDs does not contain ultraviolet (UV) or infrared (IR) radiation in its spectrum – it can be used in situations where other types of light cannot be, e.g. in a museum to protect old manuscripts;
- light output degrades slowly over time when compared with filament lamps;
- efficacy (lm/W) of most LEDs has currently exceeded incandescent lamps and is equivalent to mercury vapour lamps; with the rate of improvement continuing – in time, the industry expects LEDs to be more efficacious than any other type of lamp;
- they can be dimmed easily, from 0–100% output;
- lamps are capable of withstanding considerable vibration and/or impact.

As a result of the above, you are likely to record LEDs in an increasing number of different types of buildings. These might include offices, restaurants, and public houses, theatres (e.g. in staircase risers), museums (as above), sales areas and hotels.

Figure 7.12: LEDs

Identification of fluorescent control gear and ballasts

One challenge a CEA faces is identifying whether any T8 fluorescent lamps have either 'standard' or 'high frequency' ballasts as the program requires this information. The challenge is significant since most ballasts are hidden in the luminaire and not visible.

With gas discharge lamps, a starter is usually (but not always) needed to switch on the lamp. The control gear (or ballast) limits and regulates the voltage and current delivered to the lamp and is essential for proper lamp operation. The electrical input requirements vary for each type of lamp, and so each type/wattage requires gear specifically designed to drive it.

Figure 7.13: Starter for a fluorescent lamp

There are two types of ballasts that operate on alternating current: electromagnetic (EM) and electronic (EC). The EM ballast, the 'standard' ballast since fluorescent lighting was first developed, uses electromagnetic technology. The EC ballast, only relatively recently developed, uses solid-state technology.

Old style 'electromagnetic' or 'magnetic' ballasts use electromagnetic induction coils to provide the starting and operating voltages of a gas discharge lamp. Such lamps run at 50 Hertz (Hz), perceptibly flicker when they start and take a few seconds to begin running. The lamp illuminates on each half-cycle of the power source – this is why many fluorescent and neon lights visibly flicker. Since the light illuminates on half-cycles, the rate of flicker is twice the frequency of the power source, meaning the light flickers at 100 Hz. The operating flicker is usually noticeable and is useful for identification. Some people are affected in a negative way, or simply annoyed, by this flicker. EM ballasts can 'hum' slightly, or even make a considerable noise. EM ballasts are more robust than EC ballasts.

More modern EC ballasts are 'high frequency (HF) electronic' types that operate at 20,000 HZ, usually without perceptible flicker. Electronic ballasts weigh less than electromagnetic ballasts. They can be rapid start, or 'soft start' with a very brief delay, which extends the lamp life. They start at the first attempt and operate silently. Because they use solid-state circuitry, they are more efficient and therefore run cooler. They are also dimmable. Such control mechanisms provide better efficacy, i.e. reduced energy input, and heat output. In addition, unlike electromagnetic ballasts, EC ballasts can also alter the frequency of power. This means that electronic lighting ballasts greatly reduce or eliminate any annoying flicker in the lamps and do not have any significant noise or hum.

In general, T12 fluorescent lamps do not work with HF ballasts. T8s can operate with HF. T5s can only operate with HF.

Control gear power consumption for some T8, T5 and CFLs is defined in terms of efficiency by reference to an EEI (Energy Efficiency Index) based on the CELMA publication adopted by the EU (Directive 2000/55/EU). CELMA is the Federation of National Manufacturers Associations for Luminaires and Electro-technical components in the EU.

The EEI for ballasts is:

- Class A1 – dimmable electronic ballasts;
- Class A2 – electronic ballasts with reduced losses;
- Class A3 – electronic ballasts;
- Class B1 – magnetic ballasts with very low losses;
- Class B2 – magnetic ballasts with low losses;
- Class C – magnetic ballasts with moderate losses;
- Class D – magnetic ballasts with very high losses.

Thus, Class B1–D ballasts are EM, i.e. 'standard'. The others are EC, i.e. 'HF'. Ballasts in Classes C (banned in November 2005) and D (banned in May 2002) should no longer be available. Many types of ballast are stamped with the efficiency rating, e.g. 'EEI B1' would help you identify the lamp as fitted with 'standard' ballast.

Figure 7.14: Electromagnetic (EM) ballast

Figure 7.15: Electronic (EC), high frequency, ballast

Lamp identification – quick case study

Now we have dealt with lamps and their identification, here is a photograph of two different lamps in a warehouse. You do your inspection from ground level. Further details you record during your inspection are:

- left hand lamp – warm up 2 minutes, re-strike 7 minutes;
- right hand lamp – warm up 5 minutes, re-strike 1 minute.

Go back to the descriptions and try to visually identify them. We give the answer at the end of the chapter.

Lighting Core Knowledge and Recognition

Carbon Trust lamp identification information

This government-backed organisation, established in April 2001, provides UK businesses and industry with advice on reducing carbon emissions. It publishes much useful information about lighting (and other energy efficiency issues) and we point you in the direction of their website – www.carbontrust.co.uk. You will find a particularly helpful table in their publication entitled CTV021 'Lighting – bright ideas for more efficient illumination', at appendix A – 'Lamp comparison table', which will help you appreciate the differences in efficacy, lamp life and colour rendering between the various lamp types:

Figure 7.16: Two lamps, same warehouse

Lamp comparison table

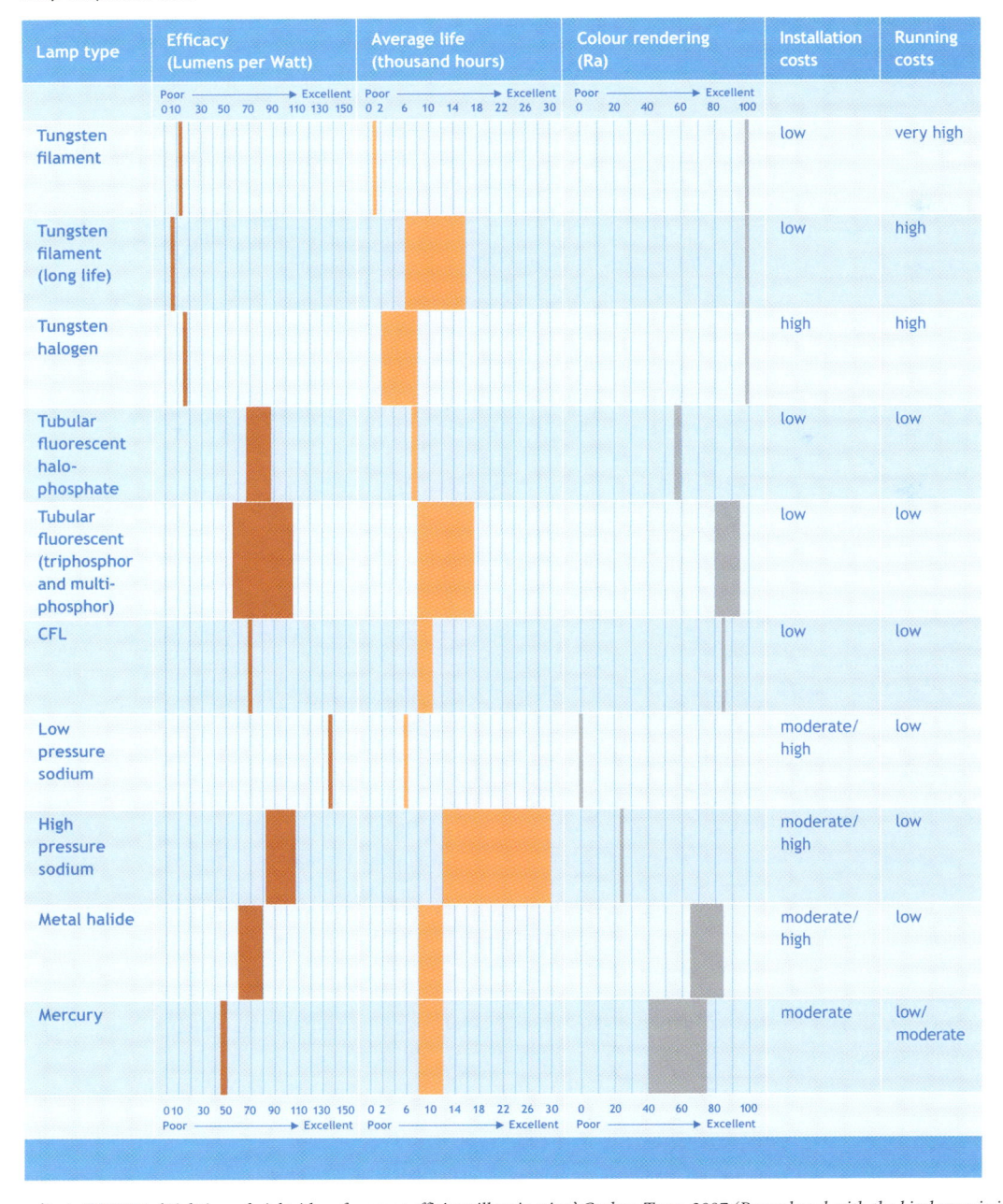

Lamp type	Efficacy (Lumens per Watt)	Average life (thousand hours)	Colour rendering (Ra)	Installation costs	Running costs
Tungsten filament				low	very high
Tungsten filament (long life)				low	high
Tungsten halogen				high	high
Tubular fluorescent halo-phosphate				low	low
Tubular fluorescent (triphosphor and multi-phosphor)				low	low
CFL				low	low
Low pressure sodium				moderate/high	low
High pressure sodium				moderate/high	low
Metal halide				moderate/high	low
Mercury				moderate	low/moderate

Figure 7.17: Appendix A, CTV021, 'Lighting – bright ideas for more efficient illumination' Carbon Trust, 2007 (Reproduced with the kind permission of the Carbon Trust)

Commercial Energy Assessor's Handbook

Another good document from the Carbon Trust is CTG010 'Display Lighting'. Figure 7.18 is Table 3 from that publication.

Table 3 *Characteristics of the key types of display lighting, and a guide to which are most appropriate for use*

Lamp type	Luminous efficacy (Lumens/Watt) Min	Max	Colour appearance (Kelvin) Min	Max	Colour rendering (Ra) Min	Max	Life (Hours) Min	Max
 Incandescent tungsten filament	6	14	2,700	2,700	100	100	1,000	1,000
The least efficient type of lighting with the shortest lifetime. Occasionally acceptable if a desired aesthetic is being pursued but in most cases should be replaced with modern alternatives.								
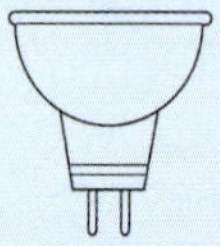 Tungsten halogen (quartz halogen)	13	26	3,000	3,000	100	100	2,000	8,000
Often used in spot lighting and display lighting. If low voltage tungsten halogen spotlights are installed there is a further saving using, for example, 35W infrared coated (IRC) bulbs instead of the standard 50W bulbs.								
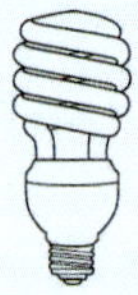 Compact fluorescent (CFL)	45	70	2,700	4,000	82	82	6,000	15,000
A wide range of attractive modern CFL bulbs are available which can be a direct replacement for standard tungsten bulbs and are also acceptable for downlights, display and feature lighting.								
38mm T12 'Standard' F/Tube	61	86	2,950	6,000	51	76	7,000	9,000
At least 7% less efficient than T8 or T5 equivalent tubes and now obsolete.								
25mm T8 'Standard' F/Tube (S/G)	53	80	2,950	6,000	51	76	8,000	9,000
At least 16% less efficient than equivalent triphosphor tubes. Lower capital cost is not cost-effective over whole life.								
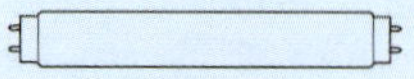 25mm T8 Full spectrum multiphosphor	52	66	3,000	6,000	95	98	15,000	20,000
Recommended when excellent colour rendering is required e.g. colour matching or medical examination but at least 25% less efficient than equivalent triphosphor lamps, so not recommended for general use.								

Figure 7.18: Table 3, CTG010 'Display Lighting' Carbon Trust, 2008 (Reproduced with the kind permission of the Carbon Trust)

Lighting Core Knowledge and Recognition

Lamp type	Luminous efficacy (Lumens/Watt) Min	Luminous efficacy (Lumens/Watt) Max	Colour appearance (Kelvin) Min	Colour appearance (Kelvin) Max	Colour rendering (Ra) Min	Colour rendering (Ra) Max	Life (Hours) Min	Life (Hours) Max
25mm T8 Triphosphor	63	100	2,700	6,500	80	85	12,000	60,000

A preferred high frequency lamp for general commercial lighting.

16mm T5 H/F Triphosphor	37.5	106	2,700	6,500	80	85	16,000	48,000

A preferred lamp for general commercial lighting. May be more efficient than T8 equivalent if optical benefits of luminaire are maximised.

Metal halide	70	107	3,000	6,000	65	96	6,000	20,000

Good quality white light for areas which require infrequent switching. Ceramic versions are more efficient.

Compact metal halide	68	100	3,000	5,900	73	83	2,000	15,000

Good quality, efficient white light with a variety of applications.

Mercury	36	58	4,000	4,000	42	49	12,000	20,000

Provides white light but metal halide lamps are preferable.

High pressure sodium	66	130	2,000	2,000	25	25	12,000	28,500

High luminous efficacy but very poor colour rendition. Deluxe model has better colour rendition than standard high pressure sodium (although still classed as poor) but at the expense of luminous efficacy.

White sodium	37	51	2,500	2,500	80	80	6,000	10,000

Much better colour rendition than standard high pressure sodium (classed as good) but at a heavy cost in luminous efficacy. Consider using ceramic metal halide instead.

Figure 7.18 (cont.)

Commercial Energy Assessor's Handbook

Lamp type	Luminous efficacy (Lumens/Watt) Min	Luminous efficacy (Lumens/Watt) Max	Colour appearance (Kelvin) Min	Colour appearance (Kelvin) Max	Colour rendering (Ra) Min	Colour rendering (Ra) Max	Life (Hours) Min	Life (Hours) Max
Light emitting diodes (LEDs)	>25	100	3,000	6,000	Too variable to state	Too variable to state	12,000	>50,000

Rapidly improving technology. The best choice for coloured effect lighting, particularly in low light level areas. Take care if specifying for white light applications as other lamp options may be more appropriate.

Colour appearance	
Description	Kelvin
Candlelight	1,500
Warm white	3,000
White	3,500
Cool white	4,000
Sunlight	6,000
Overcast sky	7,000

Colour rendering	
Description	Ra
Very poor	20-39
Poor	40-59
Moderate	60-79
Very good	80-89
Excellent	90-100
Daylight	100

Luminous efficacy	
Description	Lumens/Watt
Very poor	Less than 39
Poor	40-59
Moderate	60-79
Good	80-99
Excellent	>99

Lamp life	
Description	Hours
Low (5,000+)	5,000-10,000 hrs
Moderate (10,000+)	10,000-15,000 hrs
Long (15,000+)	15,000-50,000 hrs
Very long (50,000+)	Greater than 50,000 hrs

Figure 7.18 (cont.)

LIGHTING IN SBEM

Zone lighting energy

There are three ways in which you can specify lighting energy or efficiency in SBEM calculations. Your aim is to define how much electrical power is consumed to achieve the required lighting level. This can be defined in terms of lighting power density, which is the amount of power required for each square metre of the zone for each 100 lux required (W/m² per 100 lux). Lighting levels (design illuminance in lux) are defined for each activity type in the NCM databases. The three ways of specifying lighting power requirements are:

(1) inferring the lighting power density from a general classification of lamp type (lighting parameters not available);
(2) calculating the efficacy of the installation in terms of lumens per circuit Watt (lighting chosen but calculation not carried out);
(3) using the results of design calculations (full lighting design carried out).

Which approach you adopt depends on the available information. We suggest the second approach is the most appropriate and accurate for existing buildings and we describe this in detail. This approach is, however, not possible for certain activity types. If you use either method (2) or (3) you must include power required for any control gear.

Defining lamp type

One way to define the lighting design in SBEM calculations is to just define the lamp type and rely on SBEM to infer what the lighting power density is likely to be. As you will have seen, the efficiency of different lamp types and control gear varies considerably – both between types but also between individual manufacturers. Consequently, using this approach necessitates some broad assumptions and approximations. SBEM allows you to choose between 11 types. These types and the lighting power densities SBEM assumes are shown in Figure 7.19. The assumptions are conservative and the values do not necessarily meet current Building Regulation requirements. In practice modern lighting designs are able to achieve noticeably lower power density.

Note that whichever method of specifying lighting power you choose it is necessary to specify the lamp type. This information is not used in calculations for data entry methods (2) and (3) above, but is used in reporting.

If you do not have any information about the lamp type SBEM assumes tungsten (GLS) and, as you can see from the table in Figure 7.19, this implies very high energy demands. If you have some information about the lighting type and power density but it does not exactly fit the available categories (e.g. low voltage tungsten halogen) you should choose that with the most similar power density.

Efficiencies and efficacies of different lamps

Wherever possible you should attempt to calculate the efficacy of the lamps that light the zone, i.e. the lumens (amount of light emitted or visible output) per circuit watt (amount of power being consumed or electrical input) – we are, after all, trying to identify the likely CO_2 emissions. In practice, you will often find this is difficult because of problems related to access to the lamp and identification of the required information.

The information you require, i.e. the lumens and the wattage are sometimes, but not usually, available as a legend on the lamp – as we discussed above. In most cases you will need to contact the manufacturers, find a delivery box that holds spare lamps (manufacturers

Lamp Type	Power Density in W/m² per 100 Lux Commercial Application	Power Density in W/m² per 100 Lux Industrial Application
Tungsten lamp	28	-
Fluorescent - compact	4.6	-
T12 Fluorescent - halophosphate - low frequency ballast	5	3.9
T8 Fluorescent - halophosphate - low frequency ballast	4.4	3.4
T8 Fluorescent - halophosphate - high frequency ballast	3.8	3
T8 Fluorescent - triphosphor - high frequency ballast	3.4	2.6
Metal Halide	5.5	4.1
High Pressure Mercury	7.6	5.7
High Pressure Sodium	4.5	3.3
T5 Fluorescent - triphosphor-coated - high frequency ballast	3.3	2.6
Fluorescent (no details)	5	5
Don't know	28	28

Figure 7.19: Lamp power densities assumed by SBEM – Table 14 of iSBEM User Guide, p. 141

tend to provide more information on the box) and/or obtain the data sheet on the internet or from suppliers for the particular lamp.

For incandescent lamps, i.e. a lamp without control gear, your calculation can be relatively simple once you have obtained these two figures. Thus, for a tungsten lamp with:

- 410 lm and power of 40W, the efficacy calculation is 410 ÷ 40 = 10 lm/W;
- 700 lm and power of 60W, 700 ÷ 60 = 12 lm/W;
- 1350 lm and power of 100W, 1350 ÷ 100 = 13.5 lm/W.

However, for discharge lamps, that have ballasts, or control gear, you will also have to include the power requirement of that gear in the calculation. The challenge is discovering that figure – such a process can be long and sometimes tortuous. As a very approximate rule of thumb, older EM ballasts typically add 20% to the power required for a lamp over and above the lamp power, compared with an additional 5% required if high-frequency EC ballasts are fitted. Using such a rule of thumb is acceptable in many instances so long as you tell your client the basis of your calculation. However, for larger assessments where such an approximate rule could cause significant errors in your assessment, you need to be more accurate.

Thus, Figure 7.20 is from a Carbon Trust publication.

We have chosen the information about the T8 fluorescent lamps in the guide. This shows the energy consumed by the lamp type (in the left hand column) and total energy consumption including the control gear (in the third column). The ballast type is in the second column.

Let's say you have identified two types of 36W T8 halo-phosphate single fluorescent lamps 1200mm long and you want to calculate the figure for lumens/circuit-watt. One lamp type is fitted with older 'standard' ballast and the other with modern 'electronic' high frequency (HF) control gear.

In Figure 7.20 you can see that in the first instance the total power consumed by a single luminaire is 45W, (36W [lamp] + 9W [ballast]), i.e. the ballast consumes an extra 9W, or 25% over and above the power used by the actual lamp. The comparison figure for the modern HF powered lamp is 38W (36W [lamp] + 2W [control gear]), i.e. only 6% extra for the control gear.

We now need the amount of lumens emitted by the lamp. Figure 7.21 is a page from a manufacturer's brochure. Let us assume we have identified the lamps as being from this manufacturer. The tenth column shows the 'initial lumens (lm)'. We should use the 'initial lumens' figure, as we are conducting an 'asset rating', not an 'operational rating'. This is despite the

Lamp	Ballast Type	Single Lamp	Twin Lamp
T8 Fluorescent			
18W	Electronic	21	38
18W	Standard	26	45
18W	Standard Low Loss	19	37
36W	Electronic	38	76
36W	Standard	45	90
36W	Standard Low Loss	45	72
58W	Electronic	58	112
58W	Standard	70	140
58W	Standard Low Loss	56	112
70W	Standard	82	164
70W	Standard Low Loss	70	140

Figure 7.20 – Part of Table 1 from 'Lighting Guide 007 – Installers Guide to the assessment of energy efficient lighting installations', Carbon Trust, 2004, p.5

Catalogue	Designation ILCOS	ZVEI	Watts (W)	Nominal Dimensions (mm)	Cap	Colour Temp (K)	Colour Appearance	Colour Rendering Index/Group	Initial Lumens (lm)	Rated Life (50% Survivors)	Lumen Maintenance At Rated Life	Cat. No.
T8	FD	T26	18	L 600 x Dia 26	G13	4000	Cool white	67/2B	1175	10000 hr	75%	FT218CW
T8	FD	T26	18	L 600 x Dia 26	G13	3500	White	56/3	1200	10000 hr	75%	FT218W
T8	FD	T26	30	L 900 x Dia 26	G13	4000	Cool white	67/2B	2300	10000 hr	75%	FT330CW
T8	FD	T26	30	L 900 x Dia 26	G13	3500	White	56/3	2300	10000 hr	75%	FT330W
T8	FD	T26	36	L 1200 x Dia 26	G13	4000	Cool white	67/2B	2850	10000 hr	75%	FT436CW
T8	FD	T26	36	L 1200 x Dia 26	G13	3500	White	56/3	2900	10000 hr	75%	FT436W
T8	FD	T26	58	L 1500 x Dia 26	G13	4000	Cool white	67/2B	4600	10000 hr	75%	FT558CW
T8	FD	T26	58	L 1500 x Dia 26	G13	3500	White	56/3	4800	10000 hr	75%	FT558W
T8	FD	T26	70	L 1800 x Dia 26	G13	4000	Cool white	67/2B	5800	10000 hr	75%	FT670CW
T8	FD	T26	70	L 1800 x Dia 26	G13	3500	White	56/3	5800	10000 hr	75%	FT670W

Figure 7.21: Part of manufacturer's table (Crompton)

fact that the lumens emitted are likely to be less after the lamps have been used for some time due to deterioration. Indeed, the deterioration for halo-phosphate lamps is greater and more rapid then for tri-phosphor lamps.

From the table you can identify which lamp you identified on-site from the legend in the right hand column, which you have recorded in your site notes – 'FT436CW' (i.e. FT ['fluorescent tube'] 4 [4000 'Kelvin'] 36 ['36 Watts'] CW ['cool white']. You can therefore confirm with certainty a lumens output of 2850 lm.

For the less efficient luminaire (45W power consumption), your calculation for efficacy is therefore (lumens ÷ total watts in the circuit), i.e. 2850 ÷ 45 = 63 lm/W.

For the more efficient luminaire (38W power consumption), your calculation is 2850 ÷ 38 = 75 lm/W.

The two figures of 63 and 75 lm/W are the figures you would enter into the program for each of the appropriate zones where you use the 'lighting chosen but calculation not carried out' (in iSBEM) button. Your equivalent figures for incandescent lamps do not have to include control gear. You must also confirm the lamp type for each zone in the entry field.

As discussed, the *BRE User Guide* at Figure 7.19 confirms the power densities, measured in watts/m^2/100 lux for the different general building types (commercial and industrial) and lamp types the program assumes for the zone. You can identify the *significant* differences in power consumed (over 1000%) the program assumes from that table and hence why it is important to attempt to identify the correct lamp type.

You should note that in existing buildings you must use the total power based on the above types of calculation. You should not use power indicated by the LOR (light output ratio) as you would thereby use false (low) data. You should use the LOR calculation for certain uses for new-build calculations – see above under 'Building Regulations'.

In existing buildings you ultimately base your calculation on the total energy the lamp, or lamp and ballast, consume; not the amount of light emitted by the fitting, which could be significantly less due to the effect of the luminaire, particularly the diffuser – that calculation is the 'LOR'.

Electronic ballasts can even reduce the power consumption of ballast/lamp circuits to less than the rated power of the lamp at 50 Hz. Thus, one manufacturer gives an example for the relative power consumption of a 'T8 36W 840' lamp fitted with different types of EEI control gear as follows:

A1 ≤ 19 W
A2 ≤ 36
A3 ≤ 38
B1 ≤ 41
B2 ≤ 43
C ≤ 45
D ≤ 45

Where possible, we think you should always try to obtain such exact information about the lumens, lamp and control gear direct from the manufacturer.

'Full lighting design carried out'

You use this SBEM option when provided with a design by a lighting engineer you believe you can rely on. The design must give you the two required items of information for that zone, i.e. total wattage (power) and designed illuminance (luminous flux). In practice you will generally only have this information for a new building. Indeed, you must have this information for a 'new-build' assessment.

In existing buildings you may have such information available to you. You will need to judge whether you can rely on this information, based on how accurate the evidence is. In reality, you may find lighting designs are not always followed by electrical contractors, e.g. not installing specified lamps. Furthermore, we have seen major differences in actual fitted lighting systems when compared with the system specified for other reasons such as:

- occupiers quickly alter or adapt a system because the design fails to suit their requirements;
- failed lamps are replaced to different specifications.

You should therefore only accept such information in an existing building if you have thoroughly checked the entire system and compared it with the design.

You will need to identify the lamp types for the program so it can generate appropriate recommendations.

Air extracting luminaires

Heat emitted from a lamp is very wasteful and can contribute to overheating and increased zone cooling demands. Air handling luminaires can be used in air conditioned rooms with false ceilings and incorporate air passages through which air extracted from the room passes before entering the main extract ducts. This reduces the amount of heat from the lamp that directly enters the room. Air conditioning energy demands can be slightly reduced. Coincidentally, this helps cool and therefore increase the lamp life.

Accurate identification of these lamps is problematic in our experience; usually caused by lack of access. As the luminaires are the means by which air is extracted from the room you would not expect to find any extract grilles.

LIGHTING CONTROLS

General

Lighting controls should be designed to ensure the building and the occupants make maximum use of natural day-lighting and only use artificial lighting systems when required. Lighting controls in buildings should provide for flexible and efficacious control either by simple 'on-off' procedures, or by 'dimming' of the lamps.

Manual switching, i.e. a light switch, is perhaps the simplest type. Alternatively, occupiers/users can achieve a similar degree of control by remote infrared transmitters. That type of system requires a receiver at the luminaire.

Manual lighting controls and remote switching provide good flexibility and empower the users. However, research has shown such controls suffer from a fundamental flaw; i.e. occupiers turn lights on when natural light falls to an unacceptable level, but seem to find it difficult to turn them off again if the opposite prevails.

The lighting industry has therefore responded to this lack of human initiative by developing alternative methods of control. These include systems such as time switching, movement/noise/occupancy detection and automatic reaction to the amount of available daylight. The latter type of system uses photocell technology.

It should be fairly obvious that if you 'dim' a lamp, you will reduce the energy the lamp uses, thereby reducing energy consumption. However, whilst tungsten filament lamps can easily dim, discharge lamps present more of a problem, sometimes requiring expensive and energy-intensive control gear. Furthermore, dimming systems can suffer from the same problem as general lamp controls, i.e. inadequate human initiative. Hence, automatic dimming controls are preferable to manual controls.

The *BSRIA Illustrated Guide to Electrical Building Services* has a good introduction to lighting controls that you may find useful (p. 45).

SBEM light controls

You should choose 'local manual switching' for the zone where:

- each of the light switches (e.g. rocker switches, push buttons, pull cords, infrared, radio, sonic, ultrasonic and telephone, handset controls) in the zone is less than 6m from all of the lamps it controls:
  - or the distance between the switch and the lamp is less than twice the height from floor level up to the lamp (if that measurement is greater than 6m);
  - or if the area of the space is less than 30m².

If none of these cases apply, you should not tick this box. The *BRE SBEM Technical Manual* suggests manual switching does not apply to certain types of area such as corridors.

If you find photoelectric controls you must identify whether they are of the simple 'on-off' 'switching', or more flexible 'dimming' type. In addition, you must decide whether they operate on 'stand alone sensors' or 'addressable systems'. An addressable system is a programmable system that allows the occupier greater flexibility. Timers can be pre-set to switch lighting on or off at pre-determined times. Both 24-hour and 7-day versions are available. The 7-day version allows for different settings at weekends or during evenings (e.g. a school which runs evening classes might need corridor lighting on a Tuesday evening, but not during the rest of the week).

For areas with natural light, astronomical timers can be used to adjust the on-off times gradually over the year to account for seasonal changes in lighting levels. This system must be programmed with the geographical location. Such a system can turn a percentage of lights on/off or dim as daylight alters.

As a rough 'rule of thumb' if *photoelectric switching* can be performed remotely from a 'switch' with a control unit then this is an addressable system.

There are a significant number of different types of lighting controls. We do not name or describe all of them.

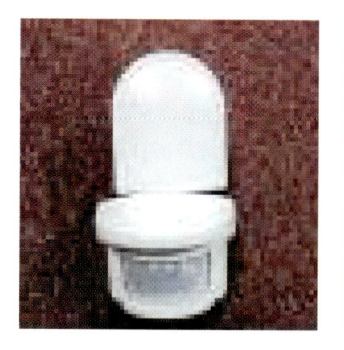

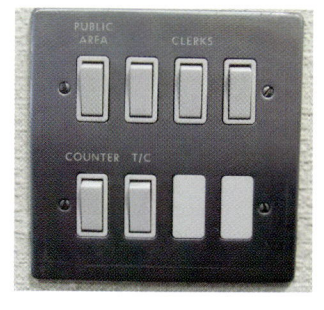

Figure 7.22: Examples of lighting controls

You should confirm if the building has occupancy sensing controls. The most common type is based on passive infrared (PIR) sensors. Occupancy switches generally have a time delay, only switching the lighting off when no movement has been detected for that period of time. The *BRE User Guide* has a useful table with guidance on identification of the different options you should consult. You only enter data for the parasitic power of the sensing device, i.e. the power consumption in W/m², if you know it; otherwise accept the default.

You can allow SBEM to automatically zone for day-lighting to reflect the existence of the either manual and/or photoelectric controls, but only if either or both are present. The *User Guide* recommends you only use this option if the building has a 'typical' arrangement of windows and/or rooflights.

OTHER LIGHTING ISSUES

Practical SBEM lighting issues

One practical issue you will find is many different types of lamps scattered 'willy-nilly' throughout the same space – that space would otherwise be a single zone but for the different lamp types, e.g. in a restaurant or public house. One method of solving this issue might be to choose the most dominant light source and allocate that to the zone. This means you ignore other lamp types in the zone and need to generate extra recommendations for those lamps in the RR to reflect their presence.

An alternative approach is to divide the space into separate zones based pro-rata on the number of lamps, with virtual boundaries between the 'sub-zones', i.e. populate each zone with a different lamp type; in effect 'move' the lamps to achieve a result. You must ensure you do this sensibly and take appropriate care with day-lighting.

Identification of lamps can be extremely problematic, particularly if they are at height. If you are uncertain, you must 'default' to the most appropriately efficacious type of lamp and confirm why you have done so in your site notes and your RR/client report. You should consider:

- close inspection if possible, or otherwise;
- find a lamp box;
- manufacturer's information;
- shape;
- colour;
- start-up and re-strike.

You can ignore any lighting service not covered by Part L of the Building Regulations, e.g. emergency escape lighting, specialist process lighting; and external lighting, i.e. lamps outside the 'building'.

You should currently identify tungsten halogen lamps as tungsten lamps. You will need to make adjustments to the RR.

You cannot choose LEDs as an option in SBEM. You must therefore choose the closest lamp type in terms of efficacy, based on the actual LED type you identify.

In all of these examples, you should follow the rules your accreditation scheme suggests you use.

Display lighting

This is defined in ADL2A as lighting intended to highlight displays or exhibits of merchandise, or lighting used in spaces for public leisure or entertainment such as dance halls, auditoria, conference halls, restaurants and cinemas. There is a separate 'activity' in the activity database for 'display area' where display lighting is used to illuminate items.

Your safe inspection of lamps

You should adopt safe procedures when you inspect lamps and luminaires closely. You must only access lamps, etc. from a safe location, i.e. off a properly situated ladder. You should only inspect any lamp with gloves and safety glasses on.

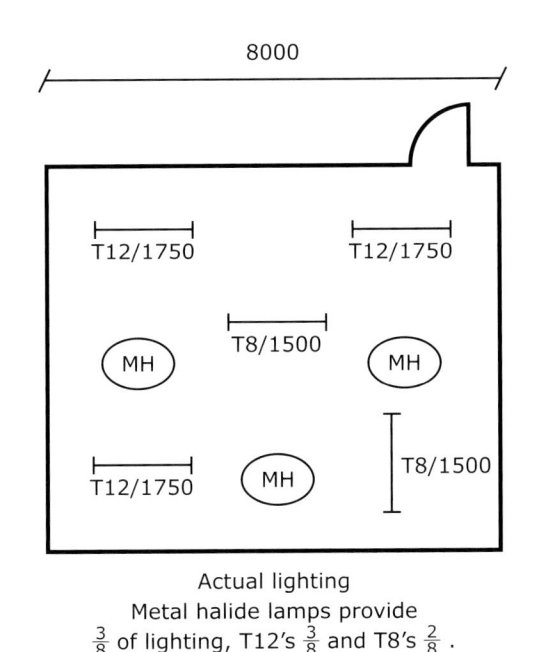

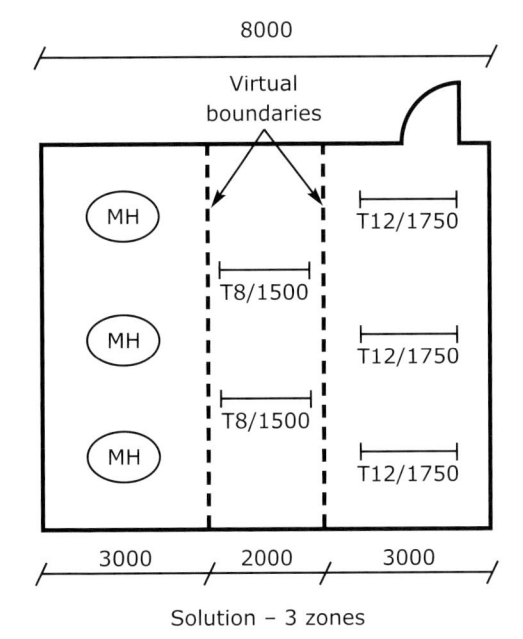

Figure 7.23: Method of allocating different lamp types in same 'zone'

You should not inspect high pressure mercury or metal halide lamps if the lamp is broken. This is because such a fault allows escape of ultraviolet radiation that can damage skin and eyes.

You should not touch any tungsten halogen lamps with your bare hands since you run the risk of contaminating the quartz envelope of the lamp with oils from your skin. As a result, a 'hot spot' can develop, leading to a possibly significant reduction in lamp life. If you do inadvertently touch a tungsten halogen lamp manufacturers suggest you remove your fingerprints using a tissue moistened with methylated spirits.

Some light fittings installed in the 1960s and 1970s, and possibly in some cases 1980s, may contain toxic chemicals called PCBs (polychlorinated biphenyls) – allegedly, amongst other issues, a hormone-disrupting compound.

We are sometimes asked during training whether you should remove any part of a luminaire to inspect the lamp, e.g. take off a diffuser on a fluorescent fitting. We believe you can remove any part of a lamp to inspect it so long as you do it with the owner's permission, safely and avoid damage. Alternatively you can ask the occupier, or any person responsible for the lighting systems, to do this for you.

The Environment Agency has determined that fluorescent tubes, sodium and metal halide lamps are now classified as hazardous waste as they contain mercury.

Damage caused by light

'Light' causes discolouration of paintwork, lifting and cracking of paint from a surface, deterioration in fabrics and distortion and cracking in surfaces such as timber.

Infrared light heats objects it falls on, the energy in ultraviolet light can alter the physical state of a material (by moving, or totally removing, electrons in the material) and visible light can cause similar changes. In certain extreme circumstances we suggest you should consider these possible effects when making recommendations in the RR, e.g. if you are instructed to report on an art gallery containing paintings, or museum with fabrics that could be affected by alterations in the lighting regime you should recommend the services of a lighting engineer.

SUMMARY

Identification of lighting systems and their controls is vital for energy efficiency. You will find much of the knowledge you need for identification will come from practice and inspection of actual systems.

Our answer to your lamp identification puzzle is: the right hand lamp is high pressure sodium; the other is metal halide.

8 Inspection and Reflection Methodologies for Existing Buildings

INTRODUCTION

In this chapter we review the methods you need to thoroughly inspect and reflect on the energy assessment of a commercial building. The main issues we address in this chapter are:

- the requirement for every job to have an 'audit trail';
- a short discussion of the practical nature of the contract between you and the client;
- your preparation before you visit the property – you can sometimes discover a considerable amount of information about the premises before you actually arrive 'on-site';
- your methodology of inspecting the premises, and how to handle the relationships with the client; or the occupiers;
- your post-inspection reflection about the data you have collected, before you begin to enter any data into the program.

FIRST CONTACT WITH YOUR CLIENT

The audit trail

Your professionalism could be called to account in the future and your file could be scrutinised. You should ensure any person who considers your job file can understand your thought and decision-making process when you carried out your inspection, entered the data and prepared your report. This is difficult to achieve and reinforces the importance for continuous review of your processes as discussed in Chapter 3. Every file should read like a well-structured book, demonstrating a complete audit trail from the note of your initial contact to the completed report and your client satisfaction questionnaire.

Figure 8.1, the 'Kings Lynn protocol' is a representation of the thought process you should embrace in order to achieve an EPC and RR 'fit for purpose'. It can act as a guide for you regarding the balance between data you receive from the client, data you collect on site and data you can accumulate from the internet and other similar sources. The amount of information will affect the way you carry out your inspection.

Initial enquiry, fees and conflicts of interest

You will receive enquiries for your services from a variety of sources including:

- solicitors – you should establish good relationships with local companies;
- commercial estate agents – acting for owners and lessees selling their freehold and leasehold interests;
- local property companies and developers – for existing and 'new-build' properties;
- individuals such as shop and factory owners;
- national and international referral companies seeking to establish a network of CEAs who can work for their retained clients.

You should try to avoid any actual or potential conflict of interest and comply with your accreditation scheme rules about this issue. In some instances, the person or company wishing to instruct you may ask for a percentage of the fee in return. At present, to the best of our knowledge, none of the accreditation scheme rules preclude such practice. However, if you are a member of any other professional body, you should check their rules of conduct – despite a relaxation of many rules in the last 20 years some professions still maintain strict (and, some might say, proper) rules of conduct and may frown on such practice.

This practice in other professions has sometimes contributed to levels of fee that are inconsistent with providing an adequate standard of service in our opinion. If a significant proportion of the fee the client pays is 'creamed off' by the introducer, your fee might be inadequate – although the client might not necessarily know this has occurred and will still, quite correctly, expect a first-class service from you. Indeed, past case law means you cannot use a low or no fee as an excuse for a low service level. We think you should

Commercial Energy Assessor's Handbook

Commercial EPC / RR 'Kings Lynn' inspection data / evidence *protocol*

START H1 First enquiry phase, following contact from prospective Client:
- Heating system above 100kW; or
- Cooling system above 12kW; or
- Other Accreditation Scheme rule confirms level 4 or 5 assessment; or
- EPC not legally required?

H2 Level 4 or 5 assessment may be required, or EPC not needed:
- Confirm why EPC not required; or
- Review your client approach; and
- Consider up-skilling Level 4 / 5 EPC required, and you are level 4 / 5 qualified?

H13 GENERAL APPROACH
- If your evidence / audit trail provides you with all of the required information, use it and input the data;
- Otherwise, use the defaults in the system;
- If you have <u>any</u> doubt, you **must** default to a more detailed inspection regime;
- Your aim is to achieve the most accurate EPC and RR possible, within acceptable Accreditation Scheme tolerances.

Route, subject to continuous reflection and review, to different level assessment, if the property is not at level assumed; or if EPC not required under Regulations

H3 Instruction confirmed, in writing, with CofE?

H4 Second enquiry phase: "*do you* **please** *have copies of all up-to-date documentation, and / or can you or somebody else confirm information regarding the following*"
- Drawings/plans/manuals/specification;
- Boiler, HWS, DH, CHP, GSHP, etc;
- Ventilation and exhaust;
- Air-conditioning and cooling;
- Lighting and controls;
- Others, e.g. H&S, SES, PV, Wind etc
- Question 100KW / 12KW again

H5 Inspection basis - RED
- Carefully carry out a **VERY** detailed inspection to assemble all of the required data;
- Are there any queries and/or uncertainties after inspection, review and reflection?

H7 Inspection basis - AMBER
- Carefully carry out a **MORE** detailed inspection than indicated in box H9
- Are there any queries and/or uncertainties after inspection, review and reflection?

H10 carefully reflect again

H6 Third enquiry phase:
- Send out specific Building Services & Fabric etc questionnaire;
- You receive a decent and full response?

H9 Inspection basis - GREEN
- Carefully carry out a **MORE** detailed inspection than indicated in box H8
- Are there any queries and/or uncertainties after inspection, review and reflection?

H8 Inspection basis - IDEAL
- Carry out very careful inspection of fabric and services etc;
- Check and confirm information provided in questionnaire; and
- Question plant/services manager
All information confirmed correct?

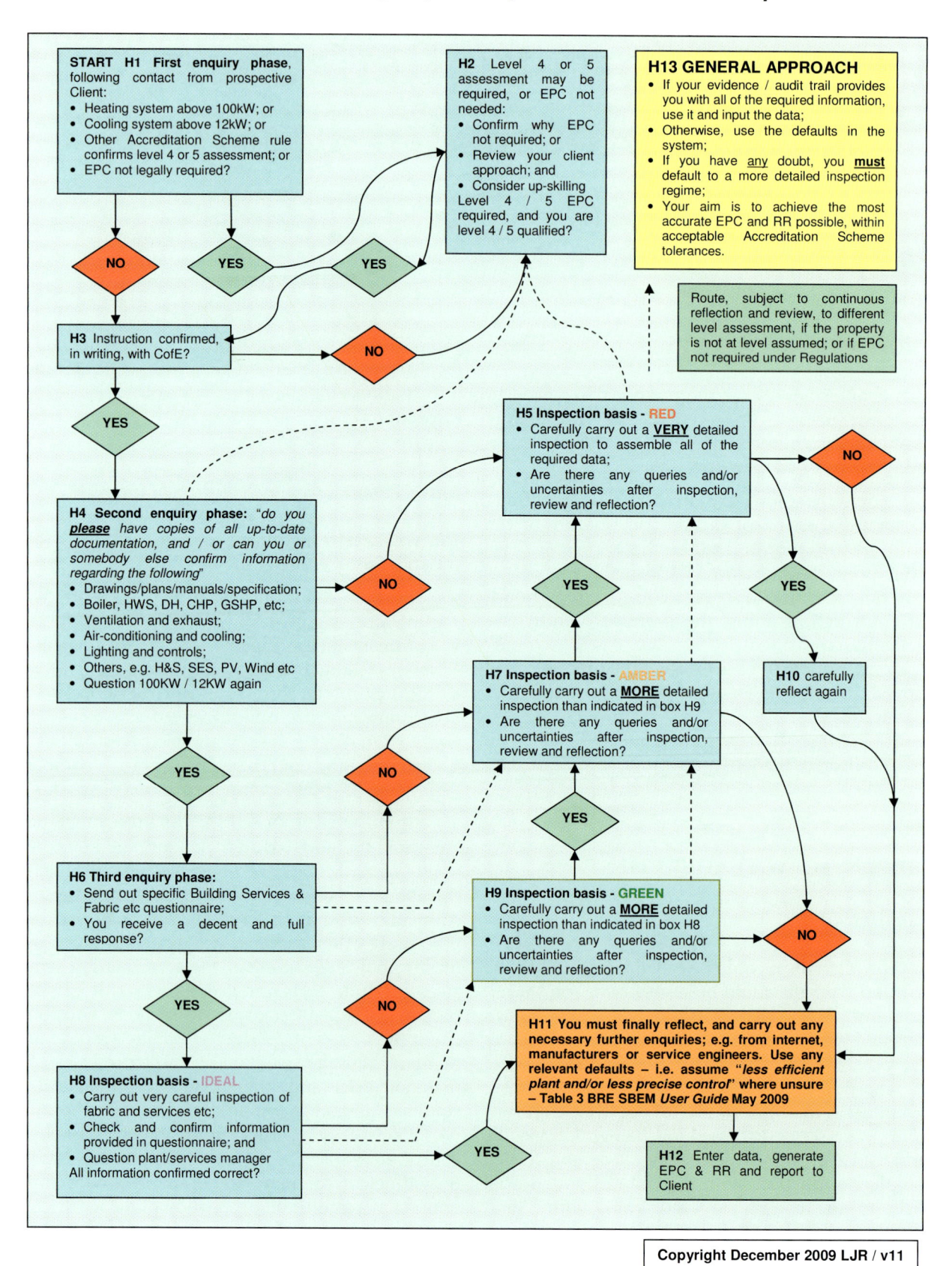

H11 You must finally reflect, and carry out any necessary further enquiries; e.g. from internet, manufacturers or service engineers. Use any relevant defaults – i.e. assume "*less efficient plant and/or less precise control*" where unsure – Table 3 BRE SBEM *User Guide* May 2009

H12 Enter data, generate EPC & RR and report to Client

Copyright December 2009 LJR / v11

Figure 8.1: Kings Lynn protocol

134

only work for an appropriate fee. If you work for anything less, the temptation to 'cut corners' is potentially in your mind.

Such outcomes are neither in the interest of the consumer, nor the professions concerned. On balance, we suggest you should try to resist this practice; unless you are sure there is no conflict of interest and you do not reduce your standard of service.

However, if your introducer requires an introductory fee we believe this represents a payment that could be seen as a potential conflict of interest. You should confirm this in the disclosure box in the report using words along the following lines:

> 'I received the instruction for this EPC and Recommendation Report from the seller following a recommendation from the selling agent to whom I will be paying an introductory fee which is a proportion of the total fee. I confirm this has not affected the way in which I have prepared my report.'

During your first contact you should attempt to establish exactly what level of service your potential client requires. Some will want 'the bits of paper' to comply; other, more discerning, clients might seek a 'value added' service for reasons discussed in Chapter 1 about how EPCs might affect the value of a property.

Client instructions

You should ensure you agree your contract with the client in the general manner, and for the specific reasons, we have described in Chapter 2. In this regard, we remind you that the EPC and the RR are valid for ten years. The person who instructs you is not likely to be the only person or legal body that relies on your report.

PRE-INSPECTION PRACTICE

General comment

Before your inspection you should try to discover as much possible information about the building's fabric and services. The NOS require you to 'assemble and collate information … [not only] … from your on-site inspection ... [but also] …from other relevant and reliable sources' (elements 5.2, 6.2 and 7.2). We now discuss some of the sources that can help you comply with the NOS.

Seller's questionnaire

In Appendix B we have included an example of a seller's questionnaire. You should send a questionnaire in every case; to the client, their agent or solicitor. The document gives your client the opportunity to inform you about their property, any changes they have carried out and specific details of the services.

You should have different questionnaires for different property types. You may find the complexity of the document in Appendix B is inappropriate for smaller types of property. The small 'corner shop' owner is likely to believe questions contained in the document are 'over the top'. Indeed, you might well provoke a reaction along the lines of: 'I'm paying *you* to do the assessment, not me!' A lengthier document is likely to be required for some large level 3, or level 4 and 5 properties.

Desk study

In many instances you will not receive a completed questionnaire, despite repeated requests. If you do receive it back, you should consider the information before you visit the property.

You may obtain further information from a number of other sources and by using a number of methods, including:

- internet searches;
- local authority website;
- land registry;
- the EPC register;
- manufacturers' information;
- the energy technology list at www.eca.gov.uk.

You should consider all of the information from the client and other sources as part of your assessment in a formal 'desk study'. Your information might include:

- aerial photographs – many internet mapping services now include images, e.g. you can sometimes see roof vents or flues that might not be visible on site; areas on adjoining properties you are unable to access;
- previous energy assessments of the property – ensure your assessment does not conflict, without good reason, with previous assessments; even if prepared on a different basis, e.g. an operational rating;
- reports on the condition and serviceability of the property's structure and services – to assist you in deciding whether there are 'condition' issues you should take into account in your recommendations in the RR;
- past planning permissions or current applications – to confirm the ages of different parts of the property;
- current planning status, in particular whether the property is:
  - listed, including the grade,
  - in a conservation area,
  - in a National park, site of special scientific interest, or area of outstanding national beauty, or even
  - a national monument;

- whether the property is in a smoke control zone – this could cause you to remove or edit recommendations automatically generated in the RR;
- Building Regulations applications and approvals – providing more detailed information about the original building/services and/or later extensions or refurbishments;
- a copy of the CDM (Construction, Design and Management) Regulations health and safety file – can provide you with information about construction and services; and for your health and safety;
- test, servicing or commissioning certificates for service installations;
- maintenance or operating log books;
- manufacturers' information – e.g. efficiencies of heating or cooling systems and controls, or 'U' and 'K$_m$' fabric values;
- guarantees, warranties and other reports – e.g. relating to boilers, chillers or other cooling systems;
- fire risk assessments and fire safety certificates – a good source of a plan as we describe below;
- an asbestos register – for your own protection and to ensure you do not disturb this material to the detriment of current and later occupiers;
- COSHH assessments – for the same reasons as above;
- schematics, e.g. of heating and/or cooling systems – these could be vital if access is poor in this regard, and it usually is;
- an access audit prepared for compliance with Part 'M' of the Building Regulations or similar – another possible source of a plan;
- building plans and/or specifications, licences for alterations and contracts – to confirm past work;
- copy of the lease – to confirm when alterations were completed;
- any other source as appropriate – e.g. old photographs, local history books.

You should not treat this list as exhaustive. You are likely to find particular local issues relevant to your practice. For this reason we believe you should consider restricting your geographical area of practice, certainly when you begin your working life as a CEA.

The concept of restricting the size of your work area is one accepted in many property-related professions, to protect you and your client. At the time of writing, we note indications that CLG is becoming concerned that energy assessors may be operating in areas of the country they are unfamiliar with, and possibly therefore generating EPCs and RRs that are incorrect.

SITE INSPECTION – A SUGGESTED PROCEDURE

Inspection – some general comments

We now outline a methodology of inspection practice for energy assessment of existing non-domestic buildings. We do not suggest this will always be the best approach, or all aspects will be appropriate, e.g. your method may vary if the building is empty. However, we hope you find some aspects of the suggested procedure helpful. We take the view that what we are suggesting is the *minimum* acceptable standard a CEA should adopt to achieve what clients require – a thorough, accurate and professional job.

The NOS require that you 'identify and record the method of construction of the property and the main materials used' and 'undertake a methodical visual inspection of all relevant aspects of the property' (elements 5.1, 6.1 and 7.1). The NOS provide further definition of this requirement, stating that in order to ensure compliance you must know and understand 'how to recognise different types of building construction, materials and services from drawings as well as buildings' and 'how to conduct the inspection in a thorough, methodical and consistent manner'.

In our view, you face a significant challenge to achieve these NOS requirements in a consistent manner:

- you will require a good quality management system that enables you to regularly reflect and review your working practices, as we have indicated in Chapter 3; and
- you should work towards developing your ability to build and hold in your imagination a three-dimensional image of the building and services – a skill all good property professionals have and one that new SBEM software programs that incorporate such an image can help you with.

At the time of writing, it seems most accreditation scheme rules will allow you to practice as a CEA using your own personal methodology, so long as that methodology complies with the accreditation scheme rules and the CLG 'Minimum Requirements'. This is notwithstanding the fact you must provide a service your accreditation scheme can properly audit to ensure compliance with scheme standards. One argument therefore might be that in order to ensure complete consistency, scheme members must all follow the same procedures and methods, even to the extent of insisting all scheme members use the same set of site inspection notes.

However, given the fact that each of us has different learning and working styles, we are firmly of the view that a methodology that recognises at least some of your personal preferences, and is 'owned' by you, is the best approach so long as your system is appropriate and complies with scheme requirements. This may not suit the internal audit processes of larger firms, but there is research to indicate that better working practices result from such a personal process, with fewer mistakes occurring.

Arrival

You should always try to arrive on time, meet and greet the client or occupier. It is rude and unprofessional to arrive late. It is also disrespectful to the client who is paying your fee.

It is a requirement of the NOS that you 'identify yourself to those present at the property before commencing the inspection' (elements 5.1, 6.1 and 7.1). You should therefore introduce yourself with a business card; and explain, possibly again, why you are there, approximately how long you expect the inspection will last, and how they can best help you to complete the inspection as quickly but nevertheless as accurately as possible. Some possible issues could be as follows:

- if the business or premises are being sold (as is likely in most instances), you should establish whether the employees/staff know this, and how you will deal with any questions from them. You might not like 'being economic with the truth', however if you inadvertently disclose the purpose of your visit and this is part of a sale process, you could be at risk of prejudicing the sale, your client's interests and even the employment rights and prospects of the staff. You should already have addressed the confidentiality issue in your client questionnaire, but there is no harm in ensuring again that you are not treading on egg shells in this regard. Thus, you could suggest, if asked, you are inspecting the property for insurance purposes – one of the writers has used this stratagem with success for many years;
- you are inspecting a shop – it might be best to inspect the public areas during a period when there are few customers to disturb;
- the owner expects a delivery of goods – you should inspect the area of the unloading bay first;
- you might even need to arrange your inspection 'out of hours', or at a weekend.

There will be many other instances where you should consider altering your usual order of inspection for the benefit of your client, and we can neither list them all nor set out what procedures you should follow. The fundamental point is you are at the property on your client's behalf and you should do all you can to accommodate their wishes within reason and without prejudicing the final results.

You should ideally set yourself up a 'base camp', with the occupier's permission; ideally in a quiet location; somewhere you will be safe, able to work, reflect and concentrate, can keep your equipment and won't interrupt business. We recommend a position near the kettle, tea bags and milk!

You should ensure your file is secure at all times – you may have personal or confidential information, so don't leave it in open view.

Information from the client

You should collect the completed 'commercial EPC questionnaire' together with all of the associated paperwork, e.g. drawings, manuals, manufacturers' information relating to heating and cooling systems. If the occupier has not completed the questionnaire, you can help them complete it there and then for you – ideally ask them to sign the completed document.

During your inspection, you will need to check the contents of the questionnaire, and the documentation and probably discuss some issues with the occupier.

You will see from the questionnaire in Appendix B that you should have specifically asked whether the client knows about health and safety risks you should be aware of so you can plan your inspection properly and safely. Now is the time, indeed the final opportunity, to clarify any particular issues relating to your health and safety before you begin your formal inspection of the premises.

In addition to the above formal enquiries, you should make verbal enquiries of site personnel (with circumspection and possibly owner's permission) and the owner/client when you require clarification of issues you find during your inspection of the property. You need to record their answers in your site notes.

Plan of the property

You should always begin your consideration of every property with a plan. This is because SBEM is 'plan-led'.

The plan you initially use can be the plan provided to you by the occupier. However, you must check it very carefully for accuracy. This is because you will find most if not all plans you are given will be incorrect in some way. Some reasons why include:

- human beings make mistakes;
- buildings are constructed by and under the supervision of human beings, usually with equipment, with similar consequences;
- at any time after around 5–10 years following the original date of construction (our suggested time is arbitrary, but based on experience) there is a risk the building and services may have been altered, extended or refurbished; and those changes may not have been recorded accurately, if at all.

For these, and other, reasons, you should always include measurement checks of the plan in your usual methodology. You should carefully measure appropriate parts of the property and compare your measurements with, say, 10% of the measurements stated on the plan(s) – if you find a significant number of inaccuracies either reject the plan because it could potentially mislead you, or carry out further checks until you are satisfied you can use the plan. You must confirm this process in your site notes, as part of your audit trail.

However, even very inaccurate plans can be very useful, because they can provide you with a general layout of the premises and interrelationship between the different spaces. Thus, if the client cannot provide you with a formal building or services plan for the property; do not dismiss out of hand any other sketch representation of the premises. A good example is a fire safety assessment plan. You will seldom find this plan is correct in a dimensional sense. However, it can often serve as the starting point for your own plan.

It is almost inevitable that you will need to prepare your own plan in most cases even if the client gives you one. This is because they will usually want you to return it, so you will not be able to write on it, or make any notes. We tend to draw our plans on sheets of A3 paper, rather than A4 size; although there are significant advances being made in electronic writing and drawing aids and paper may soon be 'out of date'.

Some CEAs, particularly if they are new to the profession, seem to find great difficulty in preparing a plan that is either accurate or helpful. If you are in this category, you may find useful a drawing aid known as 'Pyramid-Liner'. This backing material behind your sheet of paper helps guide your pen or pencil along a straight line. We wholeheartedly recommend this type of aid and within literally seconds you will find you can draw plans that suggest you have been sketching free-hand drawings for years!

In any event, we do not recommend you prepare a 'scruffy' site plan, and then later 'tidy it up', since:

- you will possibly make mistakes when transferring the information from your rough site plan onto your pristine plan prepared in the office; and
- you will waste time.

Instead, we suggest you persevere in your plan drawing and aim for perfection – you will rapidly find regular practise will work wonders.

Contents of the plan

You must prepare your plans in a methodical and careful manner. Suggestions for the general layout include the following:

- draw the 'front elevation' at the bottom of the page, so you do not confuse 'left, right, front and rear'; or
- show the 'south elevation' at the top of your page, following the SBEM envelope data entry;
- ensure you set the building out on the paper so you have enough space to include the building and/or each entire floor without running out of space;
- internally, draw the 'boxes', i.e. the floors, walls and ceilings/roofs that are the physical elements and boundaries of the rooms and spaces, (these are *not* necessarily the zones – see Chapter 9);
- if the building has more than one floor, or is attached to another building, ensure you take care to get the proportions of your drawing correct as you will need to reflect on 'adjoining conditions' and you will find this is difficult if your plan is out of proportion; even if you use a program that draws a two or three-dimensional plan of the building;
- always use your compass and record a 'north point' on *every* plan you draw – a supplementary check of where the sun happens to be can be helpful in this respect (with climate change, we should see more of it!);
- use a standard list of 'keys' as we have in Appendix B, i.e. diagrammatic indicators of objects such as radiators, central time controls and thermostatic valves, e.g. + (plus) sign for heater, - (minus) sign for cooling unit.

On your plan, you must show and record information that defines:

- horizontal and vertical measurements of surfaces (floors, walls and ceilings/roofs) in every room and space;
- construction details including thicknesses, of the main elements, i.e.:
  ○ walls, external and internal,
  ○ roofs, pitched and flat,
  ○ floors, suspended and solid,
  ○ doors, vehicular and personnel, and
  ○ windows and rooflights, including the extent of shading, 'overhangs' and 'fins';
- position and type of insulation, together with the thickness.

You also need to record the factors that define the zones, as specified in section 3.3 of the *BRE User Guide*. These details are:

- activity within each room or space;
- heating, cooling and HWS generators, outlets (emitters) and controls;
- air conditioning systems, outlets and controls;
- ventilation/extract systems and outlets;
- lighting information (lamp type, lumens, ballasts, etc.) and controls;
- day lighting – this will be implicit in the position and type of windows and rooflights.

Finally, here is some further advice on plans:

- if you are given one to use, check any scale – pdfs and photocopies can distort the scale;
- whether you use their plan or your plan, always take external and internal overall 'check' measurements and compare them with your totals for accumulated small spaces.

Order of inspection

You should complete your inspection in *at least* three 'laps', as follows:

- an initial inspection throughout the entire property, internally and externally, to:
  ○ establish your 'feel' for the building including the general layout and relationship of the different spaces and floors,
  ○ check health and safety issues and complete your written health and safety risk assessment,

- identify, record and report any existing damage to the building and services, so it is not attributed to you.

Do *not* immediately begin to 'zone' the property at this stage – you cannot zone the building until you have collected all the data and completed any other investigations. You might need to carry out some of those investigations after your inspection of the building.

Some occupiers will insist on, or offer to, accompany you on this initial tour of the property. In our experience, accepting such an offer can sometimes be very helpful, particularly:

- if there are a considerable number of people working in the premises, you will be seen 'with the boss' and this can (usually, but not always) help later when you make enquiries of personnel,
- the owner can personally point out any specific health and safety dangers,
- if the property is complicated in layout or construction; or has been extended on a number of occasions, having somebody who knows their way around can be a good introduction to the building and services,
- you will continue to develop your relationship with your client;

• a second 'lap' to prepare your plan and then logically, methodically and carefully record all relevant information. You might want to start outside to record the main construction, overall shape and also carry out 'check' dimensions. In practice, you can only collect all of this information during one 'lap' if the property is a particularly small one. For larger properties, most practitioners prefer to split this part of the procedure into a number of different 'tours' of the premises, recording each substantive part of the property independently from others, e.g.:

- dimensions, activities and construction in one lap,
- heating and cooling in the next,
- air-conditioning including ventilation, and exhaust, and finally
- lighting systems and controls.

In practice, the exact number of times you tour the property depends on how efficiently you observe and record the data you need, and the complexity of the building and services;

• lastly, a third lap, to check and confirm your information, and take photographs (see later). You should always make a final tour of the entire premises. You will be amazed, even after years of experience, at how much you can miss during your careful examination of the property; for reasons such as being distracted by people talking to you and preventing you recording details of the AHU; to the sun shining from a certain direction and preventing you seeing the particular type of brick bond or cladding detail.

Some practitioners prefer to inspect and record the property on a room-by-room, or space-by-space basis. During such an inspection they will record all of their data in one 'lap' only. You will find such a methodology requires considerable practice, sometimes over years. We take the view that whilst it is possible to collect all of the data in such manner in a small building; in a larger property it is inevitable you must return to certain parts of the building to check and confirm details, particularly if you discover very complicated HVAC or lighting systems.

Furthermore, in our experience, the more times you tour the property, the greater is the chance you will notice details you inadvertently missed during the previous tour, e.g. when you were preoccupied noting down the details of that very interesting lamp fitting!

Some CEAs record information about the property using digital recording machines. This can be an excellent and quick method of recording data – most people can talk faster than they write notes. However, you will need to ensure you follow a strict pre-prepared checklist and tick each item off as you proceed, and you should ensure the notes are typed up as soon as possible after the inspection. If you do choose this system, bear in mind that one day the technology *will* fail and when it does you may have nothing more to rely on than your memory. In such case you will have to return to the property and do the job again – the client might not be too impressed!

In residential energy assessment and surveying some practitioners now use technology that enables them to record data and generate the final report on site. The systems can send the completed report electronically within minutes of arriving on site. Such systems might develop for commercial energy assessment, although commercial assessment is currently much more complicated. We do not recommend this practice as it fails to allow you sufficient time to reflect on the data you have collected. You increase the risk of entering incorrect data and generating a false result, with the possible need to revise the report later when you realise your mistake(s).

As we discuss throughout our book, 'reflection' is a fundamental part of your decision-making process as a property professional. Hasty preparation fails to fulfil the requirement of reflection. The case of *Watts v Morrow* [1991] 4 All ER 937 is instructive; it concerned the practice of dictating a survey report (not notes for it) directly on site. In the view of the judge this produced a report that was 'strong on immediate detail but excessively, and I regret to have to say negligently, weak on reflective thought'. The defendant was found negligent by 'falling below the standard of reasonable care and skill required of the ordinary professional man exercising the same function as himself'.

At the time of writing, customers and clients are beginning to make successful claims against energy assessors. You will find such claims impossible to defend if you cannot later demonstrate:

- full and complete site notes; and
- evidence of mature professional reflection about the data you collected on site; before you entered that data into the program.

Photographs

We support the idea of recording data about the building and services using digital photography or 'moving films' – photographs are excellent. However, we are firmly of the opinion photographs should be *supplementary* to your main site notes. They are not a substitute for you recording good hand-written site notes, as some of the accreditation schemes require in their rules. We support this approach because:

- taking a photograph is too 'easy' – you can be lulled into believing you have captured all of the data in a photograph, only to realise later the information you require is frustratingly just 'out of shot'; whereas
- a hand-written note prepared at the property requires at least some reflective thought or consideration; and
- you will reflect much better at the property; rather than later, sitting at your computer.

In summary, photographs and similar recording devices are excellent tools, but only when you use them appropriately.

Extent of inspection – 'invasive' inspections

DEAs carry out a 'non-invasive' inspection. The current *BRE User Guide* (version 3.4a) in section 3.2 includes useful comment, guidance and tips for gathering information on 'new-build' properties; and existing buildings, including:

- 'Don't assume that adequate information on an existing building can be obtained easily';
- 'Any direct investigation of construction details such as wall or roof constructions and thermal bridges by opening them up should only be undertaken with the written approval of the building owner and consent of occupants'; and
- 'It can be difficult to identify systems from simple visual inspection'.

The guidance suggests discussing the required information with others such as the facilities manager. We agree with these comments. You should read section 3.2 in full. The inference of this guidance is that opening a building up might be required to establish vital construction, service and insulation details as otherwise without good information you will need to adopt 'inference procedures' and should 'err on the pessimistic side ... [or] ... err towards a less efficient plant and/or less precise control'.

We suggest you expose any part of the property with extreme care, as a 'last resort' and only after you have exhausted other attempts to discover the required data. However, if you believe 'opening ... up' is required, we suggest you:

- establish who is responsible for any extra fees or costs;
- ask permission of the building owner/occupier, or other person in authority, either:
  - *at least* verbally, in which case record the conversation on site notes, or
  - *ideally* in writing, with the permission confirmed in writing;
- prepare, and record, a health and safety risk assessment;
- carefully assess and consider whether the opening up can be accomplished without causing any damage, significant or otherwise;
- open up the relevant part of the fabric very carefully, ideally:
  - by using contractors skilled in such work, but if this is not possible,
  - carry out the task personally, using appropriate tools.

You should consider inspecting the fabric in areas that are already exposed or damaged. Areas accessed by asbestos surveyors are often useful and their location should be described in the asbestos report, or register.

Some accreditation scheme rules preclude such opening up. You should check your scheme rules. Many CEAs confirm in their Conditions of Engagement they will not carry out any type of invasive inspection.

However, we can foresee circumstances when an inspection based upon an invasive inspection might be the only certain method to arrive at the 'correct' result and might be considered a reasonable approach in all of the circumstances. You will need to consider such issues, in particular in cases where your liability could be greater because of any failure to open up.

Site inspection and reflection notes

You should record all of the information you require to accurately complete data entry on a SBEM program and advise your client in your notes. As has been noted above, and below, you will not necessarily collect all of the required information at the property.

The Minimum Requirements from CLG require that:

'Energy Assessors must make accurate and legible records of the data gathered at the premises. These records must be of sufficient detail to enable a third party to interpret the Energy Assessor's findings.' (para. 20)

The implication of this statement is clear – you must ensure you do not have to rely on your memory, or photographs, to complete your data entry. A useful exercise is to regularly place a copy of your completed site notes in front of a work colleague and ask him or her to 'interpret' the notes. You should note and act on any lack of clarity or other failings as part of your quality management system.

The Minimum Requirements further require that:

'Energy Assessors must … make available EPCs, RRs and associated site notes to their Scheme Operators.' (para. 28)

In this regard, the various accreditation schemes may have different specific requirements. However, there is now a good body of opinion to confirm what is, and what is not, acceptable as being good practice in relation to site notes. That evidence includes past case law, particularly relating to cases involving surveyors, architects, engineers and other property professionals; established practice employed by such practitioners supported by learned opinion and articles; and guidance that has been developed during the recent Home Inspector and Domestic Energy Assessor programmes.

In Appendix B, we have included a copy of the site and reflection notes we prepared for case study 2. Please turn to those pages now and look at the notes to familiarise yourself with the minimum level of inspection we believe you require for competent commercial energy assessment.

You should not believe your site notes will ever reach a state of 'perfection'. In line with the need to regularly update your professional knowledge, understanding and competence, you should subject your standard site notes to regular review and improvement in response to changing working practices, and any alterations to the iSBEM data entry requirements. We include a separate section in our site notes proforma for such review. In this way, you can record contemporaneous notes for later consideration in the more formal exercise of 'site notes review', say every month or quarter.

Indeed, we believe it is very likely if your standard site note proforma has the same appearance and layout 12 months hence; you are not engaging in the thoughtful professional reflection you should embrace as a practising CEA.

Once you have completed your inspection of the building and services, you should complete the task of entering all data on your notes. You should ideally do this at the property, since it is a simple task to note any omission and return to the part of the fabric or service installation and make your notes.

In our view, and experience, you should not leave until you have completed the site notes. If you leave the property prematurely, you could find yourself repenting at your leisure when entering your data and you might have to return. This does not reduce CO_2 emissions and reflects poorly on your personal organisation of time and efficiency.

Zoning

Once you have collected all of the data at the property you can finally begin to zone the property in accordance with the *BRE User Guide* 'rules' and any other 'conventions' from your accreditation scheme.

A common error made by new trainees and inexperienced CEAs is to walk into a room, space or an entire building and begin to zone it immediately. This is a mistake as there are so many variables to consider in the zoning process. You can only begin to zone once you have recorded *all* of the required information in your site notes and then reflected on that information. We discuss this issue further in Chapter 9.

We find it is very useful to use coloured pens to show the zones on the plan – look again at our zone plan in Appendix B.

Two different possible data collection methodologies

In Appendix B we have included the site notes for case study 2 and an example of a 'zone data sheet' to demonstrate two possible approaches to preparation of your site notes and plans. Both approaches have been developed by CEAs in practice. We can best describe those two approaches as follows:

1. One plan, or plans; and supporting site notes dependent on the size of the property. The plans include all of the information shown in case study 2. Your site and reflection notes support and underpin those plans. Once you have collected all of the necessary data, you zone the property and confirm your decisions by colouring the different zones.
2. Zone data sheets and zone identification plan. You prepare a global zone plan with the different zones coloured as before; and then separate zone data sheets to support your global plan.

We do not suggest these two different approaches are the only methodologies you can use; nor do we offer our opinion on which method we prefer. You should decide on the approach that suits you best; or arrive at your own methodology.

The advantage of the first method is you assemble all of the data and then prepare your zone plan – as we have suggested, you cannot begin to zone without all required data. You can therefore be sure your zoning process should be correct. The disadvantage is this method can be unnecessarily complicated and unwieldy for smaller properties.

The advantage of the second method is that all of the information you require about each zone is easily available to hand, on one data sheet. The disadvantage is that you can only prepare the data sheet once you have collected all of the required data and there is a danger you may decide on your zone boundaries before you are ready, in an attempt to press on with the job.

Advances in software programs and data collection may render some of these discussions void. In the meantime though, we suggest you practice and decide on your preferred methodology.

AFTER YOUR INSPECTION

Post-inspection enquiries

In many cases, following your inspection you will need to carry out further enquiries before you can enter all of the data. In our experience, issues that arise in this regard can include any of the issues we have discussed earlier in this chapter under the heading of 'desk study'.

In practice, you are likely to find that such enquiries typically include the need to research issues relating to:

- manufacturer's details of heating and cooling sources, hot water systems and storage facilities, controls and lamps, in particular efficiencies;
- confirmation of whether heating, cooling, lighting and control systems attract 'enhanced capital allowances' and assumed system efficiencies implicit in inclusion in the energy technology list;
- the local authority's planning register to check the date of any extensions or alterations you identified.

To confirm the general thrust of this chapter regarding your requirement to maintain an audit trail for each case, you would do well to consider that part of the NOS that confirms energy assessors should 'maintain internal records which are clear, complete and conform to accepted professional and statutory requirements' (elements 3.2, 4.2, 5.2, 6.2 and 7.2). You therefore need to record these details and keep a copy on your file.

SUMMARY

In this chapter, we have considered and discussed the important issues you must build into your quality management system including developing a system that demonstrates an 'audit trail', inspection practice with a structured methodology and a good overall layout plan with detailed site and reflection notes, using supplementary photographs and the need to do some pre- and post-inspection enquiries.

9 Data Entry into the Program

INTRODUCTION

In this chapter we consider issues you need to be entirely clear about after you have carried out your inspection and begin to enter data into the program. We have generally and deliberately chosen not to differentiate between the many types of programs and have instead concentrated on the issues that affect the assessment rather than any issues that might affect data entry problems specific to one, or many, types of program.

However, we discuss the entry using the order implicit in the iSBEM program for consistency. You may use other programs with different data entry fields.

In very general terms you should follow the guidance in the *BRE User Guide*, Chapters 4–7 inclusive in this book, advice from your accreditation scheme and from other sources to complete the data entry and generate the EPC. We deal with the RR in Chapter 10.

MAIN SBEM TABS

'General'

Your data entry under this tab should be reasonably self-explanatory. However, there are some common errors new entrants make and here are a few:

- ensure you enter the 'purpose of the analysis' – e.g. are you carrying out a 'new-build' ('Part L') assessment, or assessing an existing building ('EPC Wales' or similar);
- decide 'weather location' based on the closest geographical location in relation to the 14 regional centres available to you in the UK – 10 in England, one in Wales (Cardiff), one for Northern Ireland (Belfast) and two in Scotland (Edinburgh and Glasgow);
- decide on whether the analysis is level 3 or 4 – ignore the option for level 5 as SBEM cannot assess that type of building and it is not clear therefore why this option is in the program;
- information you enter for 'building address', 'City' and 'Postal Code' will define the name of the project on the final EPC and RR – you need to take care here not to upset your client by simple misspellings, etc.;
- tick the 'special conservation status' box if the building is listed, in a conservation area or an ancient monument (see Chapter 4);
- you need to disclose any 'related party' issues. Your accreditation scheme rules should provide further guidance, e.g. if you are a CEA preparing an assessment for a property your firm has instructions to sell you should confirm this information – such confirmation will appear on the two required documents; e.g. 'I am an employee/partner/director of the firm currently instructed to sell the freehold interest in the property, i.e. Braxton Hicks & Co';
- 'building types' – the NCM database currently contains a list of 29 building types a CEA will need. If you encounter a type not on the list, you should contact your accreditation scheme for advice. The list includes 'airport terminals' to 'workshops/ maintenance depot', and 'prisons' to 'launderette'.

'Project database'

General comment

This part of the program defines the different types of construction you have found in the building and from other sources. At this stage you are not defining the size or scope of the building or the services – you are simply describing the types of material components the property is constructed from, not how many or what size.

Before we discuss the five different types of construction that define the external and internal envelopes, it is important we consider the different methods of how to define those constructions. Let us summarise those issues for you here:

- 'Import one from the library' – use this button when you are reasonably certain about the type of construction in an existing property based on your evidence trail, or, e.g. you may have seen the insulation and the construction in detail and carried out a calculation to confirm the figures;
- 'Help with inference procedures' – this button is generally based on the appropriate date of the Building Regulations at the time of construction or refurbishment and intended by BRE to be used for existing buildings. You should use it:

- ○ when you have inadequate, indeed poor, information about the property. This is the case with many existing buildings, and
- ○ you will need to confirm the date of construction of the element, but
- ○ the number of available options is limited – if you cannot find an exact 'match' you should choose the closest option in terms of 'U' and 'K$_m$' values;
- 'Introduce my own values' – you are only likely to use this button when you are reasonably certain about the construction. In practice we believe you are unlikely to use this button very often for existing buildings, given the difficulty of knowing exactly how the building is put together, particularly if it is of some age. Thus, you are probably only likely to use this button if you are doing an 'on-construction' assessment for a new building where you have been provided with information based on 'proposed' drawings and/or a specification or similar document.

Your choices under this tab will define the levels of thermal insulation and thermal capacity for the fabric of the building you will use later in the data entry process to describe the building elements. You should attempt to ensure your choices are reasonable in all of the circumstances and you take into account the available evidence.

This information is not as important as that information concerning the services that condition and/or otherwise populate the spaces within the building. However, you should be prepared to justify your choices at a later date.

We now discuss some helpful tips about the different types of construction.

Construction for walls

You are likely to need to define at least one external and one internal wall type. Note we are talking about the 'description' of walls here. The data entry of a wall's dimension, which we discuss later, is made gross, i.e. including the areas of windows and doors in the wall. You should not confuse 'description' with 'dimension'. The same point applies to the other constructions.

Construction for roofs

Here you describe the different types of roof construction.

Construction for floors

These are likely to be:

- an external floor; or
- an internal intermediate floor or ceiling.

Construction for doors

Here you describe the door types, not the individual doors.

Glazing

As with the doors, do not describe the individual windows, merely the different types of window.

'Geometry'

Global building infiltration

You need to set the rate here, i.e. how 'leaky' you believe the building is. Our advice is you should adopt the higher, default, figure for older buildings, i.e. 25m³/h/m² at 50pa, unless you have firm information you can rely on. That information should be the results of a pressure test by an appropriately qualified person. If the results are older than 12 months we suggest you treat them with considerable caution.

You should *not* 'make up' your own figures or approximations without good evidence or following guidance from your accreditation scheme. Different infiltration parameters apply to some 'on construction' assessments and you must take care you do not use these for existing buildings.

Activity

SBEM works by standardising the building type and activity in order to make comparisons between the different types of building. During the zoning process (see later), you are required to divide the building into areas where different 'activities' occur. SBEM standardises activities in terms of:

- heating and cooling temperatures;
- occupation density and duration;
- hot water demand;
- internal heat gains from equipment assumed to be used in the area;
- fresh air supply; and
- natural and artificial lighting levels.

The above assumptions are standardised for each activity and in relation to each building type. BRE have obtained information for each activity from various industry publications, recognised organisations such as CIBSE, or have calculated figures based on their extensive knowledge.

There are currently 64 different types of activity in the SBEM NCM database. You will find full details in the NCM database you should download from BRE. The database applies the activities to each of the 29 building types in the program. The heating, cooling and other parameters can differ each time you apply a particular activity to a building type. This means the 64 activity types are allocated to the 29 building types to provide a list of 491 different types of activity. However, each of the building types has only some of the activities allocated to it.

As an example, the most usual types of activity you are likely to encounter in one of the most common types of building, i.e. 'office', are set out in Table 9.1, with information about the assumed density of occupation, water demand and our comments.

Table 9.1: Some different SBEM activity types, assumptions and definitions, based on NCM database

Building type/activity	NCM database definition	People density (person/m²)	HWS requirements (litres/person/day/m²)	Our comment and any other specific issues
Office – cellular office	Enclosed office space, commonly of low density	0.07	0.21	Usually considered to be an area or room where no more than six people can be satisfactorily accommodated to work Relatively low level of energy assumed for equipment use of 10W/m²
Office – open plan office	Shared office space commonly of higher density than a cellular office. For very high density with a correspondingly high IT load, refer to 'high density IT work space'	0.11	0.33	Any larger space than a cellular office and in use as an office Slightly higher level of energy assumed for equipment use of 15W/m² when compared with cellular office
Office – circulation (corridors and stairways)	For all circulation areas such as corridors and stairways	0.11	0	Generally also taken to include space occupied by lift shafts
Office – common room/staff room/lounge	An area for meeting in a non-work capacity. May contain some hot drink facilities	0.11	0	Likely to contain space for a number of easy chairs and a table Low level of energy assumed for equipment use of only 5W/m²
Office – tea making	Areas used for making hot drinks, often containing a refrigerator with transient occupancy. For larger areas containing seating and a small hot drinks making area refer to 'common room/staff room/lounge'	0.11	0	Note the difference between this activity and the one described above – this will typically be a small kitchen Relatively low level of energy assumed for equipment use of 10W/m², but higher than a staff room You might believe the HWS assumption, i.e. '0', is a nonsense or mistake for the room where occupiers brew tea, etc. but the nil figure serves to emphasise that HWS allocation depends on the likelihood of people occupying the space, rather than use
Office – high density IT work space	High density desk based work space with correspondingly dense IT	0.20	0.60	Note energy assumed for equipment use of 30W/m² Compare it with 50W/m² for 'office – IT equipment'; 15W/m² for open plan office and 10W/m² for cellular office

Table 9.1: Some different SBEM activity types, assumptions and definitions, based on NCM database (cont.)

Building type/activity	NCM database definition	People density (person/m²)	HWS requirements (litres/person/day/m²)	Our comment and any other specific issues
Office – IT equipment	An area dedicated to IT equipment such as printers, faxes and copiers with transient occupancy (not 24 hours). For areas which have 24 hour gains from equipment select from the 'miscellaneous 24 hour activities' building type either IT equipment (low-medium gains) or data centre (high gains). For areas with IT equipment and desk based staff, use one of the office activities	0.11	0	High level of energy assumed for equipment use of 50W/m²
Office – meeting room	An area specifically used for people to have meetings, not for everyday desk working. For everyday desk working areas refer to the appropriate office category	0.20	0.06	Low level of energy assumed for equipment use of 5W/m²
Office – plant room	Areas containing the main HVAC equipment for the building, e.g. boilers/air conditioning plant	0.11	0	High level of energy assumed for equipment use of 50W/m²
Office – storage area	Areas for un-chilled storage with low transient occupancy	0.11	0	Very low level of energy assumed for equipment use of only 2W/m²
Office – reception	The area in a building which is used for entry from the outside or other building storeys	0.11	0.03	Relatively low level of energy assumed for equipment use of 5W/m²

There are many more definitions. You should make careful note of the differences between the various activities in the *BRE User Guide*, and make your decisions about the activity accordingly. Note that while many of the assumptions made in the program in respect of each activity about issues such as density of occupancy, hot water demand, etc. remain constant no matter what the building type; this is not always so.

If you tend to focus on particular types of building you should carefully study the NCM database for those building types. You might wish to prepare a similar table to Table 9.1 above and use it for reference during your process of decision-making in relation to zoning. Preparation of such a table will also help you increase your knowledge about the activity types available for your particular area of practice.

As you carry out more assessments on this building type you could perhaps add other building/activity types to the list. You should review your list whenever BRE issue a new version of SBEM – you cannot assume all of the background assumptions will remain constant over time. If you look carefully at the various sub-tabs in the NCM database you will note the vast amount of information BRE has entered, and the various sources. BRE are likely to update this information whenever the sources of information indicate a revision is required.

Building types and activities – an example

You do not have to adhere strictly to the building type. Thus, if you decide an area you have identified in the building has a certain type of activity, but cannot find the 'activity' under the 'building' type, you should alter the building type to one that has the type of activity you require – 'activity' is one of the most important, if not the most important, issues in the program.

Let us assume you have inspected an office building with a classroom, i.e. a teaching area for life-long learning. You search the types of activity available to you under 'office' – 'classroom' activity is not available. You might consider adopting the activity of 'meeting room'. However, you could also consider choosing a different building type. As an example, if you consult the NCM database, you might look at a building type such as 'primary school' that includes an activity of 'classroom'.

This is a relatively easy course of action to adopt, since 'primary school' is close to 'office' in the database; and an activity such as 'classroom' is what you seek. Indeed, you might feel entirely justified in adopting a different type of building in order to find a more appropriate activity when you see the differences between the assumptions the program will make – look at Table 9.1 again and note the differences.

However, whilst we intend the examples we have shown in that figure to encourage you to consider this issue, and take action when appropriate, it should also sound a note of caution. When you decide to use another activity and/or building type you should ensure you use the database appropriately. In our view, our quick, lazy, use of 'primary school – classroom' we have demonstrated would be inappropriate, for reasons that should become clear if you look at Table 9.2.

A more appropriate building and activity would be 'university – classroom', or even 'prison – classroom'. If you decided to choose 'primary school – classroom', you can see that some of the assumptions are the same or very close; but some are significantly different. Instead of the more appropriate assumptions in the two right-hand columns of Table 9.2, assumptions about occupation density and HWS demand would be incorrect to a significant extent. However, assumptions the program would make about heating, cooling and other matters would be 'correct', or nearly so.

The combination of greater density and HWS demand assumed for school children would cause a significantly increased demand on the hot water generator, with a consequent effect on CO_2 emissions and the final rating. We would have done better to have remained with 'office – meeting room', rather than alter to 'primary school – classroom'.

It is a tall order to remember all 491 of the activities currently spread across the different building types in the database. However, you will gradually develop the required knowledge as a result of practice. You should only alter the activity and/or building types:

- with great care; ideally
- based on a reasonable knowledge of the NCM database; and
- using common sense.

For more information about what we have discussed in this part of the chapter, consult the *SBEM Technical Manual*, and in particular:

- Table 1: List of building types; and
- Table 2: List of Activity areas with definitions.

Building type and activities – a case study

At this stage in the chapter and having been introduced to the NCM database and hopefully opened it up and looked at it, you may be wondering if you really do need to consider such issues; or a change in career! Perhaps an example might help.

While writing this book, one of us completed the data entry on a industrial building and received a shock. The building was a mixture of 1970s and 1980s construction, TUFA in the region of 5,000m². The HVAC was old and not on the ECA list. HWS was from direct electricity and there were inefficient T12 lamps. The result, at '42 – band B', was a considerable surprise as he anticipated a much lower result for a building with relatively poor insulation and inefficient services.

Like any good CEA, he checked his data entry and found it correct as far as he could see, but during a telephone conversation with another CEA remembered he had chosen 'industrial process' for the activity in the main space. He had believed such an activity justified by the heavy duty equipment left at

Table 9.2: Examples of some different building/activity types, based on NCM database

NCM database assumptions	'Office' – 'meeting room'	'Primary school' – 'classroom'	'University' – 'classroom'	'Prison' – 'classroom'
People density – person/m²	0.20	0.55	0.20	0.20
Cooling set point/°C	24	23	23	23
Heating set point/°C	22	18	20	20
Metabolic rate – W/person	120	140	140	140
Occupancy latent gain – %	39	50	50	50
HWS requirements – l/day/m	0.06	1.50	0.15	0.15
Ventilation requirements, outdoor air – l/s/p	10	5	5	5
Lux	300	300	300	300
Equipment – W/m²	5	5	5	5

the property by the outgoing tenant. He reflected and altered the activity in the main space to 'speculative industrial'. He could have considered using 'workshop – small scale'; however the building is more than 'small scale' workshop use. The revised result – '144 – band F'!

He considered this result to be much closer to what he believed to be a 'correct' figure, based on his experience. The information in Table 9.3 may help to explain the considerable difference between the ratings using the different assumptions. Note, in particular, the differences in assumptions SBEM makes about lighting and equipment.

This demonstrates how you need to continually reflect and revise your approach to each job. We discuss this issue further in Chapter 10. The particular 'learning outcomes' for this CEA were:

- use professional judgment to assess the result – 42/'B' did not sit well with the description of the actual building and its services;
- remember to check the impact of the chosen activities on the rating outputs;
- discuss any issues you are unsure about with other CEAs and/or your accreditation scheme

Choosing the activity

You should choose your activity on the basis of the available evidence. For 'new-build' properties you will need to rely on drawings, specifications and other documents. You might need to discuss the matter with the designer – record (i.e. write down), your discussions.

For existing buildings you must carefully note how the spaces are being used on the day of your inspection. Your photographs will help considerably, i.e. recording furniture, equipment and other items. If the building is empty, you may need to make a decision based on the previous use, if you know it; or even what seems reasonable in all the circumstances.

Zones and zoning

SBEM requires that you separate each floor in a building into 'zones' in order to reflect any differences in the following:

- heating, ventilation and air conditioning systems;
- type of activity;
- built in lighting systems; and
- natural lighting levels from windows and rooflights.

You should by now realise why you are required to divide the building into different zones; i.e. so the program can allocate different levels of CO_2 emitting fuels used by the fixed building services to each of those areas (or 'zones') based on the different activities.

If you find that any of the defining factors are different anywhere in the area that is being reviewed, a different zone should be applied to that area. The *BRE User Guide* and *Technical Manual* succinctly describe the process of decision making for dividing each floor of the building into zones. Zoning is *absolutely fundamental* to the SBEM process. You should therefore break off from reading this chapter and read BRE's process for zoning in the current version of the *User Guide*.

The rule to remember is if any of the factors that define a zone are different (or as BRE state 'significantly different'), you should apply a different zone to that area of floor. In some cases it may be necessary to create a 'virtual' zone to accommodate some of the zone criteria. For example, a space may contain different lamp types. One approach is to select the most dominant form of lighting type, but an alternative is to create a separate, 'virtual', zone within the space and assign one of the lighting types to it – see the basement in case study 2 in Appendix B.

You should note the rules relating to day-lighting, as they will help you appreciate the considerable effect of natural lighting on energy use in a building. However, current advice from BRE is you are generally not required to use the day-lighting criteria for zones and

Table 9.3: Some further SBEM building/activity types based on NCM database

NCM database assumptions	'Industrial'/ 'industrial process'	'Industrial'/ 'speculative industrial'	'Workshop'/ 'workshop small scale'
People density – person/m²	0.02	0.01	0.07
Cooling set point/°C	–	25	–
Heating set point/°C	13	18	18
Metabolic rate – W/person	250	180	180
Occupancy latent gain – %	73	62	73
HWS requirements – l/day/m	0.05	0.05	0.21
Ventilation requirements, outdoor air – l/s/p	10	10	10
Number of lux	1000	300	500
Equipment – W/m²	50	2	5

most programs will zone the building using the day-lighting rules for you.

You should take special care with zoning. In particular, you will find if you or your data gatherer do not carry out a very careful inspection *and* record the data as carefully, your zoning process can easily be incorrect.

Zone height and corners

In older versions of the program CEAs were required to count the number of external corners in the building. Corners are important because they are repeating thermal bridges – see Chapter 4. In iSBEM you will find the program requests data entry for zone height under the headings of 'zone height (global)' and 'flr-to-flr height'.

We understand the program uses the zone height measurement to calculate the thermal transmittance along the length of the thermal bridges at corners of the building and junctions between internal walls and external walls. This latter junction will typically occur where an internal wall meets an external wall; or where a party wall between the building you are assessing meets the external wall. We show a typical plan view of this instance in Figure 9.1.

We are aware that different training organisations and accreditation schemes have varying conventions for zone height. We believe the important issue is the length of the thermal bridge. You should note this may not always be the greatest height of the envelope concerned (look at Figure 9.2). You should follow the convention of your particular scheme. We show some examples of vertical representations of zone height in Figure 9.2.

Envelopes

Once you have defined the zones in a *horizontal* sense, i.e. at floor level, you must begin to consider each zone *vertically*. The program requires that you describe the 'envelopes' that surround each zone. For this purpose, try to imagine each zone as a 'cardboard box' with (for descriptive purpose, but not always) six surfaces as follows:

- four walls;
- one floor and one ceiling; or
- a roof – some buildings with floors above may not have any roof.

Thus, most buildings will comprise a series of such 'boxes' stacked in adjoining positions, and in the case of buildings with more than one storey, above other zones. The boxes will sometimes be shaped irregularly. You should ideally be able to visualise such relationships. Indeed, we believe the need to be 'spatially aware' is usually a requirement for any energy assessor. Some software programs are able to provide such visualisation on the computer screen – this facility is a considerable advantage.

When you decide where your envelopes are, you allocate types of construction to them, from the 'project database'; i.e. the three types of construction above. You then add any relevant doors, windows and rooflights that sit within the envelope. If any envelope does incorporate such openings, you should note the following order of data entry:

- enter the gross area of that envelope, i.e. the wall, floor or roof;
- then enter the area of the door, window or rooflight – that area will 'cancel out' the area of the underlying envelope it sits in.

In practice you will find that some envelopes are 'virtual', i.e. they do not exist. A typical example would be a large open plan office that you divide into two separate zones for any reason, e.g. one end of the room is heated by electric storage heaters and the other end by radiators. Each zone will have three physical vertical envelopes (walls), together with the horizontal envelopes (floor and ceiling); and another vertical

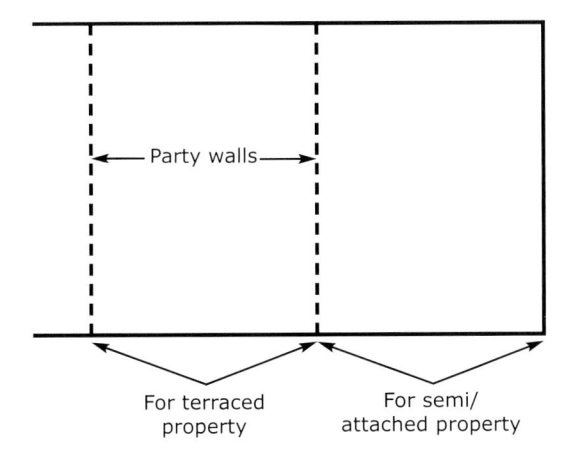

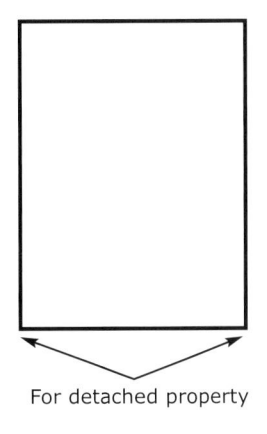

Figure 9.1: Plan view of corners intended to be captured by entering zone height

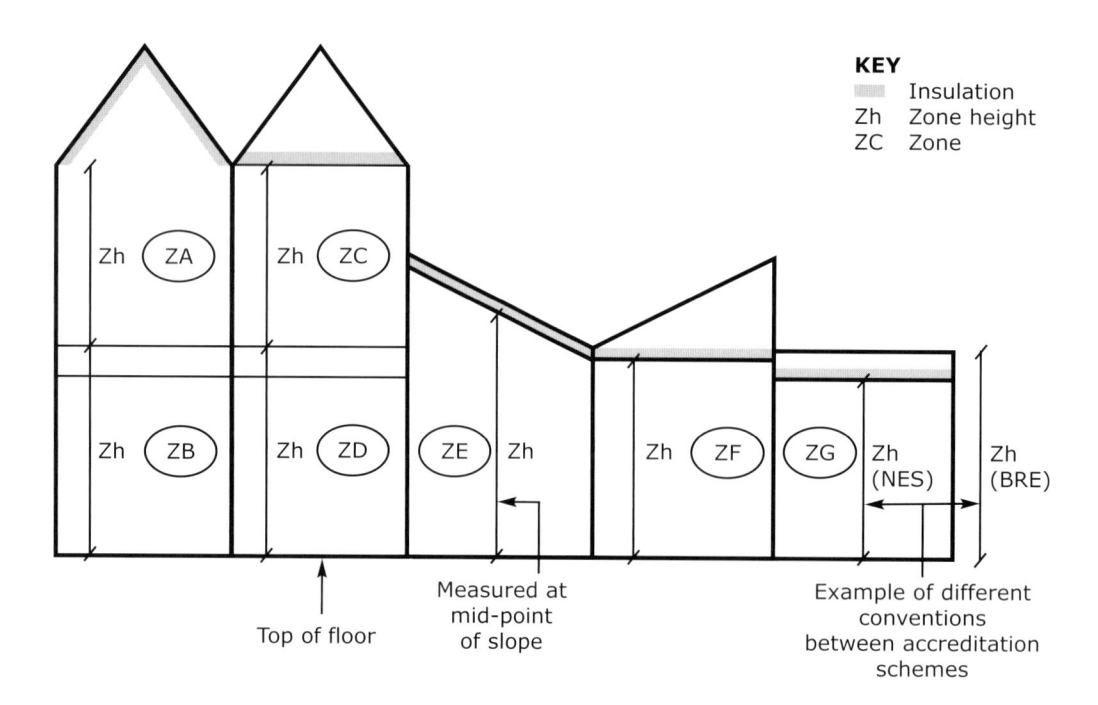

Figure 9.2: Examples of vertical representation of how to calculate zone height

envelope common to each zone that is not defined by a physical boundary.

In such a case, SBEM assumes no transfer of heat, cooling, or light occurs in either direction across this virtual envelope.

Adjoining conditions

One of the fundamental premises of SBEM is that each zone is 'conditioned' (or not) in some way. The conditioning applied to the zone is dependent on the issues indicated elsewhere. The five conditions are:

- exterior;
- strongly ventilated space;
- unheated adjoining space;
- conditioned adjoining space; and
- underground.

We describe these types of adjoining condition with some examples in Table 9.4.

Specific issues relating to conditioning

As already indicated, you must confirm in the program the type of conditioning that is, or is likely to be, applied on the other side of the envelope – either outside the building or in the adjoining space. In order to make an informed decision regarding this issue you will need to carefully inspect the entire premises that are the subject of the assessment, internally and externally.

This should be reasonably easy in properties that have simple layout, usage, construction and conditioning, e.g. the typical corner shop. However, you may encounter difficulties in certain buildings, such as:

- buildings where processes such as manufacturing mean that access on the day of inspection might be impossible;
- areas where there might be issues relating to confidentiality or security;
- spaces where stored items such as stock, machinery or other stored or fixed items mean that an inspection is not possible; and
- rooms where, for whatever reason 'the door is locked and the key isn't available' at the time of your visit.

In all such instances, and similar cases, you should endeavour to carry out an inspection of the area concerned. This will mean you should attempt to establish a relationship with the occupier that will help to facilitate access to all necessary parts of the property. Experience will teach you how to deal with certain types of property/occupiers. However, you would be wise to ensure that you make clear to the client/occupier that you do require reasonable access to the entire property and the services in the following ways:

- verbally to the client/occupier when discussing the instruction;
- in your 'conditions of engagement';
- in the letter confirming the appointment; and
- again upon your arrival at the property.

In all cases, you must adopt a prudent, but reasonable approach that reflects the circumstances. In our view you *must* make all reasonable attempts to inspect all of the necessary areas. Inspection of such areas may be vital to the final EPC rating – a judge or a prospective

Table 9.4: Discussion of adjoining conditions

No.	Adjoining condition	Examples
1.	Exterior	This is a condition outside the building defined by air (or water); examples are: – on the external side of an outer wall, or roof; – beneath an exposed floor, such as over a vehicle or pedestrian access way
2.	Strongly ventilated space	Example of this type of condition is: – a space with openings that cannot be closed with a capacity for the supply of fresh air and extract of inside air, determined according to section 5.3 of NEN 1087, of at least $3 \times 10\text{-}3$ m^3/s per m^2 useable area, e.g. a car park. The definition in the *User Guide* is very specific
3.	Unheated adjoining space	Any space where there is neither heating, nor is heating likely to be provided
4.	Conditioned adjoining space	An area on the other side of a zone envelope, within the subject building or in an adjoining building (see later), where 'conditioning' is applied
5.	Underground	Examples include: – a space where there is a floor (solid, ground-bearing or suspended) immediately or very closely above the soil the building sits on; – on the other side of a wall to a basement below ground level, or at ground floor level where part of an outer wall (i.e. envelope) is below ground level, such as on a sloping site

occupier will quite correctly take a dim view of a CEA who does not make such reasonable attempts, and produces a misleading energy assessment. This may require you to return to the property on another occasion, e.g. when the occupier has found the key or has made arrangements for the secure area to be temporarily cleared for the inspection.

However, it may be that in some circumstances you cannot carry out an inspection of the area in question. In all such instances, you must record any limitations to your inspection in your site notes, supplemented with a photograph (but taken with care and circumspection, and *not* of the secure area!).

In addition, you should note significant restrictions in the program, e.g. 'user notes' and EPC audit in SBEM (see Chapter 10); and in the accompanying client report if you provide one. In this way, any person who might rely on the EPC, and to whom therefore you have a duty of care, will be aware of the limitations of the advice contained in the report.

However, you will usually have no automatic legal right of entry into properties that adjoin the property that is the subject of the assessment. You should realise therefore that when you are carrying out an inspection of a property that is attached in some way to an adjoining building, there will necessarily be envelopes (typically floors and walls) beyond which you may not be able to properly assess the nature of the adjoining condition.

In such circumstances, you will still be required to enter information regarding the adjoining condition. You will therefore need to carry out as complete and careful inspection of the building, the adjoining buildings, and the surroundings as is possible, bearing in mind the physical facts at the property on the day of the inspection. You must record any restrictions in your site notes. Supporting photographs could be particularly important in this respect.

However, reg. 50 of the 2007 Regulations (SI 2007/991) provides that:

'It shall be the duty of every person with an interest in, or in occupation of, the building to ... allow ... access to any energy assessor appointed ... as is reasonably necessary to inspect the building (and) cooperate with the responsible person so far as is reasonably necessary to enable him to [prepare the EPC and RR],'

Thus, we believe while you cannot reasonably demand access into another building that is not involved in the transaction, a landlord of a block (and in some cases a tenant if they have an interest in the 'building' in question) must give you such access.

You should otherwise conduct your external inspection only from within the boundaries of the property that is the subject of the energy assessment and from any public land or highway.

Specific issues you would be wise to bear in mind in relation to this general issue include:

- at party walls, make particular note of the vertical alignment between the subject property and the adjoining properties – noting whether the floor and roof levels are consistent across the boundary line; e.g. if the property is on a slope or the roof on the adjoining property is lower than the subject property, perhaps part of the envelope in the building you are inspecting has an adjoining condition to 'conditioned adjoining space', and part to 'exterior' – see Figure 9.3;
- in most instances, it is likely you will be able to make a reasonable assumption regarding what the adjoining condition is in the adjacent property, based on your observations of that property. This might be difficult sometimes though, such as:

○ where the assessor cannot see the adjoining building. However, an 'aerial photo' screen-shot could help, although some of the images are dated and/or indistinct;

○ where the assessor can see the adjoining property, but the use is not clear. In such circumstances, careful questioning of the occupier of the subject property will usually be helpful.

You should take care when recording sensitive information regarding the adjoining property, in particular when taking photographs – this can provoke sometimes extreme reactions and leave an unhappy client, occupier and/or estate agent having to deal with annoyed neighbours. This could result in a formal complaint to the accreditation scheme and a reduction in the number of recommendations for work from the agent who must deal with the resultant problems.

Certainly, under no circumstances should you commit a trespass onto an adjoining property when carrying out your inspection.

Indirect conditioning

An indirectly conditioned space is one in which any heating, cooling or ventilation is provided indirectly from an adjoining space due to a 'high level of interaction' between the spaces. An example of this would be a corridor next to offices which is not directly heated but is conditioned through the movement of heat and air from the adjacent offices due to the repeated opening and closing of the interconnecting doors. You should consider an indirectly conditioned space to be conditioned by the HVAC system that supplies the most important surrounding area. You should label envelopes between a (directly) conditioned space and an indirectly conditioned space as adjacent to a 'conditioned adjoining space' and NOT to an 'unheated adjoining space'.

This rule will be subject to the particular circumstances you observe and record on site. An example of a space that is unlikely to be indirectly conditioned and one that is likely to be directly conditioned might be:

- a store room adjoining an office where the occupants do not use, or are unlikely to use, the interconnecting doors on a frequent basis; whereas
- a door between an office and the tea making area is more likely to be in frequent use (certainly in this country).

Envelope measurement conventions

You are required to measure buildings internally and enter data into the program in the *horizontal plane* as follows:

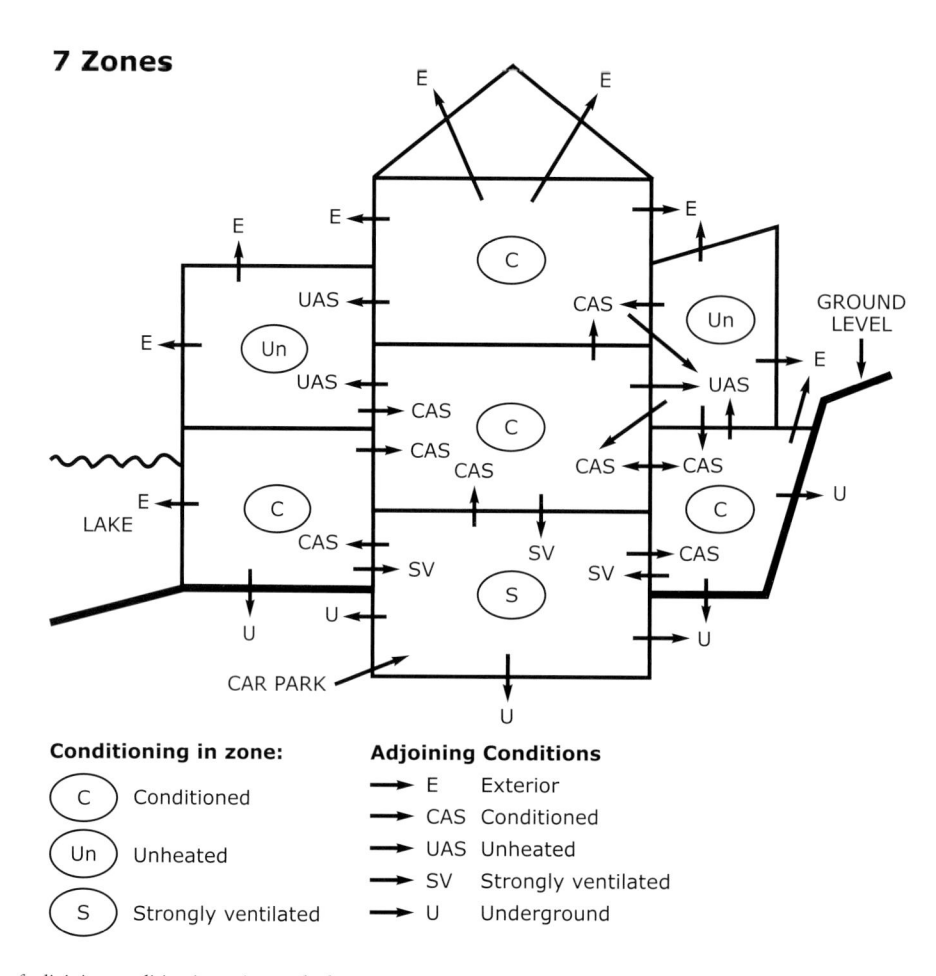

Figure 9.3: Example of adjoining condition issues in attached property

- from and up to the internal face of external and party walls (so, treat party walls as if they are external walls); and
- to the centre line of internal walls.

In addition, you should measure the building and enter data in the *vertical plane*:

- internally from the top of the lower floor slab to the top surface of the floor slab or structure above; and
- on the top floor measure to the underside of the insulation, i.e. to the ceiling.

You must input data relating to the surface area, construction, direction in which it faces to the nearest 22.5° (unless it is 'horizontal'). For vertical envelopes, the size entered is 'gross', i.e. including doors and windows. You ignore internal doors and glazing. You do not enter data for 'virtual' (i.e. non-existent) envelopes.

Thermal capacities within merged areas

In some cases you may merge many areas with the same zoning criteria, as described earlier in this chapter. If you follow this procedure you will not capture and reflect the thermal capacity of those constructions. The current *User Guide* suggests you should add together the surface areas of two or more internal walls (between merged areas) with the same construction and orientation and enter them as one envelope (assigned to the zone resulting from your merging exercise).

However, if the internal walls are partitions of lightweight construction, e.g. timber stud partitions with a 'very small thermal mass' (*BRE User Guide*, section 3.3), you need not use this practice as their thermal capacity should not cause any significant effects on the calculation if they are omitted from the data entry. You should avoid the phrase 'thermal mass' and instead use the term 'thermal capacity' – see Chapter 4. See case study 2 in Appendix B for an example.

External windows and doors

You must enter relevant information relating to these parts of the building fabric, including the surface area. The area of the window and/or door you must consider includes the frame, i.e. measurement is horizontally to masonry or similar reveals, and vertically from the underside of the lintel and to the underside of the sill, or threshold in the case of a door.

Each window has a 'surface area ratio', i.e. the relationship between the surface areas of the glass and the frame compared with the size of the opening. Entering a rooflight normally provokes a default of 1.3. You can accept this if it seems reasonable, otherwise you should calculate the figure and enter the correct amount. Clearly, a figure of 1.3 is inappropriate if the rooflight is a flush 'velux' type construction or typical industrial type, in which case you should normally enter a figure of '1'. See Appendix F for an example.

'U values' of vehicular entrance doors

Some versions of SBEM currently default under 'inference procedures' to an unreasonably low 'U value' (1.5W/m²K) for an uninsulated vehicle entrance door. We suggest you 'import one from the library' or adopt a more appropriate figure for this construction.

High usage entrance doors

This is a door likely to experience 'high traffic volumes' (ADL2A, para. 99). Such a door should usually have automatic closers, be protected by and with a lobby.

Fins and overhangs

You must confirm any glazing that is partially obscured by such building details. We have shown an example of how to calculate the extent to which the amount of light entering through the window is obscured by these projections from the external face of the building, or adjoining buildings in Appendix H.

This issue generally only tends to be significant if fins and overhangs are present on the south facing elevations.

Glazed 'doors'

You should treat doors that are more than 50% glazed as windows. The area you must consider should include the total surface area measured to the reveals.

'Building services'

Please look at Chapters 5 and 6 for details of how to enter data in this part of the program.

'Ratings and building navigation'

We discuss these parts of the program in Chapter 10.

SUMMARY

In this chapter we have discussed many of the important issues that sometimes confuse new entrants and qualified CEAs alike when they enter data into the program. We draw your attention in particular to the need to take care when choosing your activity types and ensure you have sufficient data recorded in your site notes to ensure you can enter the correct adjoining condition.

In the final chapter we deal with how you generate a result and how you can ensure you RR is 'fit for purpose'.

10 Ratings, Recommendations and Reporting

INTRODUCTION

In this chapter we consider perhaps the most important parts of the entire EPC report process including:

- how to understand and interpret the results page in SBEM;
- checks you should make before, and after, you calculate the EPC rating to ensure you have generated the 'correct' result;
- the documents, i.e. EPC, RR etc. generated by the program;
- how you should use the results page as part of your audit process, in particular use of the 'EPC Audit';
- recommendations, including how and why to edit, remove or add to them;
- how you can ensure you pass the information contained in the EPC and RR onto your client, or subsequent clients, to their ultimate benefit.

SBEM RESULTS AND DEFINITIONS

Main practical purpose of SBEM

So far as the EPBD and the UK national Regulations are concerned you carry out your inspection, and enter all of the data, for the sole purpose of providing your client with the two documents that are 'required' under the Regulations. These are the Energy Performance Certificate (EPC) and the Recommendation Report (RR). You will generate both of those documents direct from the 'Results' page of SBEM. Hence, it is appropriate that you fully understand the page.

Example of the results page

Figure 10.1 shows the iSBEM results page for a practice assessment.

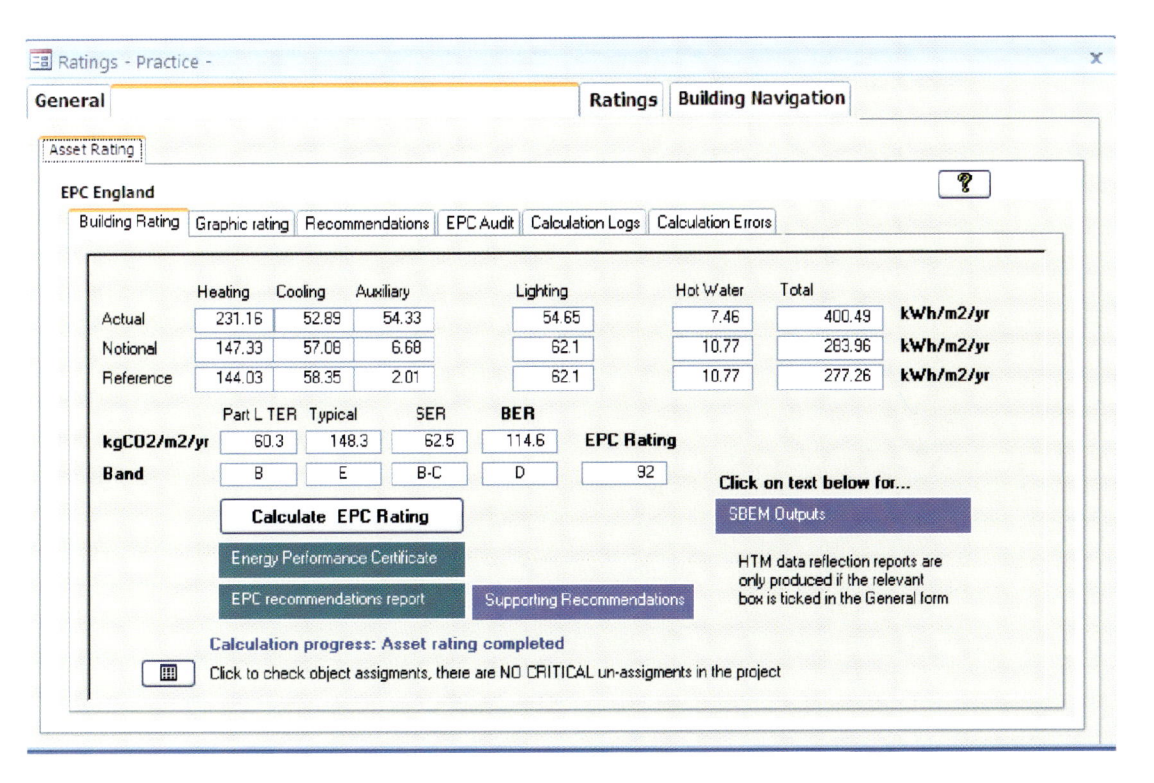

Figure 10.1: Practice assessment results page

155

Commercial Energy Assessor's Handbook

Figure 10.1: Practice assessment results page (cont.)

Figure 10.1: Practice assessment results page (cont.)

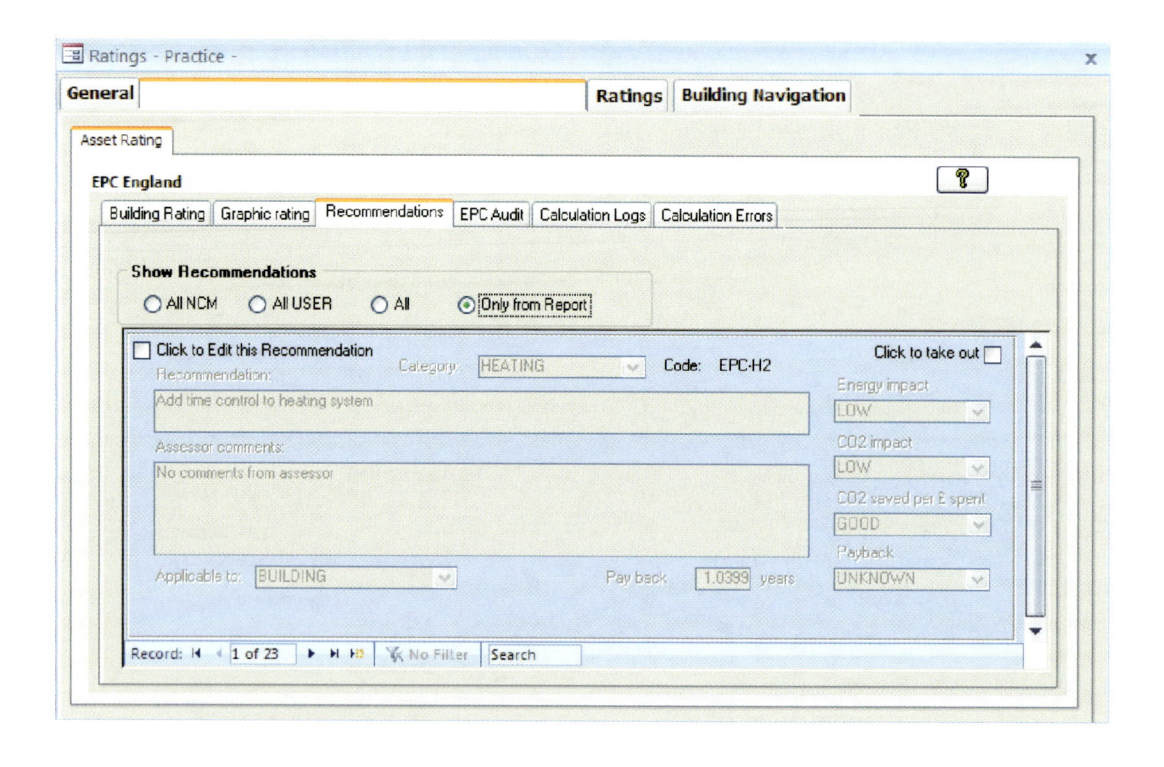

It is important you properly understand the information on this page since:

- some informed clients are likely to require an explanation for reasons we discuss in Chapter 3; and
- if you properly understand this page, you may be able to spot mistakes in your data entry.

Different SBEM building types

Before we consider the results the program displays, we need to understand the differences between, and descriptions of, the four building types that appear on the results page. The buildings are the:

- 'actual building';
- 'notional building';
- 'reference building'; and
- 'typical building'.

The 'actual' building is the building you have assessed and the results are based on your data entries.

The 'notional' building generally:
- has the same characteristics as the actual building, i.e. the same:
  ○ geometry,
  ○ activity uses, and
  ○ orientation;
- is situated in the same location as the actual building, i.e. the building is exposed to the same weather;
- is assumed to have a fixed percentage of glazing on *every* wall and roof (dependent on the building type); and
- has fabric, glazing and HVAC efficiencies that generally comply with 2002 Building Regulations Part L requirements.

The 'reference' building is a building of the same type as the notional building, i.e. it complies with the 2002 Building Regulations Part L, but with 'fixed parameters' to aid comparison, irrespective of what is in the actual building:

- heating and the hot water service are *always* assumed to be gas fired, even if gas is not available;
- fixed heating is generally assumed in every space, based on the SBEM database assumptions (even if the space is not heated in the actual building); and
- fixed cooling of every space is also generally assumed in the same way; *but*
- if any space in the actual building is not conditioned, the same space in the reference building will be unconditioned.

The 'typical' building is a building based on the compliance standards in Part L of the 1995 Building Regulations. The CO_2 emission from the typical building is the 'typical emission rate' (TYR).

Other SBEM descriptions

'BER' is the 'building emission rate'. This is the total energy consumption of the 'actual' building and its services, expressed in terms of CO_2 emissions.

'TER' is the 'target emission rate'. This is the current Building Regulations 2006 CO_2 emissions target, based on the 'notional building'. This is achieved by subjecting the notional building' to improvements to bring it up to 2006 Building Regulations Part L compliance, i.e. the 'TER'.

Those improvements comprise:

- an improvement factor in general energy efficiency of between 15–20%; and
- a 10% allowance for 'LZC', i.e. low and zero carbon technologies – this is also used for approval of new buildings, to encourage introduction of LZC.

These improvement assumptions are based on the same assumptions that underpin Table 1 in ADL2A to the Building Regulations (reproduced in Chapter 5) and the *LZC Strategic Guide* from CLG.

'SER' is the 'standard emission rate'. This is defined by applying an 'improvement factor' of 23.5% to the emissions from the 'reference building'.

There are generally a number of other tabs and sections on the final page of the different versions of the program. We will discuss them, in the next section of this chapter.

Interpretation of the results

The information on the results page is set out in the top three lines of figures as energy consumption figures. These figures are described in kWh/m² of the floor space. In addition, this page shows the associated CO_2 emissions in kg CO_2/m², on the next two lines.

The energy figures are shown split over consumption and building types. The bandings and ratings are based on CO_2 emissions. For EPC purposes, SBEM compares the CO_2 emissions for the actual building with the equivalent emissions for the 'reference' building.

The 'asset rating', i.e. the EPC rating, is the ratio of the CO_2 emissions from the actual building (i.e. the BER) to the CO_2 emissions from the reference building (i.e. the SER). It is expressed as a percentage, rounded to the nearest figure. The actual rating is derived from the calculation BER x 50 ÷ SER. This calculation derives from the fact that a rating of 50, i.e. the point where band B drops down to band C, is the SER. Thus, a building that complies with Part L of the 2006 Building Regulations, with gas fired heating, will have an asset rating of 50.

For the practice assessment, you can see that the BER is 114.6, and the SER is 62.5. Thus, the actual calculation is 114.6 x 50 ÷ 62.5 = 92 (actually 91.68), i.e. the EPC rating.

Commercial Energy Assessor's Handbook

Figure 10.2 shows the results page again, but this time we have annotated the page to summarise the main points we have outlined above for you, together with more information relating to further issues that we will now deal with under the next heading:

THE ENERGY PERFORMANCE CERTIFICATE

The EPC document presents the asset rating of the building. Important points you should note about this document are:

- the rating bands have equal widths; thus each band is 25 EPC points wide, except band G, which captures any EPC rating over 150;
- the lower the number, the better the banding of the building.

You should also note the ratings in the lower right hand corner; i.e. a rating for the same building 'if newly built' and 'typical of the existing stock'. You should use these figures to help you assess whether your figure is 'in the correct ball park'.

Graphic rating

The 'graphic rating' is a graphical presentation of the EPC rating for display purposes. The same graph comprises part of the EPC.

Figure 10.3 shows the graphic rating page from the practice assessment. We have annotated it for your reference purposes:

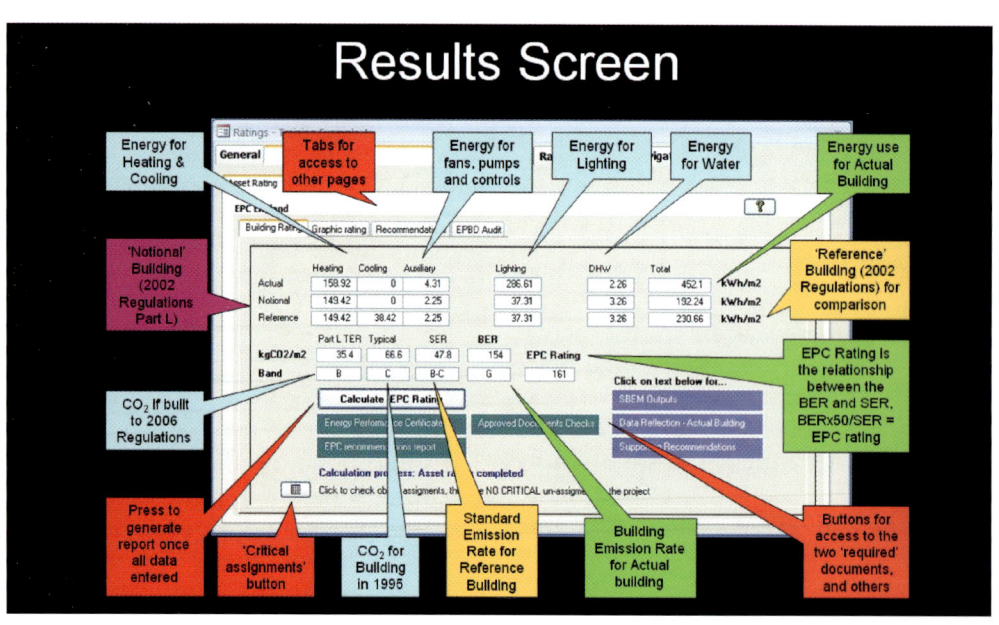

Figure 10.2: Practice assessment results page, with annotations

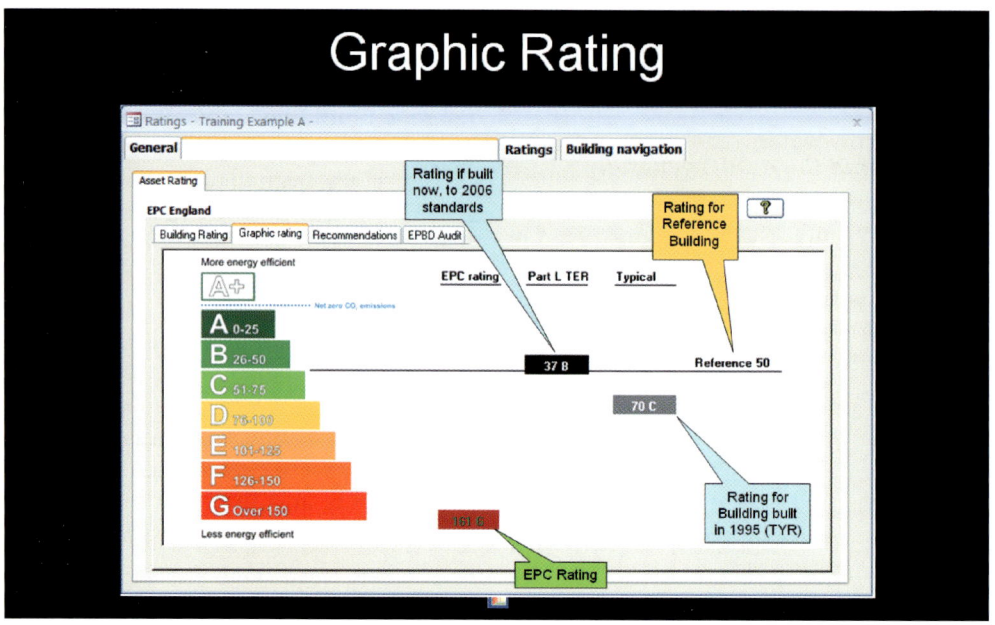

Figure 10.3: Annotated graphic rating page from the practice assessment

RECOMMENDATIONS

The RR is the second of the two required documents. Clients will take action and make decisions in accordance with your recommendations. You therefore need to take care to ensure every recommendation that appears in the RR is appropriate.

Only a small number of recommendations appear in the main RR. The RR presents recommendations in terms of 'payback' with associated potential impact on CO_2 emissions:

- short term;
- medium term; and
- long term.

All recommendations appear in the 'Secondary Recommendations Report', under the 'supporting recommendations' tab in iSBEM.

The 'SBEM outputs' tab generates the 'SBEM Main Calculation Output Document'. This is a pie-chart representation of the total energy use. In our experience, clients can find this report particularly useful as it is a general representation of the total building energy use, although it sometimes appears with a watermark on it confirming 'only for illustration' and some clients find this confusing.

All of the above documents and the EPC are generated as 'pdfs' by the program. Once you have generated these documents and considered them you may decide to make changes or additions to your calculations or your list of recommendations. If you do, you must press the button for 'calculate EPC rating' again to ensure results of your alterations appear in your revised pdfs.

HOW TO USE THE RESULTS PAGE

You should use this part of the process to generate the required documents and other documents you can use; and as part of your overall audit procedures. We now discuss how to achieve both of these goals.

Reflection on the EPC and the results generally

You should spend some time considering the results page once you have generated a final result. As we shall discuss, the program flags up any 'objects' you have not allocated, but there is no 'fail-safe' in the iSBEM software to indicate when you have made an error.

Audit checks before calculation

As you carefully go through each item of data entry you should take care to 'tick off' those items in your site notes as you make the entries. However, we are all human and can make mistakes. On large (and small) properties, issues such as ringing telephones, other interruptions and even simple boredom can break your concentration and cause incorrect data entry – and, yes, the authors have all experienced these phenomena!

You should therefore adopt a routine of audit of your work at all stages during data entry, particularly at the final stage. We have included some checks on the site notes you will see in case study 2 in Appendix B. We hope by this stage in the book you understand why we include those checks, but let us remind ourselves:

- they act as a prompt for when we forget the correct procedure; and
- the notes/ticks/strike-throughs we record using pen on-site and later, provide an audit trail confirming our careful and professional practice.

We can summarise the steps you should take when you reach the results page as follows:

- return to your 'project database' entries and ensure each of your entries for walls to windows is correct (it is easy to click on the wrong year, etc.) – take particular care you have correct 'U' and 'K$_m$' values;
- next, consider your 'geometry' entries, particularly:
  - building infiltration rate,
  - each zone area,
  - zone height, HVAC allocation and activity for each area,
  - ensure you have added the doors and windows – they are easy to forget;
- then, check your services entries, paying particular attention to:
  - HVAC efficiencies, i.e. have you altered them as may be required, e.g. by reference to inclusion on the energy technology list,
  - M&T,
  - HVAC controls,
  - HWS storage amounts and dead-legs,
  - mechanical ventilation or exhaust in each zone – also easy to forget,
  - lamp details and lighting controls – defaults in some programs may not allow themselves to be altered with ease;
- finally, in iSBEM, and in other programs, you will know if you have not allocated all of your HVAC and other systems as you will see a warning about 'critical unassigned objects' or similar – click on the tab and a list of those 'objects' appears so you can return to the appropriate page and either delete the object, or allocate it to a zone.

Once you have gone through this, or a similar process, you are ready to calculate your result. However, we always make a 'guesstimate' about the result we expect before we do so. You should have some idea of what you expect to see when you press the calculate button. Over the months/years of your career as a CEA you will develop a 'feel' for the results the program is likely to generate – a 'tone of EPC rating' that certain types of building are more likely to generate. Table 10.1 shows some examples of typical results we have noted from our own and others' actual real assessments.

Table 10.1: Example schedule of typical results

No.	Building age and type	Construction and insulation	Heating/cooling/HWS/lighting	Rating and band	Comments
1.	Pre-1900 town centre mid-terrace office, converted from former house	Solid brick external walls, no insulation; single glazing, roof insulated to 300mm depth	Gas fired 'wet' radiator system, simple cooling system to part, HWS from boiler, tungsten and fluorescent lamps + CFLs	80–100 D	You are unlikely to be able to improve this rating significantly Good result due in part to protection from adjoining buildings
2.	1950s detached factory or workshop	Solid or cavity brick walls, asbestos cement sheet roof, single glazing, no insulation	Radiant heaters to work space, electric heaters to office, toilets and kitchen, HWS from instant over-sink generator, fluorescent and tungsten lamps	125 + F or G	Take care with your activity choices, re-model with different activities if unsure
3.	1950-1960s detached community building, e.g. sports building or scouts hut	Timber frame walls, felted or asbestos cement sheet roof, timber or solid floor, single glazing, no insulation	Electric heating and instant HWS, tungsten and fluorescent lamps	100–120 E	Much scope for improvement, or sometimes (careful) demolition!
4.	Mid-1980s detached workshop with integral offices/toilets	Steel or concrete frame, profile metal cladding and roof, insulated floor, insulated walls and roof	New forced convection air heater (ECA), instant electric HWS, mix of fluorescent and sodium lamps	75–85 C	Would be worse if original heating
5.	1980s shop in mall with conditioned space all around	Concrete frame and display windows	Electric heating in shop, mechanical supply ventilation	80–100 D	No exterior conditioning
6.	Post-1995 offices	Concrete and steel frame and curtain walling	Fan coil heating and cooling, thermal wheel, T8 + CFLs	100–120 E	Inefficient HVAC
7.	Post-2000 offices, converted from old barns	Solid stone walls insulated, slate roofs, double glazing	Condensing combi (ECA), tri-phosphor T8 and/or T5 lamps with modern ballasts	40–50 B	Heavily insulated and efficient HVAC
8.	Detached two-storey post-2002 offices	Steel frame brick/block and curtain wall cladding; double glazing 50 + zones	Split air-con, 2 No 80kW boilers to under-floor heating to part, mechanical ventilation. CFL and T8 lighting	55–65 C	Reasonably efficient HVAC
9.	Small workshop 2002, mid-terrace	Brick, insulated cladding to walls and roof	Gas condensing combi (ECA) and seasonal efficiency of 90.1, T8 lamps, SES on roof	20–25 A	Adjoining workshop with inefficient boiler, no SES and T8 lamps achieved 55/C

We do not suggest that these figuires are necessarily 'correct' or should be held up as firm benchmarks. We have no doubt you will generate results that are different to those in the table, depending on the different data you enter. However, we do suggest you should keep a similar record based on your results and for your particular locality and the usual types of building you inspect. In time, based on those results, you will also develop a 'gut feeling' for assessment that will help you to critically analyse the results you see displayed.

Audit checks after calculation

You should *not* assume that the result the program displays is necessarily correct. You should instead take time to consider whether the result is indeed correct.

Thus, we believe once the results appear on your screen, the first question to ask is 'do the BER and EPC ratings seem "about right" for this particular building type and activity?'. If the result is significantly different from the correct 'tone'; and even if it isn't, the next stage in your audit process is to look at the different fuel consumption figures – do they seem consistent with the data you believe you have entered and your other past assessments?

There are other actions we recommend you should take before you are satisfied you have generated the correct result:

- consider the results for the TYR and 'new' building and compare these with the 'actual' results – some clients will question you about significant differences;

- consider clicking on the 'unassigned objects' tab and go to the 'data summary report'. This is an excellent method of checking the data you have entered for each zone again 'at a glance', including activity. Even if you don't check each zone, you should discipline yourself to make some random checks. But, beware, you cannot check lighting here;
- look at the 'data reflection – actual building' report (in iSBEM you will need to click on the appropriate box under the 'system configuration' tab in 'General' to ensure this appears for you). At first glance, and certainly if you are new to commercial energy assessment, you might be somewhat daunted by what appears, for it is all of the information about the building in one document. However, we suggest you persevere and you will soon find this page can be very helpful – and it includes your lighting information.

None of these audit checks are sufficient on their own, but each of them conribute to a general consideration of whether you have achieved the correct rating. In some cases when the result seems 'simply wrong' and you are unsure about the activity type, you should return to that part of the program, alter the activity and run your results again, in extreme instances more than once or twice.

Recommendations tab

The 'Recommendations' tab lists generic recommendations that SBEM automatically produces. You can accept and/or modify the recommendations and associated impacts. This process will automatically provide you with an audit trail. You must have good justification for altering recommendations the program has automatically generated. You should record such justification somewhere in your paper file and/or the program.

You should always take particularly great care when considering this part of the process. We have heard of some assessors who do not even look at, let alone consider, this particular aspect the program. We believe this is a potentially costly and silly mistake, for the following reasons:

- some of the recommendations are entirely inappropriate for the building, usually as a result of current software 'glitches'. Thus, at the time of writing, a recommendation for cavity wall insulation is normally generated, even if you enter data confirming all walls are 'solid';
- a competent CEA should consider any reasonably anticipated recommendation, even if the program has not generated it – this is implicit in the NOS and although SBEM is the 'National Calculation Methodology', it would be unreasonable to assume it accommodates *every* possible building type, services and lighting variation – you should make allowance for each particular building;
- many of the recommendations appear more than once (there is a reason for this, as we shall see) – you should remove recommendations that appear two or even three times, unless you have good reason not to do so, as otherwise clients could be confused;
- your careful editing of the recommendations is a good way to demonstrate competence, knowledge and understanding to the people who will use your professional services. You can thereby generate repeat business from wise commercial agents, solicitors and discerning clients – we have heard one very well-respected property professional confirm they removed from their panel of recommended CEAs any assessor who accepts all of the recommendations without careful and reflective thought, particularly when the accepted recommendation is entirely inappropriate for the building;
- the RR is one of the two 'required documents' and is ultimately entirely your responsibility.

We believe that it follows therefore that any competent CEA, i.e. you, should:

- carefully consider, reflect on and edit each of the recommendations based on your inspection of the building and services, your knowledge, understanding and competence, and confirm you have done so and your reasons in your site notes and/or the 'assessors comments' section in the program;
- remove any recommendation, on the same basis;
- add to the list of recommendations if the program fails to generate a recommendation you believe should appear for the particular property you have inspected.

With regard to the last point, we do not believe that a CEA will be able to hide behind an argument that is based on the following (faulty) logic:

> 'I relied on the NCM/SBEM written by BRE for the government to generate all of the appropriate recommendations, and it's not my fault that it didn't do so.'

You must accept it is your name that appears on the report and on the Register. It is *your* report, not the government's – *your* responsibility to achieve an accurate rating and recommendations appropriate for the property that clients can rely on in operating and using their building.

The above requirements are significant and potentially onerous, but we believe they represent the basic professional minimum a CEA is responsible for. It follows that such requirements will take time, indeed in some cases a considerable amount of time. However, this is a necessity in our view and should not be considered as being in any way an 'added value' service.

We believe the steps we have set out are implicitly suggested as being required in the BRE program and in the guidance, if a CEA wishes to provide a good, professional service. The NOS state you must:

> 'check the recommendations generated and make any necessary amendments … delete recommendations that are inappropriate providing your reasons.'

If assessors fail to properly edit, remove and/or add to the list of recommendations on a regular basis, in our view this will result in RRs that are not fit for purpose that in time could mean:

- CEAs will be sued for incorrect advice, unnecessarily, thereby contributing to a public perception of incompetence that could attach to all assessors;
- clients' and other users' opinions of the commercial EPC process in general, and CEAs in particular, could take on a negative hue, leading to public dissatisfaction with the entire process, and a failure to take action on the basis of the RRs.

Other matters you should consider are as follows:

- Have you noted a significant condition issue in the building that means you should warn the occupier to take particular care before following any recommendation, e.g. you noted condensation to the base of an external wall you believe may be caused by a particularly cold thermal bridge at this point and a recommendation to improve the insulation to the wall could make the problem worse?
- Might a recommendation to install any improvement cause a structural problem that is obvious to you – e.g. installing a solar array on the roof might cause overloading of a roof structure that is clearly distressed?
- Will any recommendation require planning permission, building regulations approval or similar – if so, you should warn the occupier? You should not delete such a recommendation, since the local planning policy might alter in the next 10 years.

You may think it is inequitable that CEAs should have to report these issues; we confirm we have some sympathy with such a view. However, existing domestic energy assessment practice has established DEAs must consider and report on 'condition' and associated matters; and this means CEAs must also do so in our opinion.

Figure 10.4 is a representation of one of the iSBEM recommendations pages for the practice assessment, with annotations. Other programs have different representations of this page.

Note the 'radio buttons' across the middle of the page, indicated as:

- 'All NCM';
- 'All USER';
- 'All'; and
- 'Only from Report'.

You can scroll through all of the NCM recommendations available to you by using the first button. At the time of writing, different parts of the building's fabric and/or services can each be the subject of the same recommendation. Thus, by way of example, a recommendation to 'add weather compensation controls' will normally appear twice; being 'applicable to' the 'building' and to the 'HVAC SYSTEM' – look at the boxes that will appear under the 'Assessors comments box'.

We suggest that you concentrate on the 'Only from Report' button. This button allows you to scroll

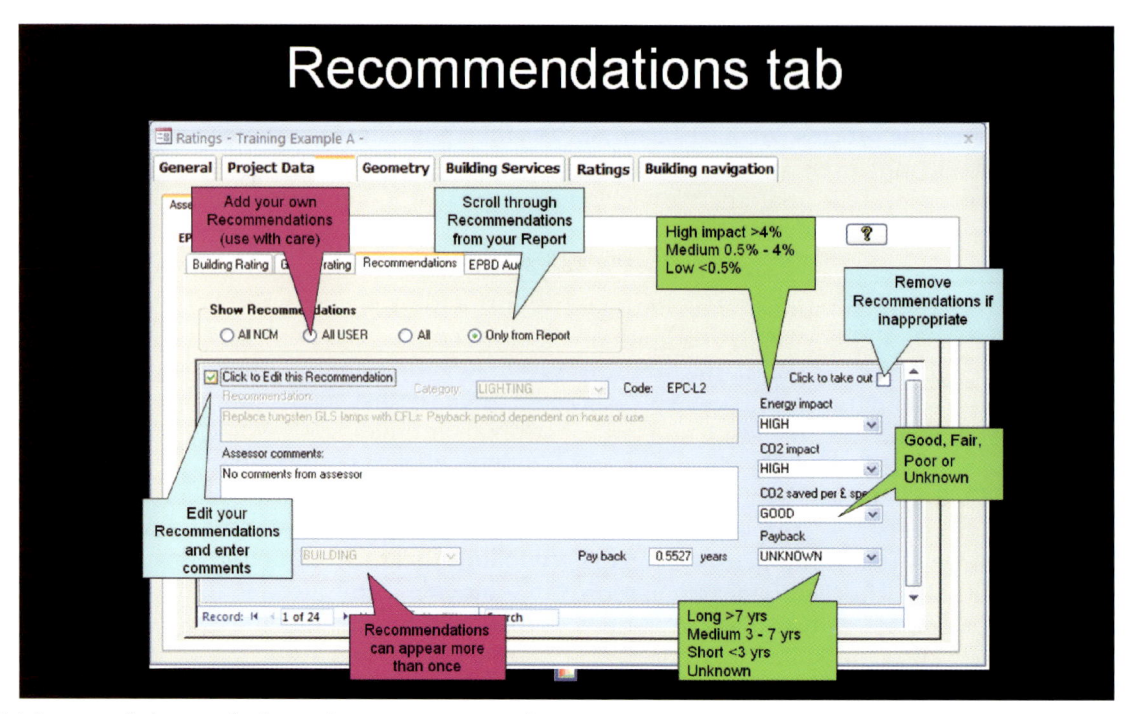

Figure 10.4: Recommendations page for the practice assessment, annotated

through each of the recommendations for your actual building.

You can add your own recommendations by using the 'All USER' button. You can use the 'All' button to view both the NCM recommendations and those you have added.

The recommendations generated by SBEM

You should look at the lists of current recommendations that can appear in the RR. They are in appendices A5.0 and 6.0 of the *BRE Technical Manual*. You should anyway read the *Technical Manual* to further develop your SBEM knowledge. You may note the strange use of the word 'reworded' amongst some of the recommendations. We suspect this has arisen due to incomplete editing of the program.

We have included four of the 42 standard recommendations in Table 10.2 and discussion where appropriate, together with paragraphs, that we sometimes include in our RR and/or in any accompanying client report – see later. We do not have space to include all recommendations and discuss them in detail. Please also bear in mind that each property and every client is different. You should *not* copy these paragraphs without being sure they are entirely appropriate for your particular case.

Developing your own recommendations

In addition to the above recommendations, after reflection you may wish to add some of your own recommendations to your reports. We have developed one recommendation that a number of other assessors use and we refer to this later in the chapter.

THE EPC AUDIT

We treat the EPC audit as a vital part of the process – we now discuss why.

Table 10.2: Some SBEM recommendations, with our comments

SBEM reference	SBEM paragraph	Discussion	Paragraph for possible insertion into client report, or RR
EPC-C3	Ductwork leakage is high. Inspect and seal ductwork	In many instances, the ductwork will only be accessible to the extent required for inspection by carrying out potentially expensive opening-up works, and by shutting the plant down – this might only be possible during operating 'down-time'	Inspecting and sealing the ductwork could be a lengthy and potentially costly exercise and might affect production
EPC-E6	Some loft spaces are poorly insulated – install/improve insulation (reworded)	When insulation is laid in any roof, the air temperature reduces in the void. This makes the roof void more prone to condensation. This issue is more likely if the roof has a layer of impermeable felt beneath the primary covering. The problem is significantly less if the sarking layer is a modern 'breathable' type, or the roof has good ventilation in accordance with AD 'F' of the Building Regulations.	The existing main roof has some insulation. You should lay further insulation. You should take care when doing so, to avoid stopping any ventilation, as this would increase the risk of condensation occurring. This work may need to include improvement of ventilation into the roof void to help prevent condensation occurring
EPC-L7	Introduce HF (high frequency) ballasts for fluorescent tubes: reduced number of fittings required	Chapter 7 confirms why this can make a difference	You should take professional advice before you reduce the number of lights in your fittings as this could have an effect on issues such as health and safety or the quality of work by people in the area involved
EPC-R2	Consider installing building mounted wind turbine(s)	Local occupiers can sometimes be extremely upset if an adjoining occupier indicates an intention to install such a system. If the property is in a conservation area, listed as being a building of special historical interest the local authority may object	You will need to obtain approvals for planning, building regulations, listed building and conservation area consent for such work

Recording comments on your assessment program

Most SBEM programs are likely to have facilities for you to enter comments that form part of your audit process. In the iSBEM program this is provided in the form of dialogue boxes for 'user notes' in the zones data entry pages and an EPC audit (in previous versions named the 'EPBD audit') on the ratings output screen.

It is our firm view you should make full use of these facilities to record the reasoning you have used for your data-entry decisions. Each inspection and assessment is likely to have an unusual feature that will affect the data entry. You will record this in your site notes, but it will only be during your later reflection on your notes that you will consider the way the particular feature influences your data entry.

You could make a record of your data-entry decisions in your site notes. However, if you do so, you should distinguish your later reflection note in some way from the notes you have made at the property. Alternatively, you could make a separate file note about your decisions. Whatever method you use you should also put a note on the program because:

- when reviewing the program file in the future it will be easier to refer to a comment on the file than retrieving an archived set of site notes;
- your accreditation scheme, which may also be your insurer, can see the comments on the electronic files you submit;
- the electronic program file is likely to be more durable and secure than any paper note, especially as an EPC has a potential shelf-life of 10 years.

The *BRE User Guide* specifically refers to the use of the EPC audit as an 'audit trail' of information that has been used as a source of reference for the data entry. There is a note written in distinctive, bold text on the EPC audit section of the iSBEM program stating: 'Please introduce concise supporting evidence for over-riding default values.' You will find a similar comment repeated in the *Guide*. The EPC tab in iSBEM is split into sections for 'construction', 'geometry', 'HVAC and HWS', and 'lighting'.

There are several examples of specific default values that are stated in SBEM, such as:

- infiltration (air permeability);
- global thermal bridges;
- surface area ratio and transmission factors (windows and rooflights);
- HVAC seasonal efficiency (heating, cooling and hot water service);
- specific fan power (supply and exhaust ventilation);
- electric power factor correction;
- parasitic power (photoelectric lighting control).

So, if you choose to use any other value than these defaults you should record your reason and source of information in the audit.

We believe what we have described above is the minimum standard of good practice any CEA should adopt. However, we also believe you should use the audit and user notes more broadly. First, it could be argued that any value built into the program's database is a default. In this case an inference value could be considered to be a default and you should therefore record your decision and source of information when you choose to enter your own 'U' and 'K_m' values. Similarly, when you decide to use a library definition for a building element as a 'close match' for the actual construction, you should record this in the audit. Secondly, we believe you should use the user notes and EPC audit to record any relevant matter that has affected your data-entry decisions. Here are some examples of notes we have made in practice:

- selection of specific activities, particularly those not provided in the data set for the building type;
- use of 'virtual' zones, e.g. 'reception' in a larger space;
- identification of zones where we have assumed indirect heating;
- reference particulars of plans we have used (drawing number; title; scale and name);
- existing services damaged or dilapidated and we have assumed they will be reinstated;
- limitations to the inspection, e.g. inaccessible zones or areas; visual identification of lamps located at high level using binoculars;
- identification of zones that have been merged or where we have ignored some zoning criteria, e.g. an insignificant area of solid flooring in a larger area of suspended floor;
- confirmation of where wall elements have been added to allow for thermal capacity when zones have been merged;
- confirmation of how we have calculated values, e.g. the volume of hot water storage from the dimensions of the storage vessel; the software program used to calculate 'U' and 'K_m' values;
- use of electric resistance heating where no heating is fitted or the type is unknown;
- reason for selection of the date of construction or services installation;
- evidence to support 'special conservation status';
- reasoning for the inclusion or exclusion of parts of the building from the assessment, e.g. residential accommodation; common-parts;
- confirmation of where thermal bridges have been added, e.g. valley gutters;
- reason for the selection of the adjacent conditioning, e.g. in basements; adjacent properties.

The above list is not exhaustive and as we carry out more EPCs we add to it. As you practise your profession, you should do likewise. If you decide to ignore our advice please do remember the words we have quoted above from the *User Guide*, i.e. 'audit trail' and the requirement that you must be able to justify all of the reasons why you arrived at the assessment you presented to your client.

REPORTING TO YOUR CLIENT

General comment

When you reflect on this aspect of the job you should not only consider your immediate client, but also the other professionals and advisers likely to be associated with any sale; *and* subsequent purchasers and their advisers for the period of validity of your EPC and RR – 10 years. When you prepare these documents you are potentially serving many masters or mistresses and they will not all necessarily sing from the same hymn sheet as you or the client who has paid you!

Client report

Many CEAs send their two required documents to their clients with either a letter or an accompanying explanatory report.

We have included an example of a client report in case study 2 in Appendix B. There is no legal requirement to send a supplementary report. You may regard such a practice as an unwarranted expense on your part. It is true that many clients and professionals such as commercial estate agents and solicitors do not currently ask for any explanation of the EPC or RR. Instead, they regard these documents as possible impediments to the sale rather than documents that might assist the sale. It is true that if the report is a 'bad one', i.e. the documents show the building(s) as inefficient; a seller or professionals acting for them might want to suggest the report is a 'waste of time and not worth the paper it's written on'.

However, an accompanying report, or even a short explanatory letter, is a good opportunity to inform your client and others who are likely to rely on your report over the next 10 years about some of the energy issues at the building, and circumstances that may have affected your inspection. Examples might include clarification about the extent your inspection was restricted by stock or other stored items; or you may have made some assumptions about plant or lighting efficiencies you wish subsequent users of the report to understand.

In addition, you could bring the user's attention to other issues, including:

- putting the building and the energy use into perspective to a greater extent than is explained in the documents generated by the program;
- further advice about how some of the energy savings flagged up in your RR can be practically implemented;
- confirming the documents, e.g. manufacturer's information, you relied on;
- further confirmation about which drawings and other documents, e.g. plant handbooks, you had access to and relied on.

We do not suggest such a client report will always be appropriate. We have sent out reports without such a report. However, we believe there are circumstances when one can significantly add to the quality of your service. We also take the view that as commercial building occupiers become more aware of how much money they spend on energy use in their buildings, and how much the value of their property might be affected by the service CEAs offer, many more clients and their advisers will realise the importance of a good EPC, RR and explanatory report.

If you decide to include a client report, or you wish to add any clarification about the EPC or the RR in an accompanying letter, you should make a note about the existence of the report or letter in the RR, otherwise your letter or report could be 'lost' in years to come, possibly to your detriment, or your client's or anybody who relies on your report. A paragraph we have used as an additional recommendation in our reports reads:

> 'This Energy Performance Certificate (EPC) and Recommendation Report (RR) must be read and considered together with our accompanying Energy Report dated (insert date). The Energy Report describes and explains the EPC and RR in much greater detail, in particular in relation to issues such as legal, planning and other matters.'

We certainly believe you should at least send every client a letter or standardised document that explains issues such as:

- your confirmation of instructions and the terms under which the EPC and recommendations report are provided;
- the status and purpose of the EPC and RR and how the rating in the EPC may be changed if your client, or others, implement the recommendations;
- the status of the EPC and RR, the underpinning legislation and the status of the documentation, including the existence and purpose of the Central Register.

We believe these requirements are either explicitly or implicitly required in the NOS. it makes good sense to explain them to your client anyway. If you decide not to send an accompanying letter you can include them in the RR as an additional recommendation.

SUMMARY

This chapter has brought us to the conclusion of your data entry process and we have considered the main purpose of your profession – advising on the current asset rating and how your clients, subsequent owners and other occupiers can improve the energy efficiency of their building(s) and services.

Appendix A – Case Study 1 – Level 3 Assessment of The Old Forge, Main Road, Burnham Deepdale, Norfolk PE31 1ET

INTRODUCTION

For our first full practical example of how commercial energy assessment works in practice we chose a building that does not require an EPC or RR. We have often been asked to advise solicitors about such issues. Some clients will thank you for telling them the fee for such an inspection is much less than a full job, so we think this appendix is helpful! We have not included a completed client questionnaire – you should refer to Appendix B for this.

We chose this property as it has some interesting practical and legal issues. The appendix comprises:

- a brief description of the building;
- photographs – these set the scene for you and represent our visual inspection of the property;
- two pages from our site inspection and reflection notes, including our plans of the property – see Appendix B for an example of a full set of site notes;
- copy of our client report; and
- discussion.

Our intention for the case study is to demonstrate how you should use your knowledge and understanding in practice.

DESCRIPTION OF THE BUILDING

The building is a detached building at least 100 years old, with walls of solid masonry and a pitched clay tile roof. The floor is earth. The doors are solid timber and windows are in timber with single glazing. The building has a main electrical supply. There is internal lighting. The space comprises a single area used as a blacksmith's forge.

The building was empty at the time of our visit. It is situated on a working farm in the area of outstanding natural beauty that is the north Norfolk coast.

PHOTOGRAPHS

Figure A1: Front (west) elevation

Figure A2: Rear (east) elevation

Figure A3: Left hand (north) elevation

Figure A4: View internally showing floor

Figure A5: View of the forge

CLIENT REPORT

Here are some paragraphs from our report to the Client:

'Further to your instructions in respect of the above as you know I have carried out an inspection of your building(s) on 1 April 2009.

Thank you for returning my signed Conditions of Engagement and thereby confirming my instructions. Thank you also for giving me access to your property on the day of my inspection.

As my original letter of engagement to you confirmed, I am a Non-domestic Energy Assessor registered with the NHER Accreditation Scheme, registration number SAVA004349. You can check my details at www.ndepcregister.com.

As requested, I have sent a copy of this report to your Solicitor, Cameron Green Esq of Kenneth Bush Solicitors in Kings Lynn.

As I have explained to you, in general terms any owner who sells or rents a "non-domestic" property is required by the Government to provide an Energy Performance Certificate and Recommendation Report at the earliest opportunity in order to comply with the current legislation that applies to energy performance. However, some types of building are exempt from the Regulations as confirmed in *The Energy Performance of Buildings (Certificates and Inspections) (England and Wales) Regulations 2007 (SI 2007/991)* and associated Regulations.

I carried out a careful inspection of your building. The building is a detached building used as a forge, with a "total useful (gross internal) floor area" of 43.99m². It has no fixed heating and is unlikely to be provided with such heating – the internal space is heated by the forge process.

In the light of my inspection and my subsequent reflection I am of the opinion the building does not require production of the two documents specified in the Regulations, i.e. an Energy Performance Certificate and Recommendation Report are not needed as the building is exempt from the Regulations.

I am of such view for the following reason(s), with reference to the paragraph numbers in the Regulations I have named above:

- the building does not use, and is unlikely to use in my opinion, energy to condition the indoor climate [Regulation 2(b)];
- the building(s) is a workshop with low energy demand [Regulation 4(1)]; and
- the building is a stand-alone building with a total useful floor area of less than 50m² and is not a dwelling [Regulation 4(1)].

Appendix A

PAGES FROM SITE NOTES

RT2.D0671   COMMERCIAL EPC SITE / REFLECTION NOTES   File / page reference *2001 / 03*

REFLECTION ISSUES

EPC not required for whole or part of property?

Building(s) constructed with **walls and roof**	**Yes** / no
Building(s) used solely or primarily as place(s) of **worship**	Yes / **no**
Temporary building(s) with a planned time of use of two years or less	Yes / **no**
Industrial sites, workshops and non-residential agricultural buildings with **low energy demand**	**Yes** / no
Stand-alone building(s) with a total useful floor area of less than 50m² which are not dwellings	**Yes** / no
Building(s) will be **demolished** (owner must believe on reasonable grounds new owner will demolish)	Yes / **no**
Has, or intention of having, **fixed services** and ability to include fixed services to condition climate	Yes / **no**
Construction of the building(s) is **not complete**	Yes / **no**

Part of building(s) to which above issue(s) apply: *All – used as forge, no fixed heating/conditioning – low energy demand*

Documentary evidence available from inspection, client, internet or other source

Previous energy assessments Yes / **no** – EPC ☐ DEC ☐   Survey report Yes / **no**   Services report Yes / **no**

Planning permissions Yes / **no**   Building Regulations Yes / **no**   CDM health and safety file Yes / **no**

Commissioning certificates for service installations Yes / **no**   Test / service certificates **Yes** / no

Maintenance or operating log books Yes / **no**   Manufacturers' information Yes / **no**   Air-con report Yes / **no**

Guarantees / warranties Yes / **no**   Fire risk assessments Yes / **no**   ACM register **Yes** / no

COSHH Yes / **no**   Schematics Yes / **no**: Heating systems ☐ Cooling systems ☐ HWS ☐ Lighting ☐

Access audit Yes / **no**   Building plans Yes / **no**   Specifications Yes / **no**   Licences for alterations Yes / **no**

Contracts Yes / **no**   Copy of the lease Yes / **no**   Client / other photographs Yes / **no**   Other Yes / **no**

Details:

None

'Existing' drawing audit

Drawings provided / seen Yes ☐ No ☒   *5720 x 7690 = 43.99m²*

Drawing number / date:

Drawing(s) checked Yes ☐ No ☐

Percentage measurements checked: 10% ☐ 20% ☐ ?____ ☐

Which measurements checked? See plan / other:

Drawings confirmed as being reliable Yes ☐ No ☐ Other ☐

Other relevant information

Figure A6: Two pages from our site notes

Commercial Energy Assessor's Handbook

RT2.D0671   COMMERCIAL EPC SITE / REFLECTION NOTES   File / page reference __2 3W/__

15 – OVERALL / ZONE PLAN

External (ext) check measurements taken   Yes / no

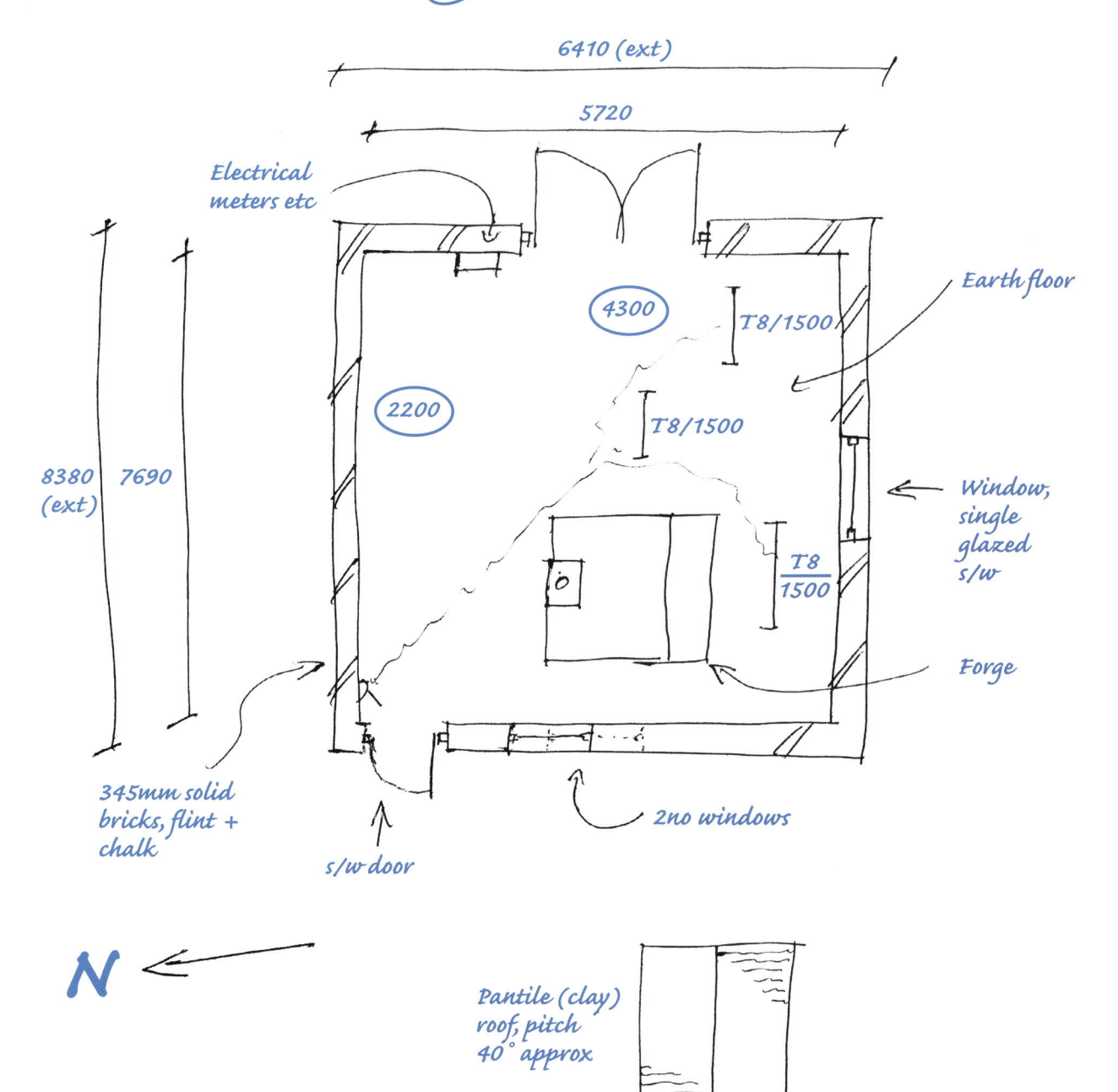

[Hand-drawn zone plan with the following annotations:]

- 6410 (ext)
- 5720
- Electrical meters etc
- 4300
- T8/1500
- Earth floor
- 2200
- T8/1500
- 8380 (ext)
- 7690
- Window, single glazed s/w
- T8/1500
- Forge
- 345mm solid bricks, flint + chalk
- s/w door
- 2no windows
- N ←
- Pantile (clay) roof, pitch 40° approx

KEY:

s/w – softwood

Continue on other sheet: Yes / No

Copyright Larry Russen June 2009

The Regulations only require one reason for a building to be exempt.

In arriving at my opinion I have had regard to:

- my experience as a practising Non-domestic Energy Assessor;
- the Regulations and other associated Regulations relating to non-domestic energy assessment I have referred to above;
- publications available from the government at www.communities.gov.uk and in particular *Improving the energy efficiency of our buildings – A guide to energy performance certificates for the construction, sale and let of non-dwellings*, 2nd edition, July 2008.

The law relating to commercial energy performance certificates is relatively new and is subject to interpretation. However, I believe my opinion is correct and valid and is unlikely to be disputed.

Thank you for pre-payment of the fee. I have already sent you my receipted account. Thank you for your instructions.

Lastly, your opinion of my service is important to me. In this connection, I enclose a short questionnaire for you to complete. I would be very grateful if you could take the time to give me your honest views. I will treat your response with total confidence. I enclose a stamped and addressed envelope for your use. Thanks for your time in this regard.'

DISCUSSION

We knew very soon after entering the building it was likely to be exempt, for the reasons indicated in our client report. The presence of a nameplate externally confirming it is a forge was a good indication and when we saw the forge internally this was good confirmation of the use.

We found no heating in the building, other than from the forge. We came to the conclusion it was the 'industrial activity' in the space that conditioned the air and had particular regard to the July 2008 CLG guidance (section 1.5, p. 11 'Situations where an EPC is not required' and 'Glossary of terms' on p. 39):

> 'Activities that would be covered include foundries, forging … where … the air in the space is not fully heated or cooled … the intention here is to include buildings that are conditioned primarily, or entirely, by the industrial process that is taking place in the space … you will need to judge each case on the individual merits of each building.'

We did judge the building on its merits and concluded this was justification by itself for an EPC and RR not being required.

However, to be doubly certain, we measured the building as we suspected it was less than 50m². Confirmation of a gross internal floor area less than 50m² provided the third reason for exemption.

If we had discovered the building had a TUFA of more than 50m² we might have contacted the local authority planning department to confirm the existing use for planning purposes, i.e. to confirm a use class consistent with a forge and provide us with formal justification for our decision.

You will note we were very careful to complete some of our inspection notes (we completed other sections, in particular the health and safety section) to record the appropriate details and take some photos. We hold this information on file as confirmation of our conclusions in our report to the client. We might be required to justify such an opinion in the future and must therefore retain the evidence.

After our careful inspection, consideration of the matter, readings of the Regulations and associated documentation and discussion with colleagues we were confident in our opinion and wrote the report you have read. However, if we had been uncertain we might have suggested the client take legal advice, although at present many solicitors do not have knowledge of the Regulations and tend to defer to CEAs. Or, we could have included one of the following two paragraphs in our report:

- 'I have also spoken to the enforcing authority for non-domestic energy assessment, the local Trading Standards Officer at Norfolk County Council. I discussed the matter with (text) and she confirmed she was of the same view as I in my interpretation of the Regulations' – we would record this discussion in our file notes, or with an email to her as an audit trail;
- 'The law relating to commercial energy performance certificates is relatively new and is subject to interpretation. However, I believe my opinion is correct and valid and is unlikely to be disputed. If you want "official" confirmation and/or clarification of my opinion, you can contact the local enforcing authority for non-domestic energy assessment, the Trading Standards Officer at Norfolk County Council on 01553 669269'.

In the event of finding alternative reasons for exemption, in accordance with the issues we have discussed in Chapter 3, other bullet points for our report might include the following:

- the building(s) is/are not a roofed construction having walls [Regulation 2];
- building(s) is/are used primarily or solely as (a) place(s) of worship [Regulation 4(1)];
- the building(s) is/are a temporary building(s) with a planned time of use of two years or less [Regulation 4(1)];
- construction of the building(s) is not complete [Regulation 4(2)];

- I understand the building(s) will be demolished by you or the subsequent purchaser/tenant [Regulation 7(2)] – guidance from the Government suggests that an application for planning permission is sufficient grounds for a belief that demolition will take place, but a mere stated intention is insufficient. I have seen a copy of the planning application, dated (text), reference number (text).

Appendix B – Case Study 2 – Level 3 Assessment of The Riverside Restaurant, King Street, Kings Lynn, Norfolk PE30 1ET

INTRODUCTION

For this case study we have prepared a complete EPC and RR on an actual property that presented some interesting assessment challenges. The appendix comprises:

- a brief description of the property;
- completed client questionnaire;
- photographs – these set the scene and represent your site inspection and observations;
- copy of our site inspection and reflection notes, including zone plans;
- copies of the:
  - EPC,
  - RR,
  - SBEM Main Calculation Output Document,
  - Secondary Recommendation Report;
- extracts from our client report;
- discussion – in the same order as our site notes, SBEM outputs and client report, emphasising issues of particular importance.

Our intention for the case study is to demonstrate how you should use your knowledge and understanding in practice; and for you to try to generate an SBEM report if you haven't yet done so.

BRIEF DESCRIPTION

End-terrace two-storey medieval building used as a restaurant situated in the conservation area of this historic Hanseatic port, listed Grade 2. The property includes a detached outbuilding in use as a laundry for the restaurant.

COMPLETED CLIENT QUESTIONNAIRE

Our client completed this before the inspection.

Commercial Energy Assessor's Handbook

RT2.D0666 - COMMERCIAL EPC CLIENT'S QUESTIONNAIRE TYPE 'B'

Commercial Energy Performance Certificate

Client's questionnaire – Type 'B'

Hello.

You have asked us to prepare a commercial Energy Performance Certificate (EPC) on your property. Thank you very much for those instructions.

We will of course carry out a very careful inspection of your property in order to collect and record all of the necessary information that is required in order to prepare an accurate EPC. However, some parts of buildings, in particular the services, are covered or unexposed, and this means that we cannot see sometimes vital parts of the property. This can adversely affect the accuracy of your EPC. It has been shown that an accurate EPC can sometimes affect the value of a property to a significant extent.

Some additional information can therefore only be discovered by talking to you, or other people who know about your property; or by looking at documentation such as drawings or manufacturers' information. We will be able to provide you with a better service if we have as much of that additional information as possible. We would therefore be grateful if you could please take some time completing this questionnaire, and return it in the self-addressed and pre-paid envelope.

We accept that some of your answers may be approximate, or guessed. However, if that is the case, please indicate as such; and please be as precise as is possible.

You will note that in many of the instances below we ask you if you have documentary evidence available to confirm the details that we require; e.g. copies of design drawings, specifications, handbooks, or manufacturers' information. If you do have this information, or know where it can be found, we would be grateful if you could tell us on this questionnaire. In that event, we would also generally appreciate either seeing a copy of the relevant documentation (at least); or being provided with a copy for our file (preferable).

Please tick the appropriate box, or write the answer to our question in the space provided. The form may look very complicated; and indeed, there are many questions. Once you begin completing it however, you will find that it is unlikely that all of the questions relate to your property.

Some of the questions are of a technical nature, and you might not know the answer. Please do not worry about that; simply complete as much of the questionnaire as you can by writing your answer, circling 'yes' or 'no', or ticking the appropriate box, below.

Alternatively if you have a professional who acts for you, e.g. an Architect, Chartered Surveyor, or Heating / Lighting Engineer; you could ask them to complete the relevant part of the form. We will be happy to discuss any details with that person if you like – just write their contact details in the appropriate box.

If you have a particularly large property with complicated and / or many different types of system; you or your professional advisors may find that this questionnaire has insufficient space to write all of the information. In that case, please use a continuation sheet, or ask us for another questionnaire.

Figure B1: Client questionnaire

Appendix B

RT2.D0666 - COMMERCIAL EPC CLIENT'S QUESTIONNAIRE TYPE 'B'

Lastly, we would ideally prefer you to complete this questionnaire and return it before our inspection. However, if you would prefer to wait until our visit, and fill the questionnaire out with our assessor present, then please let us know before the inspection. We must stress though that it can sometimes be very helpful for us to have the information requested <u>before</u> the inspection, so that we can consider that information and assess the time we will require for our visit.

Thank you for your time. Please call us if you or your professional advisors need any help in completing this questionnaire.

Additional information or explanation	Our question	Your answer, and any other information you believe is relevant.
General information		
Please confirm the exact postal address of the property, including the postcode	Address	*27 KING STREET* *KING'S LYNN* Post code *PE30 1ET*
An EPC is normally required for a legal reason	Reason for your EPC	Sale of freehold ☐ leasehold ☐ other ☒ (please confirm): *ASSESSMENT OF ENERGY*
If you do not occupy the property, how long have you owned the property	Length of occupation or ownership	Occupation ☒ Ownership ☒ Both ☒ *2½* years
A simple description of the property is sufficient	Property type	Shop ☐ Factory ☐ Warehouse ☐ Restaurant ☐ Office ☒ Other ☐:
Please tick one of the boxes	Detachment and original nature	Detached ☐ Semi-Detached ☐ Terraced ☐ End-Terraced ☒ Other ☐ Details: Purpose Built ☐ Converted ☒ *(WAS WAREHOUSE)*
This can be a difficult one, but the age of the property can sometimes make a big difference to the final result. Please be as specific as you can	Age	Pre 1965 ☒ 1965–1976 ☐ 1977–1982 ☐ 1983–1990 ☐ 1991 – 1995 ☐ 1996–2002 ☐ 2003–2006 ☐ post 2006 ☐ OR, if you have more specific information: Exact year: *1478* Approximate year: —

Commercial Energy Assessor's Handbook

RT2.D0666 - COMMERCIAL EPC CLIENT'S QUESTIONNAIRE TYPE 'B'

Only general information required – more detailed information requested later	Services	Present in property: Don't know ☐ Main electricity ☒ Main water ☒ Main gas ☒ Main drainage ☒ LPG ☐ Oil ☐ Other ☐ Details: If LPG, oil or similar, please state storage location: _____
This information is important; for health and safety, and also EPC purposes	Location of main service points	Please confirm location of: Electricity meter(s) __LOBBY__, or Don't know ☐ Electricity fuseboard(s) __LOBBY__, or Don't know ☐ Gas meter(s) __BASEMENT__, or Don't know ☐ Main water stop valve (internal) __KITCHEN__, or Don't know ☐ Main water stop valve (external) __FRONT DOOR__, or Don't know ☐ Other _____, or Don't know ☐
If you answer 'yes', please answer the more specific questions later under the heading of planning permission and building regulations	Alterations	Any alterations to the building or the services, such as heating Yes ☐ No ☒ Don't know ☐
It is helpful to us if you can indicate the size of your property, so we can estimate the length of time we will need for our inspection. This can help to reduce the risk of any inconvenience to you. If you are not sure, please don't worry	Property size	m² sq.ft Don't know ☒ How do you know this information: Estimate ☐ Based on drawings ☐ Agents measurements ☐ Other ☐
It is also helpful to us if you can indicate the type and number of rooms and spaces in your property, so	Property accommodation	Details:

RT2.D0666   Page 3 of 16   Quality Manual Record Form   June 2009   3

Appendix B

RT2.D0666 - COMMERCIAL EPC CLIENT'S QUESTIONNAIRE TYPE 'B'

we can estimate the length of time we will need for our inspection. This can help to reduce the risk of any inconvenience to you.		
This can help us to consider access arrangements, and matters such as health and safety	Storeys	Please confirm number of storeys Don't know ☐ 1 ☐ 2 ☒ 3 ☐ 4 ☐ 5 ☐ 6 ☐ More ☐ If you have answered 'more', how many:
It can be easy for the assessor to miss the presence of a cellar, particularly if the access is covered, e.g. by floor coverings	Cellar or basement	Present in the property Yes ☒ No ☐ Don't know ☐ If yes, where is it: And, how is it accessed: *EXTERNALLY*
Planning and Building Regulations		
If your property has special planning status, we need to know as this can sometimes affect the EPC and energy assessment	Listed buildings and Conservation Areas	Property 'listed' as being a building of special architectural or historical interest Yes ☒ No ☐ Don't know ☐ if you have answered 'yes', please confirm: Grade 1 ☐ 2* ☐ 2 ☒ *(BUT NOT SURE)* Property situated in a Conservation Area Yes ☒ No ☐ Don't know ☐
If you have answered yes to any of the questions about extensions or alterations etc; or if the property is new, or relatively new, you may have had to obtain necessary approvals from the local council.	Planning permissions, Listed buildings, or Conservation Area consent	Date(s) of any permissions: *NONE* Details of work: Local authority planning reference numbers:

RT2.D0666 - COMMERCIAL EPC CLIENT'S QUESTIONNAIRE TYPE 'B'

	Building Regulations	Date(s) of any building regulations approvals: *NONE* Details of work: Local authority building regulations reference numbers:
Any alterations to any parts of the property can have a significant effect on the energy rating	Alterations to the structure, use, size, shape or arrangement of the building	Alterations *NONE* Yes ☐ No ☐ Don't know ☐ Date: Details of the work: Introduction of additional insulation Yes ☐ No ☐ Don't know ☒ Date: Details of the work: Where: Other: Yes ☐ No ☐ Don't know ☐ Date: Details of the work: Where:
Any alterations or replacements to any of the services in the property can have a significant effect on the energy rating	Alterations or replacements to any of the services in the building	Heating Yes ☐ No ☒ Don't know ☐ Date: Details of the work: Where: Ventilation Yes ☐ No ☒ Don't know ☐ Date: Details of the work Where: Air-conditioning Yes ☐ No ☒ Don't know ☐ Date: Details of the work:

Appendix B

RT2.D0666 - COMMERCIAL EPC CLIENT'S QUESTIONNAIRE TYPE 'B'

		Where: Lighting Yes ☐ No ☒ Don't know ☐ Date: Details of the work: Where: Other: Yes ☐ No ☒ Don't know ☐ Date: Details of the work Where:
Drawings etc If you have specific details and information about the original property, or any alterations. Old photographs can also be very helpful. Please confirm whether you have any of the documentation and that you can provide a copy of the document(s) to the assessor		
Drawings might be of the original building, or an extension, or alteration	Drawings	Yes ☐ No ☒ Copy returned herewith ☐ Copy available at the property ☐ Copy available elsewhere ☐ Details:
A specification usually describes the work in detail	Specifications	Yes ☐ No ☒ Copy returned herewith ☐ Copy available at the property ☐ Copy available elsewhere ☐ Details:
Old photographs can help to confirm details in areas that are not readily accessible	Photographs	Yes ☐ No ☒ Copy returned herewith ☐ Copy available at the property ☐ Copy available elsewhere ☐ Details:
You may have other information that might be helpful	Other	Yes ☐ No ☒ Copy returned herewith ☐ Copy available at the property ☐ Copy available elsewhere ☐ Details:
Health and safety information		

RT2.D0666 - COMMERCIAL EPC CLIENT'S QUESTIONNAIRE TYPE 'B'

It is important for the energy assessor to be sure that he / she will be able to carry out the inspection safely; and will not cause any issues for you, your work-people, or for others; whilst on site. Your answers will help the assessor to ensure that such is the case during the inspection. **Please take particular care in completing this section of the questionnaire**	Please confirm whether you have the following documentation and that you can provide a copy of the document(s) to the assessor	Health and safety statement or policy Yes ☒ No ☐ Don't know ☐ Copy returned herewith ☐ Copy available at the property ☒ Copy available elsewhere ☐ Details: Fire Safety Risk Assessment Yes ☒ No ☐ Don't know ☐ Copy returned herewith ☐ Copy available at the property ☒ Copy available elsewhere ☐ Details: Risk Assessment under the Control of Asbestos Regulations 2006 Yes ☒ No ☐ Don't know ☐ Copy returned herewith ☐ Copy available at the property ☐ Copy available elsewhere ☒ Details: *COUNCIL* Any other relevant health and safety documentation Yes ☐ No ☐ Don't know ☒ Copy returned herewith ☐ Copy available at the property ☐ Copy available elsewhere ☐ Details:
Some activities in properties mean that visitors must wear special protective equipment	Personal Protective Equipment (PPE)	Please confirm whether our assessor will require any special PPE Yes ☐ No ☐ Don't know ☒ if you have answered 'yes', please confirm: 'Hard hat' ☐ Boots or other footwear ☐ Overalls ☐ Mask ☐ Goggles / eye protection ☐ Gloves ☐ Other ☐ Details: Further specific details of any of the above:
There may be specific health and safety issues that we need to know about so that our assessor's health and safety is not compromised during the inspection. Thank you.	Please also confirm whether there are any other health and safety issues that our assessor should know about,	Working at height Yes ☐ No ☐ Don't know ☒ Details: Confined spaces Yes ☒ No ☐ Don't know ☐

RT2.D0666 - COMMERCIAL EPC CLIENT'S QUESTIONNAIRE TYPE 'B'

	or that our assessor needs to take special precautions against, and please confirm all details.	Details: *BASEMENT*	
		Lone working	Yes ☐ No ☐ Don't know ☒
		Details:	
		Contamination	Yes ☐ No ☒ Don't know ☐
		Details:	
		Dangerous substances	Yes ☐ No ☒ Don't know ☐
		Details:	
		Other	Yes ☐ No ☒ Don't know ☐
		Details:	
Our Inspection			
We are always happy to try to carry out the inspection at a time that is convenient to you	Inspection time	When would be best for you? Morning ☐ Afternoon ☐ Out of hours ☒ Other ☐:	
It will be helpful to know whether our assessor can have full access to your property, or are there any parts that we will not be able to gain access to	Access	Full access available to entire property If you have answered 'no', please let us know which parts are not accessible:	Yes ☒ No ☐
Photographs are normally one of the essential methods an assessor uses to record details of your property	Photographs	Please confirm whether you are happy for our assessor to record images of your property	Yes ☒ No ☐
		Sensitive areas where you would prefer that our assessor does not take any photographs Please state where:	Yes ☐ No ☒

RT2.D0666 - COMMERCIAL EPC CLIENT'S QUESTIONNAIRE TYPE 'B'

If you employ other professionals, or maintenance personnel, to look after any part of the property, particularly the services such as heating, they can sometimes provide vital information that can help to ensure the accuracy of the EPC	Other professionals and maintenance personnel	Architect Telephone: Email: Address:	Yes ☐ No ☒
		Surveyor Telephone: Email: Address:	Yes ☐ No ☒
		Heating / Lighting Engineer Telephone: Email: Address:	Yes ☐ No ☒
		Maintenance company Telephone: Email: Address:	Yes ☐ No ☒
		Other Telephone: Email: Address:	Yes ☐ No ☒
		Are you happy for us to contact them direct	Yes ☐ No ☐
Construction			
A significant amount of heat escapes through the outer walls; so what they are made from, and whether they are insulated, can have a major impact on the EPC	External walls	What are the outside walls made from Don't know ☐ Brick ☒ Concrete ☐ Steel ☐ Glass ☐ Asbestos ☐ Other ☐ Details: Insulation Yes ☐ No ☒ Don't know ☐ Details including thickness:	

RT2.D0666 - COMMERCIAL EPC CLIENT'S QUESTIONNAIRE TYPE 'B'

		Documentary information available: building log book ☐ handbook ☐ schematic drawing ☐ plan ☐ manufacturer's information ☐ specification ☐ guarantee ☐ none ☐ other ☐ Copy returned herewith ☐ Copy available at the property ☐ Copy available elsewhere ☐ Details:
Similarly, a significant amount of heat escapes through the roofs; so what they are made from, and whether they are insulated, can have a major impact on the EPC	Roof coverings	What are the roofs made from Don't know ☐ Tile ☒ Slate ☐ Steel ☐ Asbestos ☐ Felt ☐ Asphalt ☐ Other ☐ Details: Insulation Yes ☐ No ☐ Don't know ☒ Details including thickness: Documentary information available: building log book ☐ handbook ☐ schematic drawing ☐ plan ☐ manufacturer's information ☐ specification ☐ guarantee ☐ none ☐ other ☐ Copy returned herewith ☐ Copy available at the property ☐ Copy available elsewhere ☐ Details:
The exact construction of the windows, and the type of glazing, can also have a significant effect on your EPC	Windows and glazing	What are the windows made from Don't know ☐ Frames: Aluminium ☐ Hardwood ☐ Metal ☐ PVC ☐ Softwood ☒ Steel ☐ Don't know ☐ Other ☐ Details: Glazing: Double ☐ Quadruple ☐ Roof-light ☐ Single ☒ Triple ☐ Other/special ☐ Don't know ☐ Documentary information available: building log book ☐ handbook ☐ schematic drawing ☐ plan ☐ manufacturer's information ☐ specification ☐ guarantee ☐ none ☒ other ☐

RT2.D0666 - COMMERCIAL EPC CLIENT'S QUESTIONNAIRE TYPE 'B'

		Copy returned herewith ☐ Copy available at the property ☐ Copy available elsewhere ☐ Details:
Ground floors have generally had to be insulated in all new buildings constructed after around 1995, as a significant amount of heat can escape through them	Floors	What are the floors made from Don't know ☐ Solid ☒ Suspended ☒ Concrete ☐ Timber ☐ Other ☐ Details: Insulation Yes ☐ No ☒ Don't know ☐ Details including thickness: Documentary information available: building log book ☐ handbook ☐ schematic drawing ☐ plan ☐ manufacturer's information ☐ specification ☐ guarantee ☐ none ☒ other ☐ Copy returned herewith ☐ Copy available at the property ☐ Copy available elsewhere ☐ Details:
Heating		
The type of heating system in a property governs how much carbon dioxide is generated during the heating process. Please indicate the type of heating you have. Please also confirm further details about documentation you may have, as required	Heating	Heating present in property Yes ☒ No ☐ Don't know ☐ If you have answered 'yes" please confirm type: *GAS FIRED BOILER + RADIATORS* Was it installed on Yes ☐ No ☒ Don't know ☐ or after 1989 Details: *AROUND 1980* Documentary information available: commissioning certificate ☐ current test certificate ☐ maintenance manual ☐ operating manual ☐ log book ☐ handbook ☐ schematic drawing ☐ plan ☐ manufacturer's information ☐ specification ☐ guarantee ☐ none ☒ other ☐ Copy returned herewith ☐ Copy available at the property ☐ Copy available elsewhere ☐ Details:
This is usually the source of the carbon dioxide emissions	Fuel type	Don't know ☐ Natural gas ☒ LPG ☐ Biogas ☐ Oil ☐ Coal ☐ Biomass ☐ Grid Supplied Electricity ☐ Anthracite

Appendix B

RT2.D0666 - COMMERCIAL EPC CLIENT'S QUESTIONNAIRE TYPE 'B'

		☐ Waste heat ☐ Smokeless Fuel (including coke) ☐ Dual Fuel Appliances (Mineral & Wood) ☐ Other ☐ Details:
Air conditioning		
This might be a simple air conditioning unit that discharges through a wall	Simple air-conditioning	System in the property　　　　Yes ☐ No ☐ Don't know ☒ Split or multi-split system ☐　　　　Don't know ☐ Single room cooling system ☐ Other ☐ Details (e.g. manufacturer's name): Documentary information available: commissioning certificate ☐ current test certificate ☐ maintenance manual ☐ operating manual ☐ log book ☐ handbook ☐ schematic drawing ☐ plan ☐ manufacturer's information ☐ specification ☐ guarantee ☐ none ☐ other ☐ Copy returned herewith ☐　Copy available at the property ☐ Copy available elsewhere ☐ Details:
The capacity, or output, will contribute to the operating costs and carbon dioxide emissions	Size of the system	What is the total effective rated output of the air-conditioning system in the building?　　Below 12kW ☐ Between 12 – 250 kW ☐ Greater than 250 kW ☐ Output unknown ☐
It is a legal requirement that certain sizes of air-conditioning systems must be regularly inspected and tested	Air-conditioning inspection	Has an air conditioning inspection been commissioned for compliance with Energy Performance of Buildings regulations? Yes – inspection completed ☐ Yes – inspection commissioned ☐ No inspection completed or commissioned ☐ Don't know ☐
Ventilation & exhaust This might be simple natural ventilation by opening windows, or mechanical extract ventilation, e.g. from toilets		Present in property　　　Yes ☒ No ☐ Don't know ☐ If you have answered 'yes" please confirm type: Natural ☒　Mechanical ☒　Other ☐　Details: *TOILETS + KITCHEN* Documentary information available: commissioning certificate ☐ current test certificate ☐ maintenance manual ☐ operating

RT2.D0666 - COMMERCIAL EPC CLIENT'S QUESTIONNAIRE TYPE 'B'

		manual ☐ log book ☐ handbook ☐ schematic drawing ☐ plan ☐ manufacturer's information ☐ specification ☐ guarantee ☐ none ☒ other ☐ Copy returned herewith ☐ Copy available at the property ☐ Copy available elsewhere ☐ Details:
Hot water system		
This is the type of hot water supply your property has	Type, age and fuel	Do you know details of the hot water supply system Yes ☒ No ☐ Don't know ☐ if your answer is 'yes', confirm type: Dedicated boiler ☐ Stand-alone water heater ☐ Instantaneous only ☐ Instantaneous combi ☐ Heat pump ☐ Other ☐ Details: **FROM MAIN BOILER VIA CYLINDER** Is it older than 1989 Yes ☐ No ☐ Don't know ☒ Do you know the fuel type Yes ☒ No ☐ Don't know ☐ Natural gas ☒ LPG ☐ Biogas ☐ Oil ☐ Grid Supplied Electricity ☒ Other ☐ Details: Is the system a storage system Yes ☒ No ☐ Don't know ☐ If you have answered 'yes', please confirm the storage volume in litres _____ Don't know ☒ Does the system have secondary circulation? Yes ☐ No ☐ Don't know ☒ Documentary information available: commissioning certificate ☐ current test certificate ☐ maintenance manual ☐ operating manual ☐ log book ☐ handbook ☐ schematic drawing ☐ plan ☐ manufacturer's information ☐ specification ☐ guarantee ☐ none ☒ other ☐ Copy returned herewith ☐ Copy available at the property ☐ Copy available elsewhere ☐ Details:

Appendix B

RT2.D0666 - COMMERCIAL EPC CLIENT'S QUESTIONNAIRE TYPE 'B'

Renewable energy systems		
For this type of system, which helps to heat hot water, your property will have a 'solar collector' (dish), probably on the roof.	Solar thermal hot water (SES)	Present at property Yes ☐ No ☒ Don't know ☐ If your answer is 'yes': Name of the system:_____, or Don't know ☐ Documentary information available: commissioning certificate ☐ current test certificate ☐ maintenance manual ☐ operating manual ☐ log book ☐ handbook ☐ schematic drawing ☐ plan ☐ manufacturer's information ☐ specification ☐ guarantee ☐ none ☐ other ☐ Copy returned herewith ☐ Copy available at the property ☐ Copy available elsewhere ☐ Details:
For this type of system, which helps to supply electricity, your property will also have a 'solar collector' (dish), probably on the roof.	Solar photo-voltaics (PVS)	Present at property Yes ☐ No ☒ Don't know ☐ If your answer is 'yes': Name of the system:_____, or Don't know ☐ Documentary information available: commissioning certificate ☐ current test certificate ☐ maintenance manual ☐ operating manual ☐ log book ☐ handbook ☐ schematic drawing ☐ plan ☐ manufacturer's information ☐ specification ☐ guarantee ☐ none ☐ other ☐ Copy returned herewith ☐ Copy available at the property ☐ Copy available elsewhere ☐ Details:
This information is likely to be available from the original manufacturer, or installer, particularly the power output figure	Wind turbines	Present at property Yes ☐ No ☒ Don't know ☐ If your answer is 'yes', please confirm: If your answer is 'yes': Name of the system:_____, or Don't know ☐

RT2.D0666 - COMMERCIAL EPC CLIENT'S QUESTIONNAIRE TYPE 'B'

		Documentary information available: commissioning certificate ☐ current test certificate ☐ maintenance manual ☐ operating manual ☐ log book ☐ handbook ☐ schematic drawing ☐ plan ☐ manufacturer's information ☐ specification ☐ guarantee ☐ none ☐ other ☐ Copy returned herewith ☐ Copy available at the property ☐ Copy available elsewhere ☐ Details:
Lighting		
You can only answer this question if you have very detailed information about your lighting systems	Lighting – full survey	Full lighting design carried out: Yes ☐ No ☐ Don't know ☒ Total wattage: _____ (W) Don't know ☐ Total design illuminance: _____ (Lux) Don't know ☐ Documentary information available: commissioning certificate ☐ current test certificate ☐ maintenance manual ☐ operating manual ☐ log book ☐ handbook ☐ schematic drawing ☐ plan ☐ manufacturer's information ☐ specification ☐ guarantee ☐ none ☐ other ☐ Copy returned herewith ☐ Copy available at the property ☐ Copy available elsewhere ☐ Details:
The type of lighting installed in the property usually has the most significant effect on the energy efficiency of any part of the property or services. It is therefore very important that you or your advisors take particular care with this part of the questionnaire. 'Metal halide' is more commonly known as low voltage 'tungsten halogen'	Lighting types	Please confirm lighting types: Display lighting ☐ Metal halide ☐ Fluorescent tubes ☒ High pressure sodium ☐ High pressure mercury ☐ Compact fluorescent ☒ Don't know ☐ Other ☒ Details: Please confirm where: *HALOGEN IN RESTAURANT* Documentary information available: commissioning certificate ☐ current test certificate ☐ maintenance manual ☐ operating manual ☐ log book ☐ handbook ☐ schematic drawing ☐ plan ☐ manufacturer's information ☐ specification ☐ guarantee ☐ none ☒ other ☐ Copy returned herewith ☐ Copy available at the property ☐ Copy available elsewhere ☐ Details:

Appendix B

RT2.D0666 - COMMERCIAL EPC CLIENT'S QUESTIONNAIRE TYPE 'B'

This is generally the normal light switch we are all used to	Lighting controls	Manual switching (i.e. usual light switches) present in property — Yes ☒ No ☐ Don't know ☐ If you have answered 'yes', please confirm where: **MAIN SWITCH IN EAST KITCHEN** Time switching for display lighting — Yes ☐ No ☐ Don't know ☐ If you have answered 'yes', please confirm where: Photoelectric controls — Yes ☐ No ☐ Don't know ☒ If you have answered 'yes', please confirm where: Occupancy sensing — Yes ☐ No ☐ Don't know ☒ If you have answered 'yes', please confirm where: Documentary information available: commissioning certificate ☐ current test certificate ☐ maintenance manual ☐ operating manual ☐ log book ☐ handbook ☐ schematic drawing ☐ plan ☐ manufacturer's information ☐ specification ☐ guarantee ☐ none ☒ other ☐ Copy returned herewith ☐ Copy available at the property ☐ Copy available elsewhere ☐ Details:
Other relevant information		
You may believe that you have further information that might be relevant to the energy efficiency of your property. Please state the information here.	Other matter as appropriate	Your information (please go on to additional sheet if required): **DIFFICULT TO ALTER DUE TO LISTED STATUS**

Questionnaire completed by   Signature _[signature]_

Print name **PAT ISBILL**

Date **6th JUNE 2009**

Please confirm Owner ☒ Occupier ☒ Agent ☐ Representative ☐ Other ☐ Details _____

(CLIENT HAS WORKED AT RESTAURANT FOR 22 YEARS)

Commercial Energy Assessor's Handbook

PHOTOGRAPHS

Figure B2: Centre and left end of front elevation

Figure B3: Detached outbuilding

Figure B4: Centre of front (south) elevation

Figure B5: Right hand end of front elevation

Figure B6: Main steps and void beneath terrace

Figure B7: Rear elevation of main building, accessible from adjoining pub garden – note kitchen exhaust

Appendix B

Figure B8: Basement looking west

Figure B9: Boiler and right hand side of basement

Figure B10: Restaurant area looking east

Figure B11: Lamp to north-west corner of restaurant

Figure B12: Lobby to front of left hand kitchen, looking west

Figure B13: Left hand kitchen with exhausts

Figure B14: Preparation area to right of left hand kitchen

Figure B15: Heating/hot water programmer

Appendix B

Figure B16: Lighting control panel in zone G

Figure B17: Right hand kitchen looking west

Figure B18: Hot water cylinder rear left corner of right hand kitchen

Figure B19: Office, with window [see site notes later, p. 1954 / 31]

Commercial Energy Assessor's Handbook

Figure B20: Lobby to the lower ground floor, with stairs

Figure B21: Same lobby, looking to front right hand corner

Figure B22: From lower ground floor lobby into ladies toilet

Figure B23: Lamp in ladies toilet

Figure B24: Roof space over right hand kitchen and office

Figure B25: River Great Ouse, and west gable wall – a great place to watch the fishing boats go by in the summer

SITE INSPECTION AND REFLECTION NOTES

Figure B26 is a copy of our site notes, pp. 1–35 excluding the zone plans and geometry data entry sheets and data reflection notes, but with a blank example of a geometry data sheet for you to use.

Appendix B

RT2.D0671   COMMERCIAL EPC SITE / REFLECTION NOTES   File / page reference __1954__ / __01__

1 – General information, and health & safety

Name of CEA:	*LARRY RUSSEN*	**Licence Number:**	*SAVA 00*	
Inspection Date:	*18th August 2009*			
Level of assessment:	**(3)**	4	5	
Address of property:	*RIVERSIDE RESTAURANT, 27 KING STREET, KING'S LYNN*			
		Post Code:	*PE30 1ET*	
Appointment Time:	*10:00*	**Arrival Time:** *10:00*	**Depart Time:** *14:00*	
Seller's Questionnaire completed:	Yes ✗ No ☐	**Viewing Arrangements:**	Keys / (direct) / meeting (owner) / tenant / agent / caretaker / other:	
Weather location and notes:	Belfast / Birmingham / Cardiff / Edinburgh / Glasgow / Leeds / London / Manchester / Newcastle / (Norwich) / Nottingham / Plymouth / Southampton / Swindon			
Weather:	Fine ✗ Windy ☐ Sunny ☐ Showers ☐ Cloudy ☐ Heavy Rain ☐ Frost ☐ Snow ☐ Fog ☐ Other:	**Comments on limitations / weather:** *Staff + stock*		
Occupation:	Owner ☐ Tenanted ☐ Vacant ✗	**Tenure:** FH / (LH)		
Furnished, stock, plant & machinery:	Yes ✗ No ☐ Part-furnished ☐	Other: *Stock + Furniture*		
Floor Coverings:	Full ✗ None ☐ Part Carpeted ✗ Other:			
Date of Construction:	*1450 (local knowledge)*	**Alterations & Date:**	*"Minor refurbishment" 1980*	
Property Type:	Shop ☐ Factory ☐ Warehouse ☐ Restaurant ✗ Office ☐ Other:			
Detachment:	Detached ☐ Semi-Detached ☐ Terraced ☐ End-Terraced ☐ Purpose Built ✗ Converted ☐			

ACCOMMODATION						
Type of rooms / spaces	*Storage*	*Toilets*	*Hall/Lobby*	*Restaurant*	*Kit/Prep*	*Office*
Basement / Lower Ground Floor	2	2	1			
Ground Floor	1	–	1	1	3	1
First Floor						
Second Floor						
Third Floor						
Fourth Floor						
Fifth Floor						
Total Number of rooms / spaces	3	2	2	1	3	1

CONSTRUCTION		
	Walls / frame:	*Solid brick and stone*
	Roof:	*Plain clay tiles, pitched*
	Floors:	*Solid to basement + lower G.F. Suspended above*
	Doors / windows:	*Softwood, single glazed*
SERVICES	Heating	(<100kW) / >100kW Gas ✗ Electric ✗ Water ✗
	Cooling	<12kW / >12kW
	Renewable systems	SES *No* PVS *No* Wind *No*

Copyright Larry Russen June 2009

Figure B26: Copy of partial site notes [pp. 1–35]

RT2.D0671   COMMERCIAL EPC SITE / REFLECTION NOTES   File / page reference __1954__/__02__

HEALTH & SAFETY RISK ASSESSMENT

Outside	N/A	Hazard	Risk High	Risk Med	Risk Low	Notes (Action required or taken inc. special PPE / equipment needs)
Burglary	✓	Ground floor windows and door protected. Alarm serviced				
Intruders / walk in crime		Trespassers. Theft of valuables / documents			✓	Keep equipment in office
Arson		Location vulnerable – shutters / curtains in place			✓	
Vandalism		Location vulnerable – shutters / curtains in place			✓	
Access	✓	Expected persons present. Children or juveniles alone				

On site		Hazard	Risk High	Risk Med	Risk Low	Notes (Action required or taken inc. special PPE / equipment needs)
ACM		Asbestos register seen — No			✓	Take care in cellar
Empty property	✓	Unauthorised occupancy				
Floors and timber	✓	Missing or rotten flooring. Unsafe windows				
Structure of building	✓	Roof (e.g. tiles / slates / sheets loose). Guard rails / ladder access loose. Walls bulging / loose. Frame distorted				
Electrics	✓	Unsafe sockets / switches / cables. Power off				
Other utilities	✓	Water. Gas. Oil. Solid fuel				
Weather	✓	Ice, snow, or too hot. Rain. Fog. Windy				
Working at height		Roof access. Access to walls. Inspection pits. Unguarded edges. Access elsewhere. High winds			✓	Warn staff about ladder. Take care with sea defence wall
Ladder usage		Loft access. Borrowed / loft ladder safe? Own ladder serviced			✓	As above
Own equipment works well	✓	Torch discharged. Equipment assembled safely				
Other people		Interference by others including children. Vagrants			✓	Brief owner + staff
Animals	✓	Un-tethered / aggressive dogs others. Pigeons / rats / bats / other vermin / fleas				

Copyright Larry Russen June 2009

Appendix B

RT2.D0671 COMMERCIAL EPC SITE / REFLECTION NOTES   File / page reference ___1954_/_03_

Fibres	Loft insulation Mineral wool			/	Wear mask in roof
Hazardous Materials	Flammable liquids/materials Toxic substances/waste ACMs			/	Take care around cookers + ovens
Site/gardens/grounds	Debris Holes / wells / cesspits Animals, reptiles and insects	/			
Hazardous activities by occupiers	Hot processes Machinery / conveyors Cutting Welding Vehicle movements Maintenance Noise Grease / oil on floors Chemicals Radio-active Other:			/	See above. Wear boots. Minor risk from traffic at entrance
Confined spaces	Service ducts Plant rooms Refrigeration equipment	/			

Personal-lone worker	Hazard	Risk			Notes
		High	Med	Low	(Action required or taken inc. special PPE / equipment needs)
Working alone safely	Existing / known threats in area Reporting procedure clear including mobile phone contact Large site or complicated property			/	Office know where I am
Builders present	Hard hat not worn Falling materials Untidy workplace	/			
Drug & alcohol problems	Litter and broken glass Syringes and drug cooking equipment, e.g. spoons/cans	/			
Violence, Kidnap & abuse	Known dangers	/			
Travelling safely	Overlong journey Phone contact unavailable Safe parking unavailable	/			By foot
Other(s)					

BASIC EQUIPMENT: Hard hat (Yes)/no   (Basement and roof)   Boots (Yes)/no   (Basement)   High-visibility jacket Yes/(no)   (No traffic movement)

Commercial Energy Assessor's Handbook

RT2.D0671 COMMERCIAL EPC SITE / REFLECTION NOTES File / page reference _1954_ / _04_

REFLECTION ISSUES
EPC not required for whole or part of property?
Building(s) constructed with **walls and roof** — **Yes** / no
Building(s) used solely or primarily as place(s) of **worship** — Yes / **no**
Temporary building(s) with a planned time of use of two years or less — Yes / **no**
Industrial sites, workshops and non-residential agricultural buildings with **low energy demand** — Yes / **no**
Stand-alone building(s) with a total useful floor area of less than 50m² which are not dwellings — **Yes** / no
Building(s) will be **demolished** (owner must believe on reasonable grounds new owner will demolish) — Yes / **no**
Has, or intention of having, **fixed services** and ability to include fixed services to condition climate — **Yes** / no
Construction of the building(s) is **not complete** — Yes / **no**
Part of building(s) to which above issue(s) apply: *Laundry less than 50m² – see sketch (15.74m² TUFA)*
Documentary evidence available from inspection, client, internet or other source
Previous energy assessments Yes / **no** EPC ☐ DEC ☐   Survey report Yes / **no**   Services report Yes / **no**
Planning permissions Yes / **no**   Building Regulations Yes / **no**   CDM health and safety file Yes / **no**
Commissioning certificates for service installations Yes / **no**   Test / service certificates Yes / **no**
Maintenance or operating log books Yes / **no**   Manufacturers' information Yes / **no**   Air-con report Yes / **no**
Guarantees / warranties Yes / **no**   Fire risk assessments **Yes** / no   ACM register **Yes** / no
COSHH Yes / **no**   Schematics Yes / **no**: Heating systems ☐   Cooling systems ☐   HWS ☐   Lighting ☐
Access audit Yes / **no**   Building plans Yes / **no**   Specifications Yes / **no**   Licences for alterations Yes / **no**
Contracts Yes / **no**   Copy of the lease Yes / **no**   Client / other photographs Yes / **no**   Other Yes / no
Details: *ACM register "available", but not seen*
'Existing' drawing audit
Drawings provided / seen Yes ☐ No ☒
Drawing number / date:
Drawing(s) checked Yes ☐ No ☐ *N/A*
Percentage measurements checked: 10% ☐ 20% ? ___ ☐ *N/A*
Which measurements checked? See plan / other: *N/A*
Drawings confirmed as being reliable Yes ☐ No ☐ Other ☐ *N/A*
Other relevant information

Copyright Larry Russen June 2009

Appendix B

RT2.D0671 COMMERCIAL EPC SITE / REFLECTION NOTES   File / page reference 1954 / 05

Condition and other issues
Condensation ☐ Water penetration ☐ Potential problem arising from recommendations ☐ Planning issue: Listed ☒ Conservation Area ☒ Other ☒ None ☐ Where: *Areas concealed behind dry linings and false ceiling to lower ground floor, right-hand side*
Other issues
Conflict of interest Yes / (no) Own firm selling ☐ Involved in lease ☐ Past client ☐ Employee of firm ☐ Personal ☐ Air conditioning system inspection report required Yes ☐ No ☒   Client warned Yes ☐ No ☐ N/A ☒ Other ☐
Personal site notes – matters for review in later editions
More space for ACM's in risk assessment. *"N/A" for drawing audit* *Alter "Continue on other sheet when required" to "Continued Yes/No"*

DESKTOP STUDY			
Date	Subject	Contact / reference	Information / result
17/8	'Aerial photos' (Yes) no		*Adjoining river. Exposed.*
17/8	Local authority planning www.west-norfolk.gov.uk follow planning applications property search postcode and 'search' (Yes) no	*On-line*	*Listed grade 2 and in conservation area*
17/8	Local authority building control Yes / (no)		
17/8	HVAC / lighting engineer Yes / (no)		
17/8	Enhanced Capital Allowances www.eca.gov.uk	*On-line*	CHP Yes / (no) Zone controls for heat / vent / air-con Yes / (no) Boiler Yes / (no) Warm air and radiant heaters Yes / (no) Lighting Yes / (no) Space heating heat pumps Yes / (no) Solar thermal Yes / (no) Refrigeration plant / chiller Yes / (no) Automatic M & T Yes / (no) Air to air energy recovery Yes / (no)

Copyright Larry Russen June 2009

RT2.D0671   COMMERCIAL EPC SITE / REFLECTION NOTES   File / page reference _1954 / 06_

17/8	Client report		
	Part does not require EPC	Yes	Warn client ✓
	Part(s) not inspected	No	Went to Globe Hotel garden ✓
	Insulation	Yes	Not sure about most of roof ✓
	Conservation Area	Yes	In report ✓
	Listed	Yes	In report ✓
	Condition issue	No	

SCHEDULE OF PHOTOGRAPHS			
Number	Photo description	Number	Photo description
1	LH end south elevation	32	Lamps in gents toilet
2	Outbuilding	33	Lamps in ladies toilet
3	Centre south elevation	34	Lamps in basement
4	Rth end south elevation	35	Window 5
5	Front (south) elevation	36	Window 6
6	Front (south), LH end	37	Window 11
7	Rear (north) elevation		
8	Basement looking west		
9	Boiler		
10	Restaurant looking west		
11	Lobby to front of kitchen		
12	LH kitchen		
13	Down stairs from landing		
14	Rth kitchen		
15	Office + W12		
16	HWC Rth kitchen		
17	Lobby lower GF		
18	Lobby lower GF		
19	Lobby into ladies toilet		
20	Store, lower GF		
21	Lamp LHs restaurant		
22	Roof space Rth		
23	Gas meter		
24	River		
25	Electricity meter		
26	Extracts in LH kitchen		
27	Lamps in LH kitchen		
28	Lamps in restaurant		
29	Lamps in Rth kitchen		
30	Lamps in office		
31	Lamps in lobby		

Continue on other sheet: Yes / (No)

LH – left hand

Rth – right hand

W – window

Copyright Larry Russen June 2009

Appendix B

RT2.D0671   COMMERCIAL EPC SITE / REFLECTION NOTES   File / page reference __1954 / 07__

2 – Schedule of wall construction types

	Construction type, thickness, insulation and any metal	SBEM inference procedure construction, or other	Where?	Adjoining condition	Age	Evidence for identification, source, position seen, limitations and notes
1	**Material:** (Brick) / block / (stone) / tile hung / in-situ / pre-cast concrete / steel / concrete / timber frame / steel / ACM / glass-fibre cladding **Thickness (mm)** *350 – 550* **Bond:** SB / EB / FB / Other: *Garden wall* (External) / internal Involves metal Yes /(no) cladding *No evidence of insulation*	CC - Cast in-situ concrete (including no-fines) CW - cavity wall bricks blocks FCW - framed/curtain walling MC - metal cladding system PCCP - pre-cast concrete panels (may be faced) (SB) - solid brick or block wall on in-situ concrete WB – weather boarding and tile-hung walls (t.f.) Other: 'Import one from library' _____ 'Introduce own values' _____	(Entire building) / other:	(E) SVS UAS CAS U	*Mostly very old*	(Visual,) (bond,) (measurement,) planning / building regulations approval, plans, specification, cutting open (where): _____ verbal (who): _____ other: _____ limitations: none / _____ *No access direct to western elevation*
2	**Material:** Brick / block / stone / tile hung / in-situ / pre-cast concrete / steel / concrete / (timber frame) / steel / ACM / glass-fibre cladding **Thickness (mm)** *75* **Bond:** SB / EB / FB / Other: _____ (External) / internal Involves metal Yes /(no) cladding	CC / CW / FCW / MC / PCCP / SB / WB / Other: _____ 'Import one from library' *'U' 1.2* *'Km' 11.7* 'Introduce own values' _____ *No insulation*	*Old doors south wall*	(E) SVS UAS CAS U	*20–30 yrs*	(Visual,) bond, measurement, planning / building regulations approval, plans, specification, cutting open (where): _____ verbal (who): _____ other: _____ (limitations:) none / _____
3	**Material:** (Brick) / block / stone / tile hung / in-situ / pre-cast concrete / steel / concrete / timber frame / steel / ACM / glass-fibre cladding **Thickness (mm)** *As 1* **Bond:** SB / EB / FB / Other: _____ (External) / internal Involves metal Yes /(no) cladding	CC / CW / FCW / MC / PCCP / (SB) / WB / Other: _____ 'Import one from library' _____ 'Introduce own values' _____	*Basement wall below G/L*	E SVS UAS CAS (U)	*As 1*	(Visual,) (bond,) (measurement,) planning / building regulations approval, plans, specification, cutting open (where): _____ verbal (who): _____ other: _____ (limitations:) none / _____
4	**Material:** (Brick) / block / stone / tile hung / in-situ / pre-cast concrete / steel / concrete / timber frame / steel / ACM / glass-fibre cladding **Thickness (mm)** *Not sure 100+* **Bond:** (SB) / EB / FB / Other: _____ External /(internal) Involves metal Yes /(no) cladding	CC / CW / FCW / MC / PCCP / (SB) / WB / Other: _____ 'Import one from library' _____ 'Introduce own values' _____ *(Party wall)*	*Lower GF + GF*	E SVS UAS (CAS) U	*Old*	(Visual,) (bond,) measurement, planning / building regulations approval, plans, specification, cutting open (where): _____ verbal (who): _____ other: _____ (limitations:) none / _____

Copyright Larry Russen June 2009

Commercial Energy Assessor's Handbook

RT2.D0671   COMMERCIAL EPC SITE / REFLECTION NOTES   File / page reference __1954 / 08__

#	Material	Type	Location	Exposure	Age	Evidence
5	Material: Brick / **block** / stone / tile hung / in-situ / pre-cast concrete / steel / concrete / timber frame / steel / ACM / glass-fibre cladding **Thickness (mm)** *100* **Bond:** **SB** / EB / FB / Other: External / **internal** Involves metal cladding Yes / **no**	CC / CW / FCW / MC / PCCP / **SB** / WB / Other: 'Import one from library' 'Introduce own values'	*Int. walls*	E SVS UAS **CAS** U	*Modern 20–30 yrs*	**Visual**, bond, **measurement**, planning / building regulations approval, plans, specification, cutting open (where): verbal (who): other: limitations: **none** /
6	Material: Brick / block / stone / tile hung / in-situ / pre-cast concrete / steel / concrete / **timber frame** / steel / ACM / glass-fibre cladding **Thickness (mm)** *100 studwork* **Bond:** SB / EB / FB / Other: External / **internal** Involves metal cladding Yes / **no**	CC / CW / FCW / MC / PCCP / SB / WB / Other: 'Import one from library' *Timber frame 1.2/11.7* 'Introduce own values'	*Int stud walls*	E SVS UAS **CAS** U	*Modern 20–30 yrs*	**Visual**, bond, **measurement**, planning / building regulations approval, plans, specification, cutting open (where): verbal (who): other: limitations: **none** /
7	Material: Brick / block / stone / tile hung / in-situ / pre-cast concrete / steel / concrete / **timber frame** / steel / ACM / glass-fibre cladding **Thickness (mm)** *100* **Bond:** SB / EB / FB / Other: **External** / internal Involves metal cladding Yes / **no**	CC / CW / FCW / MC / PCCP / SB / WB / Other: 'Import one from library' *Timber frame as 6* 'Introduce own values'	*Rths of restaurant + lobby at high level*	**E** SVS UAS CAS U	*Modern 20–30 yrs*	**Visual**, bond, **measurement**, planning / building regulations approval, plans, specification, cutting open (where): verbal (who): other: limitations: **none** /
8	Material: Brick / block / stone / tile hung / in-situ / pre-cast concrete / steel / concrete / timber frame / steel / ACM / glass-fibre cladding **Thickness (mm)** **Bond:** SB / EB / FB / Other: External / internal Involves metal cladding Yes / no	CC / CW / FCW / MC / PCCP / SB / WB / Other: 'Import one from library' 'Introduce own values'		E SVS UAS CAS U		Visual, bond, measurement, planning / building regulations approval, plans, specification, cutting open (where): verbal (who): other: limitations: none /
9	Material: Brick / block / stone / tile hung / in-situ / pre-cast concrete / steel / concrete / timber frame / steel / ACM / glass-fibre cladding **Thickness (mm)** **Bond:** SB / EB / FB / Other: External / internal Involves metal cladding Yes / no	CC / CW / FCW / MC / PCCP / SB / WB / Other: 'Import one from library' 'Introduce own values'		E SVS UAS CAS U		Visual, bond, measurement, planning / building regulations approval, plans, specification, cutting open (where): verbal (who): other: limitations: none /

Continue on other sheet: Yes / **No**

Copyright Larry Russen June 2009

Appendix B

RT2.D0671   COMMERCIAL EPC SITE / REFLECTION NOTES   File / page reference __1954__ / __09__

3 – Schedule of roof construction types

#	Construction type, pitch, thickness (mm), insulation and any metal	SBEM inference procedure construction, or other	Where?	Adjoining condition	Age	Evidence for identification, source, position seen, limitations and other notes
1	**Covering:** Asphalt / mineral felt / felt & chippings / concrete / (clay tiles) / natural / artificial slates / stone / thatch / profiled steel / iron / ACM / other: **Structure:** (Softwood frame) / truss / steel frame / concrete frame / steel deck / concrete deck **Insulation:** (Yes) No *See section 4* **Other details:** Thickness ____ mm Pitch: __45__ ° **Involves metal cladding** Yes / (no)	FAC – flat roof, asphalt (or chippings on asphalt) on pre-cast or in-situ concrete FAM – flat roof, asphalt on metal decking on steel frame FF – flat roof chippings and roofing felt on wood-wool or metal decking MCS – metal cladding system PAC – pitched roof – asbestos cement or similar profiled cladding PMC – pitched roof coated profile metal (steel or aluminium) cladding (PTS) – pitched roof, tile, slate and similar covering RR – room in roof 'Import one from library' 'Introduce own values'	Entire building / other: *Over LH end*	(E) SVS UAS CAS U	*Old, but re-placed last 20–30 yrs?*	(Visual,) measurement, planning / building regulations approval, plans, specification, cutting open (where): verbal (who): other: pitch (estimated) / measured (where): limitations: none / *Unable to confirm insulation – ceilings tight to rafters*
2	**Covering:** Asphalt / mineral felt / felt & chippings / concrete / (clay tiles) / natural / artificial slates / stone / thatch / profiled steel / iron / ACM / other: **Structure:** (Softwood frame) / truss / steel frame / concrete frame / steel deck / concrete deck **Insulation:** (Yes) No **Other details:** Thickness ____ mm Pitch: __45__ ° **Involves metal cladding** Yes / (no)	FAC / FAM / FF / MCS / PAC / PMC / PTS / RR 'Import one from library' 'Introduce own values' *See 'U' val. Calc. 0.45/12 Km* *(100mm insulation)*	*Over Rth end*	(E) SVS UAS CAS U	*Old, but re-placed last 20–30 yrs*	(Visual,) (measurement,) planning / building regulations approval, plans, specification, cutting open (where): verbal (who): other: pitch estimated / (measured) (where): *In roof* (limitations: none) /
3	**Covering:** Asphalt / mineral felt / felt & chippings / concrete / clay tiles / natural / artificial slates / stone / thatch / profiled steel / iron / ACM / other: **Structure:** Softwood frame / truss / steel frame / concrete frame / steel deck / concrete deck **Insulation:** Yes / No **Other details:** Thickness ____ mm Pitch: ____ ° **Involves metal cladding** Yes / no	FAC / FAM / FF / MCS / PAC / PMC / PTS / RR 'Import one from library' 'Introduce own values'		E SVS UAS CAS U		Visual, measurement, planning / building regulations approval, plans, specification, cutting open (where): verbal (who): other: pitch estimated / measured (where): limitations: none /

Copyright Larry Russen June 2009

Commercial Energy Assessor's Handbook

RT2.D0671   COMMERCIAL EPC SITE / REFLECTION NOTES   File / page reference ___1___ / ____

4	**Covering:** Asphalt / mineral felt / felt & chippings / concrete / clay tiles / natural / artificial slates / stone / thatch / profiled steel / iron / ACM / other: **Structure:** Softwood frame / truss / steel frame / concrete frame / steel deck / concrete deck **Insulation:** Yes / No **Other details:** Thickness _____mm Pitch: _____° Involves metal Yes / no cladding	FAC / FAM / FF / MCS / PAC / PMC / PTS / RR 'Import one from library' 'Introduce own values'	E SVS UAS CAS U		Visual, measurement, planning / building regulations approval, plans, specification, cutting open (where): verbal (who): other: pitch estimated / measured (where): limitations: none /
5	**Covering:** Asphalt / mineral felt / felt & chippings / concrete / clay tiles / natural / artificial slates / stone / thatch / profiled steel / iron / ACM / other: **Structure:** Softwood frame / truss / steel frame / concrete frame / steel deck / concrete deck **Insulation:** Yes / No **Other details:** Thickness _____mm Pitch: _____° Involves metal Yes / no cladding	FAC / FAM / FF / MCS / PAC / PMC / PTS / RR 'Import one from library' 'Introduce own values'	E SVS UAS CAS U		Visual, measurement, planning / building regulations approval, plans, specification, cutting open (where): verbal (who): other: pitch estimated / measured (where): limitations: none /
6	**Covering:** Asphalt / mineral felt / felt & chippings / concrete / clay tiles / natural / artificial slates / stone / thatch / profiled steel / iron / ACM / other: **Structure:** Softwood frame / truss / steel frame / concrete frame / steel deck / concrete deck **Insulation:** Yes / No **Other details:** Thickness _____mm Pitch: _____° Involves metal Yes / no cladding	FAC / FAM / FF / MCS / PAC / PMC / PTS / RR 'Import one from library' 'Introduce own values'	E SVS UAS CAS U		Visual, measurement, planning / building regulations approval, plans, specification, cutting open (where): verbal (who): other: pitch estimated / measured (where): limitations: none /
7	**Covering:** Asphalt / mineral felt / felt & chippings / concrete / clay tiles / natural / artificial slates / stone / thatch / profiled steel / iron / ACM / other: **Structure:** Softwood frame / truss / steel frame / concrete frame / steel deck / concrete deck **Insulation:** Yes / No **Other details:** Thickness _____mm Pitch: _____° Involves metal Yes / no cladding	FAC / FAM / FF / MCS / PAC / PMC / PTS / RR 'Import one from library' 'Introduce own values'	E SVS UAS CAS U		Visual, measurement, planning / building regulations approval, plans, specification, cutting open (where): verbal (who): other: pitch estimated / measured (where): limitations: none /

Continue on other sheet: Yes / **No**

Copyright Larry Russen June 2009

Appendix B

RT2.D0671   COMMERCIAL EPC SITE / REFLECTION NOTES   File / page reference _1954_ / _11_

4 – Roof spaces

No.	Roof space	Construction	Insulation thickness / type	RR - relevant condition issues
(1)	Access: Yes / **No** Where: ――― H&S: ACM / fibre / ~~poor access~~ / ~~flimsy joists~~ / Other:	Structure: **Softwood frame** / truss / steel frame / concrete *(Visible in restaurant)* Sarking: Bitumen felt / breathable / None *Unsure*	Thickness: None / 25 / 50 / 75 / 100 / 150 / 200 / 250 / 300 / other *No access* Type: Mineral wool / fibreglass / rigid board	Problem: ~~Water~~ / ~~Condensation~~ / ~~Possibility of condensation~~ / ~~Rot~~ / ~~Wood-boring insect~~ / ~~Rust~~ *Over LH end-lobby, restaurant + LH kitchen + prep. room*
(2)	Access: **Yes** / No Where: *E. kitchen* H&S: ACM / **fibre** / poor access / flimsy joists / Other:	Structure: **Softwood frame** / truss / steel frame / concrete *50 x 100 @ 400 joists* Sarking: **Bitumen felt** / breathable / None	Thickness: None / 25 / 50 / 75 / **100** / 150 / 200 / 250 / 300 / other Type: Mineral wool / **fibreglass** / rigid board *Between joists*	Problem: Water / Condensation / ~~Possibility of condensation~~ / **Rot** / Wood-boring insect / ~~Rust~~ *Gaps in felt and daylight @ eaves. Over Rth kitchen*
3	Access: Yes / No Where: H&S: ACM / fibre / poor access / flimsy joists / Other:	Structure: Softwood frame / truss / steel frame / concrete Sarking: Bitumen felt / breathable / None	Thickness: None / 25 / 50 / 75 / 100 / 150 / 200 / 250 / 300 / other Type: Mineral wool / fibreglass / rigid board	Problem: Water / Condensation / Possibility of condensation / Rot / Wood-boring insect / Rust
4	Access: Yes / No Where: H&S: ACM / fibre / poor access / flimsy joists / Other:	Structure: Softwood frame / truss / steel frame / concrete Sarking: Bitumen felt / breathable / None	Thickness: None / 25 / 50 / 75 / 100 / 150 / 200 / 250 / 300 / other Type: Mineral wool / fibreglass / rigid board	Problem: Water / Condensation / Possibility of condensation / Rot / Wood-boring insect / Rust
5	Access: Yes / No Where: H&S: ACM / fibre / poor access / flimsy joists / Other:	Structure: Softwood frame / truss / steel frame / concrete Sarking: Bitumen felt / breathable / None	Thickness: None / 25 / 50 / 75 / 100 / 150 / 200 / 250 / 300 / other Type: Mineral wool / fibreglass / rigid board	Problem: Water / Condensation / Possibility of condensation / Rot / Wood-boring insect / Rust

Continue on other sheet: Yes / **No**

Copyright Larry Russen June 2009

Commercial Energy Assessor's Handbook

RT2.D0671   COMMERCIAL EPC SITE / REFLECTION NOTES   File / page reference __1954__ / __12__

5 – Schedule of floor construction types

#	Construction type, insulation, thickness (mm)	SBEM inference procedure construction, or other	Where?	Adjoining condition	Age	Evidence for identification (with full details recorded), source, position seen, limitations and other notes
1	**Ground-bearing concrete** / 'beam and block' suspended / suspended timber / floor void / other Thickness ____ mm *Unlikely to be any insulation*	**SoF - solid floor** SusF – suspended floor 'Import one from library' ~~ 'Introduce own values' ~~	**Entire building** / **ground** / upper (which) **floor** / other: ~~	E SVS UAS CAS **U**	*Old in base-ment 20–30 yrs in lower GF*	**Visual**, measurement, planning / building regulations approval, plans, specification, cutting open (where): ~~ verbal (who): ~~ other: ~~ limitations: **none** /
2	Ground-bearing concrete / 'beam and block' suspended / **suspended timber** / floor void / other Thickness __240__ mm *No insulation*	SoF / **SusF** 'Import one from library' ~~ 'Introduce own values' ~~	*Upper floor Rth side*	E SVS UAS **CAS** U	*Old*	**Visual**, **measurement**, planning / building regulations approval, plans, specification, cutting open (where): ~~ verbal (who): ~~ other: ~~ limitations: **none** /
3	Ground-bearing concrete / 'beam and block' suspended / **suspended timber** / floor void / other Thickness __200__ mm *No insulation*	SoF / **SusF** 'Import one from library' ~~ 'Introduce own values' ~~	*Upper floor above base-ment*	E SVS **UAS** CAS U	*Old*	**Visual**, **measurement**, planning / building regulations approval, plans, specification, cutting open (where): ~~ verbal (who): ~~ other: ~~ limitations: **none** /
4	Ground-bearing concrete / 'beam and block' suspended / suspended timber / floor void / other Thickness ____ mm	SoF / SusF 'Import one from library' 'Introduce own values'		E SVS UAS CAS U		Visual, measurement, planning / building regulations approval, plans, specification, cutting open (where): verbal (who): other: limitations: none /

Continue on other sheet: Yes / **No**

Copyright Larry Russen June 2009

Appendix B

RT2.D0671   COMMERCIAL EPC SITE / REFLECTION NOTES   File / page reference __1954_/_13__

6 – Schedule of doors

#	Frame construction	Insulation	Age (years)	Width mm	Height mm	Size m²	In envelope?	Type of door	Evidence and notes
1.	A – aluminium H - hardwood M – metal N – no frame P – PVC **Sw** – softwood St – steel *2 x (500 x 650)* *single glazing*	Yes / **no** Details:	30	1550	1850	2.87	A/south	**P** – personnel V – vehicle access H – high usage entrance doors	**Visual** / BS5713 / BSEN1279 / Date stamp____ / planning or building regulations approval / plans / specification Other: limitations: **none** /
2.	A / H / M / N / P **Sw** / St *200 x 200 glazing*	Yes / **no** Details:	30	1100	1850	2.04	A/south	**P** / V / H	*Visual* limitations: **none** /
3.	A / H / M / N / P **Sw** / St	Yes / **no** Details:	30	1800	2000	3.60	K/south	**P** / V / H	*Visual* limitations: **none** /
4.	A / H / M / N / P **Sw** / St	Yes / **no** Details:	30	960	1600	1.54	H/south	**P** / V / H	*Visual* limitations: **none** /
5.	A / H / M / N / P Sw / St	Yes / no Details:						P / V / H	limitations: none /
6.	A / H / M / N / P Sw / St	Yes / no Details:						P / V / H	limitations: none /
7.	A / H / M / N / P Sw / St	Yes / no Details:						P / V / H	limitations: none /
8.	A / H / M / N / P Sw / St	Yes / no Details:						P / V / H	limitations: none /
9.	A / H / M / N / P Sw / St	Yes / no Details:						P / V / H	limitations: none /
10.	A / H / M / N / P Sw / St	Yes / no Details:						P / V / H	limitations: none /
11.	A / H / M / N / P Sw / St	Yes / no Details:						P / V / H	limitations: none /

Copyright Larry Russen June 2009

Commercial Energy Assessor's Handbook

RT2.D0671   COMMERCIAL EPC SITE / REFLECTION NOTES   File / page reference __1954__ / __14__

12.	A / H / M / N / P Sw / St	Yes / no Details:						P / V / H	limitations: none /
13.	A / H / M / N / P Sw / St	Yes / no Details:						P / V / H	limitations: none /
14.	A / H / M / N / P Sw / St	Yes / no Details:						P / V / H	limitations: none /
15.	A / H / M / N / P Sw / St	Yes / no Details:						P / V / H	limitations: none /
16.	A / H / M / N / P Sw / St	Yes / no Details:						P / V / H	limitations: none /

Continue on other sheet: Yes / (No)

OVERHANGS TO WINDOWS
Overhangs to upper windows on south and north elevations

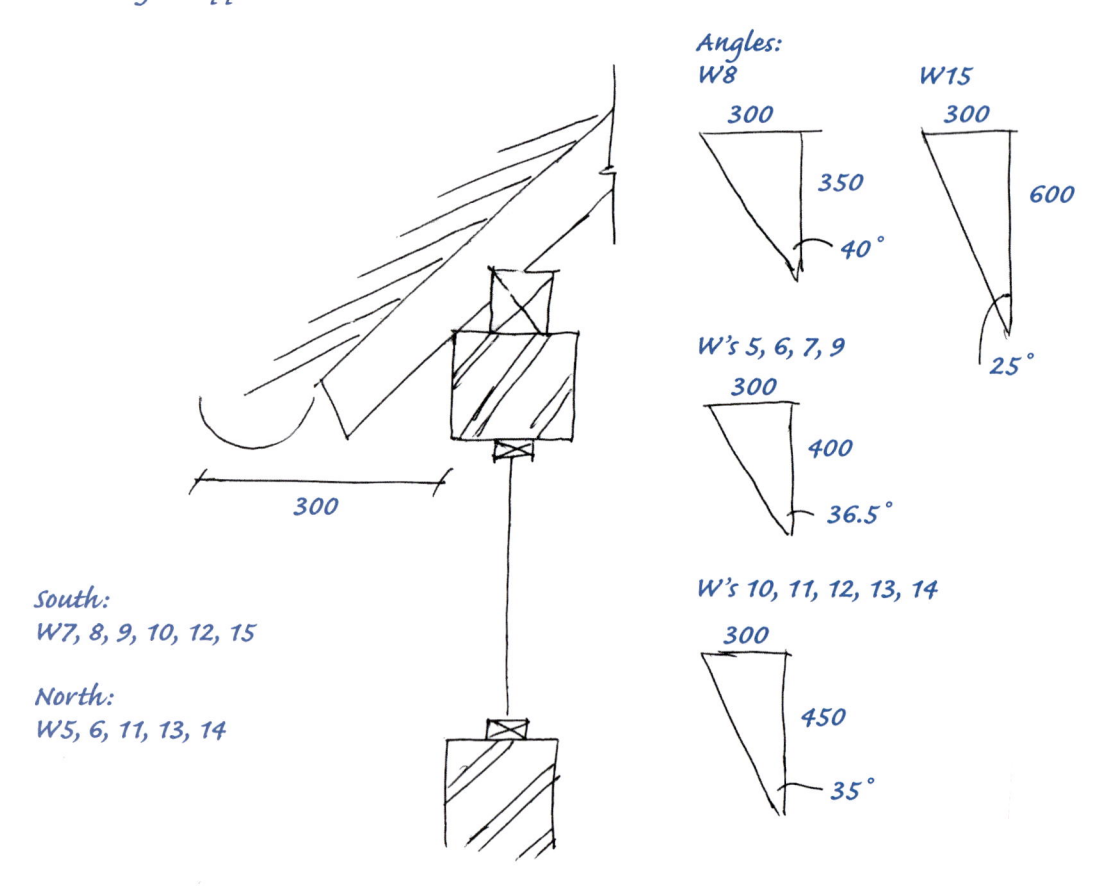

Angles:
W8   300, 350, 40°
W15   300, 600, 25°
W's 5, 6, 7, 9   300, 400, 36.5°
W's 10, 11, 12, 14   300, 450, 35°

300

South:
W7, 8, 9, 10, 12, 15

North:
W5, 6, 11, 13, 14

Adopt 55° N latitude and correction factors in BRE User Guide

Copyright Larry Russen June 2009

Appendix B

RT2.D0671   COMMERCIAL EPC SITE / REFLECTION NOTES   File / page reference ___1954_/_15_

7 – Schedule of windows, glazing and shading

	Frame construction	Glazing / coating	Fin / overhang / shading	Shading system	Width / Height mm	Size m²	Age yrs	In envelope?	Evidence and notes
1.	A – aluminium H – hardwood M – metal N – no frame P – PVC (Sw) – softwood St – steel	D – double O – other / special Qa – quadruple R – roof-light (S) – single T – triple DY – display L – reflectance, low-emissivity T – tinted (U) – uncoated, clear	Fin / Overh. 0° / 30° / 45° / 60°	EU – Ext solar protection (user moveable) EA – Ext solar protection (automatic control) O – Other (N) – None	900 / 1300	1.17	100	A/west	Visual / BS5713 / BSEN1279 / Date stamp ___ / planning or building regulations approval / plans / specification Other: limitations: (none) /
2.	A / H / M / N / P /(Sw)/ St	D /(O)/ Qa / R /(S)/ T / DY L /(T)/ U	Fin / Overh. 0° / 30° / 45° / 60°	EU / EA / (O)/ N	900 / 1300	1.17	100	A/west	Visual limitations: (none) /
3.	A / H / M / N / P /(Sw)/ St	D /(O)/ Qa / R /(S)/ T / DY L /(T)/ U	Fin / Overh. 0° / 30° / 45° / 60°	EU / EA / (O)/ N	900 / 1300	1.17	100	A/west	Visual limitations: (none) /
4.	A / H / M / N / P /(Sw)/ St	D /(O)/ Qa / R /(S)/ T / DY L /(T)/ U	Fin / Overh. 0° / 30° / 45° / 60°	EU / EA / (O)/ N	900 / 1300	1.17	100	A/west	Visual limitations: (none) /
5.	A / H / M / N / P /(Sw)/ St	D /(O)/ Qa / R /(S)/ T / DY L /(T)/ U	Fin /(Overh.) 0° / 30° / 45° / 60° 0.91	EU / EA / (O)/ N	1100 / 800	0.88	100	A/north	Visual limitations: (none) /
6.	A / H / M / N / P /(Sw)/ St	D /(O)/ Qa / R /(S)/ T / DY L /(T)/ U	Fin /(Overh.) 0° / 30° / 45° / 60° 0.91	EU / EA / (O)/ N	1100 / 800	0.88	100	A/north	Visual limitations: (none) /
7.	A / H / M / N / P /(Sw)/ St	D /(O)/ Qa / R /(S)/ T / DY L /(T)/ U	Fin /(Overh.) 0° / 30° / 45° / 60° 0.93	EU / EA / (O)/ N	1800 / 800	1.44	100	A/south	Visual limitations: (none) /
8.	A / H / M / N / P /(Sw)/ St	D /(O)/ Qa / R /(S)/ T / DY L /(T)/ U	Fin /(Overh.) 0° / 30° / 45° / 60° 0.80	EU / EA / (O)/ N	900 / 700	0.63	100	A/south	Visual limitations: (none) /
9.	A / H / M / N / P /(Sw)/ St	D /(O)/ Qa / R /(S)/ T / DY L /(T)/ U	Fin /(Overh.) 0° / 30° / 45° / 60° 0.93	EU / EA / (O)/ N	1100 / 800	0.88	100	A/south	Visual limitations: (none) /

Copyright Larry Russen June 2009

RT2.D0671   COMMERCIAL EPC SITE / REFLECTION NOTES   File / page reference _1954_ / _16_

#									
(10.)	A / H / M / N / P /(Sw)/ St	D /(O)/ Qa / R /(S)/ T / DY L / T / U	Fin /(Overh.) 0° / 30° /(45°) / 60° 0.93	EU / EA /(O)/ N	900 / 900	0.81	100	B/south	Visual limitations:(none)/
(11.)	A / H / M / N / P /(Sw)/ St	D /(O)/ Qa / R /(S)/ T / DY L / T / U	Fin /(Overh.) 0° /(30°)/ 45° / 60° 0.91	EU / EA /(O)/ N	900 / 800	0.72	100	C/north	Visual limitations:(none)/
(12.)	A / H / M / N / P /(Sw)/ St	D /(O)/ Qa / R /(S)/ T / DY L /(T)/ U	Fin /(Overh.) 0° / 30° /(45°) / 60° 0.93	EU / EA /(O)/ N	900 / 900	0.81	100	F/south	Visual limitations:(none)/
(13.)	A / H / M / N / P /(Sw)/ St	D /(O)/ Qa / R /(S)/ T / DY L /(T)/ U	Fin /(Overh.) 0° /(30°)/ 45° / 60° 0.91	EU / EA /(O)/ N	900 / 900	0.81	100	E/north	Visual limitations:(none)/
(14.)	A / H / M / N / P /(Sw)/ St	D /(O)/ Qa / R /(S)/ T / DY L /(T)/ U	Fin /(Overh.) 0° /(30°)/ 45° / 60° 0.91	EU / EA /(O)/ N	900 / 900	0.81	100	D/north	Visual limitations:(none)/
(15.)	A / H / M / N / P /(Sw)/ St	D /(O)/ Qa / R /(S)/ T / DY L /(T)/ U	Fin /(Overh.) 0° / 30° /(45°) / 60° 0.93	EU / EA /(O)/ N	900 / 1200	2.28	100	G/south	Visual limitations:(none)/
(16.)	A / H / M / N / P /(Sw)/ St	D /(O)/ Qa / R /(S)/ T / DY L /(T)/ U	Fin / Overh. 0° / 30° / 45° / 60°	EU / EA /(O)/ N	800 / 1100	0.88	100	K/south	Visual limitations:(none)/
(17.)	A / H / M / N / P /(Sw)/ St	D /(O)/ Qa / R /(S)/ T / DY L /(T)/ U	Fin / Overh. 0° / 30° / 45° / 60°	EU / EA /(O)/ N	1400 / 575	0.81	100	H/south	Visual limitations:(none)/
18.	A / H / M / N / P / Sw / St	D / O / Qa / R / S / T / DY L / T / U	Fin / Overh. 0° / 30° / 45° / 60°	EU / EA / O / N					limitations: none /
19.	A / H / M / N / P / Sw / St	D / O / Qa / R / S / T / DY L / T / U	Fin / Overh. 0° / 30° / 45° / 60°	EU / EA / O / N					limitations: none /
20.	A / H / M / N / P / Sw / St	D / O / Qa / R / S / T / DY L / T / U	Fin / Overh. 0° / 30° / 45° / 60°	EU / EA / O / N					limitations: none /
21.	A / H / M / N / P / Sw / St	D / O / Qa / R / S / T / DY L / T / U	Fin / Overh. 0° / 30° / 45° / 60°	EU / EA / O / N					limitations: none /
22.	A / H / M / N / P / Sw / St	D / O / Qa / R / S / T / DY L / T / U	Fin / Overh. 0° / 30° / 45° / 60°	EU / EA / O / N					limitations: none /

Continue on other sheet: (Yes)/ No

See pp 1954/14 for details of window overhang calculations

Copyright Larry Russen June 2009

Appendix B

RT2.D0671   COMMERCIAL EPC SITE / REFLECTION NOTES   File / page reference ___1954___ / __17__

8 – Geometrical and construction detail for the whole project

Thermal transmittance ('U' value) calculations

Ref	Detail	Calculation	Notes
1.	Roof 2	'U' value calculator / manual - 0.433 say 0.45. See attached	
2.		'U' value calculator / manual -	
3.		'U' value calculator / manual -	
4.		'U' value calculator / manual -	

Thermal capacity (Km) calculations

Ref	Detail	Calculation	Notes
1.	Roof 2	Adopt same value as roof 1, i.e. 12 – seems reasonable	
2.			
3.			

Global thermal bridges

Type of junction	Involves metal	Calculation	Accredited detail seen?
Roof - wall	Yes / **no**		Details:
Wall – ground floor	Yes / **no**		Details:
Wall – wall (corner)	Yes / **no**		Details:
Wall – floor (not ground floor)	Yes / **no**		Details:
Lintel above window or door	Yes / **no**		Details:
Sill below window	Yes / **no**		Details:
Jamb at window or door	Yes / **no**		Details:

Continue on other sheet: Yes / **No**

Copyright Larry Russen June 2009

RT2.D0671   COMMERCIAL EPC SITE / REFLECTION NOTES   File / page reference ___1954_ / _18_

9 – Building services detail for the whole project

Metering provision for lighting systems	Evidence	Fuel / energy supplies	Evidence
Is the lighting separately sub-metered? Yes / **no** / don't know	Visual, commissioning certificate, current test certificate, maintenance or operating manual, log book, schematic, building regulations approval, plans, specification, pre or post inspection questionnaire, cutting open (where): verbal (who): *Owner* other: limitations: **none** /	Location of electricity meters / switchgear noted on plan / drawing? **Yes** / no	Visual, commissioning certificate, current test certificate, maintenance or operating manual, log book, schematic, building regulations approval, plans, specification, pre or post inspection questionnaire, cutting open (where): verbal (who): other: limitations: **none** /
M&T with alarm for 'out of range' values? Yes / **no** / don't know	Visual, commissioning certificate, current test certificate, maintenance or operating manual, log book, schematic, building regulations approval, plans, specification, pre or post inspection questionnaire, cutting open (where): verbal (who): *Owner* other: limitations: **none** /	Location of gas meters noted on plan / drawing? **Yes** / no	Visual, commissioning certificate, current test certificate, maintenance or operating manual, log book, schematic, building regulations approval, plans, specification, pre or post inspection questionnaire, cutting open (where): verbal (who): other: limitations: **none** /
Lighting voltage reduction management Scheme (only for Scotland) Yes / no / don't know	Visual, commissioning certificate, current test certificate, maintenance or operating manual, log book, schematic, building regulations approval, plans, specification, pre or post inspection questionnaire, cutting open (where): verbal (who): other: limitations: none /	Location / source of LPG / Biogas / Oil / Coal / Anthracite / Smokeless fuel (including coke) / Duel fuel (mineral and wood) / Biomass / Waste heat - noted on plan / drawing Yes / no	Visual, commissioning certificate, current test certificate, maintenance or operating manual, log book, schematic, building regulations approval, plans, specification, pre or post inspection questionnaire, cutting open (where): verbal (who): other: limitations: none /
Electric power factor (EPF) **<0.9** / 0.9 – 0.95 / >0.95	**Visual**, commissioning certificate, current test certificate, maintenance or operating manual, log book, schematic, building regulations approval, plans, specification, pre or post inspection questionnaire, cutting open (where): verbal (who): other: limitations: none / *Yes*	District heating parameters – know CO² conversion factor of the DH network? Yes / no If yes, details: Primary energy conversion factor? Yes / no If yes, details:	Visual, commissioning certificate, current test certificate, maintenance or operating manual, log book, schematic, building regulations approval, plans, specification, pre or post inspection questionnaire, cutting open (where): verbal (who): other: limitations: none /

Calculation of Electric Power Factor
(Unable to calculate – inadequate information available)
kWh:
kVArh
Time period is same for both readings?   Yes / no
(if 'no', calculation based on meter readings not possible)

Continue on other sheet: Yes / **No**

Copyright Larry Russen June 2009

Commercial Energy Assessor's Handbook

Appendix B

RT2.D0671   COMMERCIAL EPC SITE / REFLECTION NOTES   File / page reference _1954_ / _19_

Air conditioning inspection	
Inspection / comments	**Notes**
Does the building have an air conditioning system? Yes / (no) / unknown Manufacturer's name and details:	
What is the total effective rated output of the air conditioning system? _____ kW (One or more air conditioning units in a building controlled by a single person are considered as a single air conditioning system for the purposes of the Regulations) Below 12kW ☐ Between 12 – 250 kW ☐ Greater than 250 kW ☐ Actual output unknown ☐	
Has an air conditioning inspection been commissioned for compliance with Energy Performance of Buildings regulations Yes – inspection completed ☐ Yes – inspection commissioned ☐ No inspection completed or commissioned ☐ Not relevant (*if the Total Effective Rated Output is 'below 12kW'*) ☐ Don't know ☐	
Any other information? Yes / (no)	

Copyright Larry Russen June 2009

215

Commercial Energy Assessor's Handbook

RT2.D0671   COMMERCIAL EPC SITE / REFLECTION NOTES   File / page reference __1954__ / __20__

10 – Heating and simple cooling

Service	Type (check <100KW)	Position located	Zone served	Other information, e.g. age, name, manufacturer, serial numbers etc	Evidence for identification, source, position seen, limitations and notes
Level 3 heating	**Central heating using water: radiators** (circled)	Restaurant, lobby lower GF lobby, toilets	A, B, J, K	HVAC system uses variable speed pumps Yes / **no** (no circled) HVAC system uses mixed mode operation strategy Yes / **no** (no circled) F – electric portable heater Spaces indirectly conditioned, and why: C + G – good inter-relationship + opening of doors	**Visual**, commissioning certificate, current test certificate, maintenance or operating manual, log book, schematic, building regulations approval, plans, specification, pre or post inspection questionnaire, cutting open (where): verbal (who): — other: — limitations: none / Access to some radiators poor due to furniture – could be more TRV's
	Central heating using water: convectors				
	Central heating using water: floor heating				
	Central heating using air distribution				
	Other room heaters: fanned				
	Other room heaters: un-fanned				
	Unflued radiant heater				
	Flued radiant heater				
	Multi-burner radiant heater				
	Flued forced-convection air heater				
	Unflued forced-convection air heater				
	Split or multi-split system				
	Single room cooling system				
	Other / hybrid:				
Level 4 heating	Single-duct VAV			HVAC system uses variable speed pumps Yes / no HVAC system uses mixed mode operation strategy Yes / no Spaces indirectly conditioned, and why:	Visual, commissioning certificate, current test certificate, maintenance or operating manual, log book, schematic, building regulations approval, plans, specification, pre or post inspection questionnaire, cutting open (where): verbal (who): other: limitations: none /
	Dual-duct VAV				
	Indoor packaged cabinet (VAV)				
	Fan-coil systems				
	Induction system				
	Constant volume system (fixed fresh air rate)				
	Constant volume system (variable fresh air rate)				
	Multizone (hot deck / cold deck)				
	Terminal reheat (constant volume)				
	Dual duct (constant volume)				
	Chilled ceilings or passive chilled beams and displacement ventilation				
	Active chilled beams				
	Water loop heat pump				
	Split or multi-split system				
	Single room cooling system				
	Other / hybrid:				
Heat source	**Low Temperature (or Pressure) Hot Water (LTHW) Boiler** (circled)	Basement	A, B, J, K	Ideal Concord 'C' 140 open flued gas fired floor mounted boiler GC41-399-80	**Visual**, commissioning certificate, current test certificate, maintenance or operating manual, log book, schematic, building regulations approval, plans,
	Medium Temperature (or Pressure) Hot Water (MTHW) Boiler				

Copyright Larry Russen June 2009

Appendix B

RT2.D0671 COMMERCIAL EPC SITE / REFLECTION NOTES   File / page reference 1954 / 21

	High Temperature (or Pressure) Hot Water (HTHW) Boiler		*Internet search found no information – very old, say 25 yrs. Not in SEDBUK*	specification, pre or post inspection questionnaire, cutting open (where):
	Direct or storage electric heater			verbal (who):
	Unflued radiant heater			other:
	Unitary radiant heater			limitations: (none) /
	Radiant heater			Thermostat (where):
	Air heater			*Zone C*
	Unflued gas warm air heater			
	Heat pump: air source			
	Heat pump: ground or water source			
	Room heater			
	Direct gas firing			
	District heating			
	Bi-valent / hybrid			
	Other:			
Efficiency	Enhanced Capital Allowances? Yes /(no) www.eca.gov.uk/etl			(Visual), commissioning certificate, current test certificate, maintenance or operating manual, log book, schematic, building regulations approval, plans, specification, pre or post inspection (questionnaire), cutting open (where):
	Was it installed on or after 1989? Yes /(no)			
	Effective heat generating seasonal efficiency			
	Generator radiant efficiency			
	Other: *Accept default – old boiler*			verbal (who): other: *ETL* limitations: none /
Fuel type	(Natural gas) LPG Biogas Oil Coal Biomass Grid Supplied Electricity Anthracite Waste heat Smokeless Fuel (including coke) Dual Fuel Appliances (Mineral & Wood) Other:		*I*	*Meter in basement on south wall*	(Visual), commissioning certificate, current test certificate, maintenance or operating manual, log book, schematic, building regulations approval, plans, specification, pre or post inspection questionnaire, cutting open (where): verbal (who): other: limitations: (none) /
CHP	Fuel type: Natural gas LPG Biogas Oil Coal Anthracite Smokeless Fuel (including coke) Dual Fuel Appliances (Mineral & Wood) Biomass Heat efficiency:_____ Electrical efficiency:_____ Building space heat supplied:___ % Building HWS supplied:___ %		CHPQA Quality Index _____ Tri-generation Yes / no System? Building cooling ___% Supplied? Chiller efficiency _____	Visual, commissioning certificate, current test certificate, maintenance or operating manual, log book, schematic, building regulations approval, plans, specification, pre or post inspection questionnaire, cutting open (where): verbal (who): other: limitations: none /

Continue on other sheet: Yes /(No)

Copyright Larry Russen June 2009

Commercial Energy Assessor's Handbook

RT2.D0671   COMMERCIAL EPC SITE / REFLECTION NOTES   File / page reference ___1954___ / __22__

11 – Chiller, HVAC ductwork, HVAC metering and controls

Service	Type (check <12KW)	Zone served	Position located	Other information, e.g. age, name, manufacturer, serial numbers etc	Evidence for identification, source, position seen, limitations and notes
Chiller type	Air cooled ☐ Water cooled ☐ Remote-condenser ☐ Heat pump (gas/oil) ☐ Heat pump (electric) ☐				Visual, commissioning certificate, current test certificate, maintenance or operating manual, log book, schematic, building regulations approval, plans, specification, pre or post inspection questionnaire, cutting open (where): verbal (who): other: limitations: none /
Pack chiller KW	Up to 100 KW ☐ 101 – 500 KW ☐ 501 – 750 KW ☐ 751 – 3.5 MW ☐		N/A		Visual, commissioning certificate, current test certificate, maintenance or operating manual, log book, schematic, building regulations approval, plans, specification, pre or post inspection questionnaire, cutting open (where): verbal (who): other: limitations: none /
Chiller efficiency	Enhanced Capital Allowances? Yes / no Generator seasonal energy efficiency ratio (SEER) Generator nominal energy efficiency ratio (EER) Mixed mode operation strategy ☐ Other:				Visual, commissioning certificate, current test certificate, maintenance or operating manual, log book, schematic, building regulations approval, plans, specification, pre or post inspection questionnaire, cutting open (where): verbal (who): other: limitations: none /
Ductwork and fanned warm air heaters	Has the ductwork been leakage tested? Yes / No Does it meet CEN classification? Yes / No Class B / Class A / Class worse than A Does the AHU meet CEN leakage standards? Yes / No Class L1 / Class L2 / Class L3 / Class worse than L3 Specific Fan Power:_____W/l/s Auxiliary Energy for fanned warm air heaters				Visual, commissioning certificate, current test certificate, maintenance or operating manual, log book, schematic, building regulations approval, plans, specification, pre or post inspection questionnaire, cutting open (where): verbal (who): other: limitations: none /

Copyright Larry Russen June 2009

Appendix B

RT2.D0671   COMMERCIAL EPC SITE / REFLECTION NOTES   File / page reference   1954 / 23

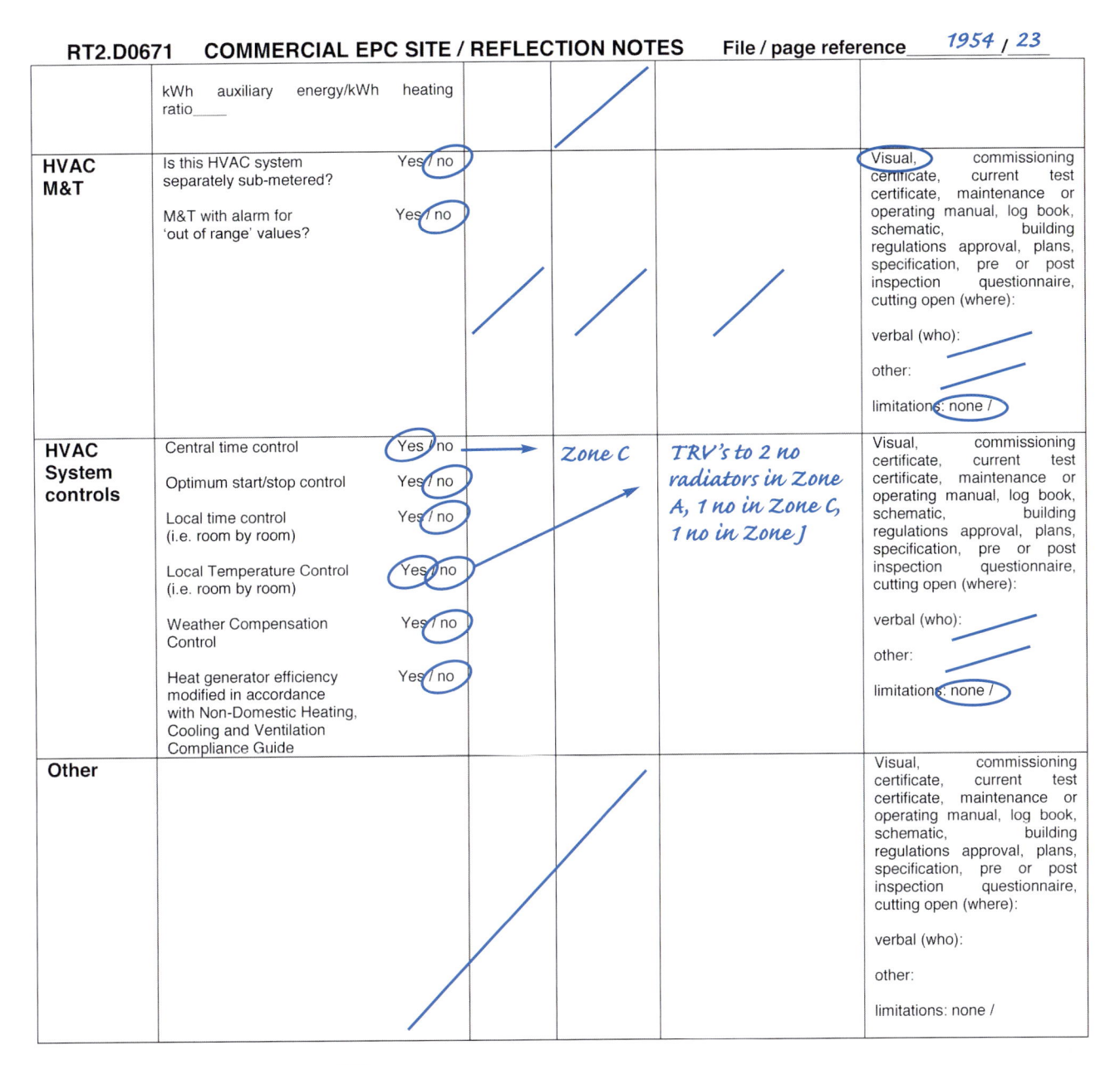

	kWh auxiliary energy/kWh heating ratio____				
HVAC M&T	Is this HVAC system separately sub-metered? M&T with alarm for 'out of range' values?	Yes / no Yes / no			Visual, commissioning certificate, current test certificate, maintenance or operating manual, log book, schematic, building regulations approval, plans, specification, pre or post inspection questionnaire, cutting open (where): verbal (who): other: limitations: none /
HVAC System controls	Central time control Optimum start/stop control Local time control (i.e. room by room) Local Temperature Control (i.e. room by room) Weather Compensation Control Heat generator efficiency modified in accordance with Non-Domestic Heating, Cooling and Ventilation Compliance Guide	Yes / no Yes / no Yes / no Yes / no Yes / no Yes / no	Zone C	TRV's to 2 no radiators in Zone A, 1 no in Zone C, 1 no in Zone J	Visual, commissioning certificate, current test certificate, maintenance or operating manual, log book, schematic, building regulations approval, plans, specification, pre or post inspection questionnaire, cutting open (where): verbal (who): other: limitations: none /
Other					Visual, commissioning certificate, current test certificate, maintenance or operating manual, log book, schematic, building regulations approval, plans, specification, pre or post inspection questionnaire, cutting open (where): verbal (who): other: limitations: none /

Continue on other sheet: Yes / No

RT2.D0671   COMMERCIAL EPC SITE / REFLECTION NOTES   File / page reference _1954_ / _24_

11 – HWS

Service	Type	Zone served	Position located	Other information, e.g. age, name, manufacturer, serial numbers etc	Evidence for identification, source, position seen, limitations and notes
HWS1 type, age and fuel	Dedicated HWS boiler / Stand-alone water heater / Instantaneous HWS only / Instantaneous combination boiler / Heat pump / **(Same as HVAC)** / Other: Is the generator later than 1989? Yes /**(no)** **(Natural gas)** / LPG / Biogas / Oil / Coal / Biomass / Grid Supplied Electricity / Waste heat / Other:				**(Visual)**, commissioning certificate, current test certificate, maintenance or operating manual, log book, schematic, building regulations approval, plans, specification, pre or post inspection questionnaire, cutting open (where): verbal (who): other: limitations: **(none)** /
HWS1 efficiency and storage	Do you know the effective heat generating seasonal efficiency Yes /**(no)** Efficiency: _____ Is the system a storage system? **(Yes)** / no Storage volume? (litres) __90__ Storage losses (MJ/month) _____ Dead-leg length measured in each appropriate zone (see plan)? **(Yes)** / no	Zone D		Deadlegs: C – 5m J – 3m	**(Visual)**, commissioning certificate, current test certificate, maintenance or operating manual, log book, schematic, building regulations approval, plans, specification, pre or post inspection questionnaire, cutting open (where): verbal (who): other: limitations: **(none)** /
HWS2 type, age and fuel	Dedicated HWS boiler / Stand-alone water heater / Instantaneous HWS only / Instantaneous combination boiler / Heat pump / Same as HVAC / Other: Is the generator later than 1989? Yes / no Natural gas / LPG / Biogas / Oil / Coal / Biomass / Grid Supplied Electricity / Waste heat / Other:				Visual, commissioning certificate, current test certificate, maintenance or operating manual, log book, schematic, building regulations approval, plans, specification, pre or post inspection questionnaire, cutting open (where): verbal (who): other: limitations: none /
HWS2 efficiency and storage	Do you know the effective heat generating seasonal efficiency Yes / no Efficiency: _____ Is the system a storage system? Yes / no Storage volume? (litres) _____ Storage losses (MJ/month) _____ Dead-leg length measured in each appropriate zone (see plan)? Yes / no				Visual, commissioning certificate, current test certificate, maintenance or operating manual, log book, schematic, building regulations approval, plans, specification, pre or post inspection questionnaire, cutting open (where): verbal (who): other: limitations: none /
HWS3 type, age and fuel	Dedicated HWS boiler / Stand-alone water heater / Instantaneous HWS only / Instantaneous combination boiler / Heat pump / Same as HVAC / Other: Is the generator later than 1989? Yes / no				Visual, commissioning certificate, current test certificate, maintenance or operating manual, log book, schematic, building regulations approval, plans, specification, pre or post inspection questionnaire, cutting open (where): verbal (who):

Copyright Larry Russen June 2009

Appendix B

RT2.D0671   COMMERCIAL EPC SITE / REFLECTION NOTES   File / page reference   1   /____

	Natural gas / LPG / Biogas / Oil / Coal / Biomass / Grid Supplied Electricity / Waste heat / Other:				other: limitations: none /
HWS3 efficiency and storage	Do you know the effective heat generating seasonal efficiency Yes / no Efficiency:_____ Is the system a storage system? Yes / no Storage volume? (litres)_____ Storage losses (MJ/month)_____ Dead-leg length measured in each appropriate zone (see plan)? Yes / no				Visual, commissioning certificate, current test certificate, maintenance or operating manual, log book, schematic, building regulations approval, plans, specification, pre or post inspection questionnaire, cutting open (where): verbal (who): other: limitations: none /
HWS4 type, age and fuel	Dedicated HWS boiler / Stand-alone water heater / Instantaneous HWS only / Instantaneous combination boiler / Heat pump / Same as HVAC / Other: Is the generator later than 1989? Yes / no Natural gas / LPG / Biogas / Oil / Coal / Biomass / Grid Supplied Electricity / Waste heat / Other:				Visual, commissioning certificate, current test certificate, maintenance or operating manual, log book, schematic, building regulations approval, plans, specification, pre or post inspection questionnaire, cutting open (where): verbal (who): other: limitations: none /
HWS4 efficiency and storage	Do you know the effective heat generating seasonal efficiency Yes / no Efficiency:_____ Is the system a storage system? Yes / no Storage volume? (litres)_____ Storage losses (MJ/month)_____ Dead-leg length measured in each appropriate zone (see plan)? Yes / no				Visual, commissioning certificate, current test certificate, maintenance or operating manual, log book, schematic, building regulations approval, plans, specification, pre or post inspection questionnaire, cutting open (where): verbal (who): other: limitations: none /
Secondary circulation system 1	Does the system have secondary circulation? Yes /(no) Circulation losses:_____ W/m Pump power:_____ KW Loop length:_____ m Time control on secondary circulation Yes / no				Visual, commissioning certificate, current test certificate, maintenance or operating manual, log book, schematic, building regulations approval, plans, specification, pre or post inspection questionnaire, cutting open (where): verbal (who): other: limitations:(none) /
Secondary circulation system 2	Does the system have secondary circulation? Yes / no Circulation losses:_____ W/m Pump power:_____ KW Loop length:_____ m Time control on secondary circulation Yes / no				Visual, commissioning certificate, current test certificate, maintenance or operating manual, log book, schematic, building regulations approval, plans, specification, pre or post inspection questionnaire, cutting open (where): verbal (who): other: limitations: none /

Continue on other sheet: Yes /(No)

Copyright Larry Russen June 2009

Commercial Energy Assessor's Handbook

RT2.D0671   COMMERCIAL EPC SITE / REFLECTION NOTES   File / page reference __1954__ / __26__

12 – RENEWABLES

Service	Type	Position located	Other information, e.g. age, name, manufacturer, serial numbers etc	Evidence for identification, source, position seen, limitations and notes
Solar thermal hot water (SES)	SES present? Yes / **no** Name (of the system) Which DHW system is it connected to? Area of the solar collector: _____ m² Orientation: N / NE / E / SE / S / SW / W / NW Inclination: 0 / 15 / 30 / 45 / 60 / 75 / 90 (90° is vertical)			**Visual,** commissioning certificate, current test certificate, maintenance or operating manual, log book, schematic, building regulations approval, plans, specification, pre or post inspection questionnaire, cutting open (where): verbal (who): other: limitations: **none** /
Solar photo-voltaics (PV)	PVS present? Yes / **no** Name of the system Type: Monocrystalline silicon ☐ Polycrystalline silicon ☐ Amorphous silicon ☐ Other thin film type:_____ ☐ Area of the solar collector: _____ m² Orientation: N / NE / E / SE / S / SW / W / NW Inclination: 0 / 15 / 30 / 45 / 60 / 75 / 90 (90° is vertical)			**Visual,** commissioning certificate, current test certificate, maintenance or operating manual, log book, schematic, building regulations approval, plans, specification, pre or post inspection questionnaire, cutting open (where): verbal (who): other: limitations: **none** /
Wind turbines	Turbine(s) present? Yes / **no** Terrain type: Smooth flat country (no obstacles) ☐ Farm land with boundary hedges ☐ Suburban or industrial area ☐ Urban with average building height less than 15m ☐ Horizontal axis ☐ Diameter of the blades:_____ m Height to the hub:_____ m Vertical axis ☐ Swept area:_____ m² Height to the geometric centre:_____ m Power output at rated wind speed:____ KW			**Visual,** commissioning certificate, current test certificate, maintenance or operating manual, log book, schematic, building regulations approval, plans, specification, pre or post inspection questionnaire, cutting open (where): verbal (who): other: limitations: **none** /
Other				

Continue on other sheet: Yes / **No**

Copyright Larry Russen June 2009

RT2.D0671   COMMERCIAL EPC SITE / REFLECTION NOTES   File / page reference __1954_ / _27_

13 – Ventilation, exhaust and heat recovery

Service	Type	Zone served	Position located	Other information, e.g. age, name, manufacturer, serial numbers etc	Evidence for identification, source, position seen, limitations and notes
De-stratification fans and air treatment	Any de-stratification fans? Yes / **no** Does the activity require high-pressure drop air treatment? Yes / **no**				**Visual**, commissioning certificate, current test certificate, maintenance or operating manual, log book, schematic, building regulations approval, plans, specification, pre or post inspection questionnaire, cutting open (where): verbal (who): other: limitations **none** /
Heat recovery	Heat recovery system Yes / **no** Type: Plate heat exchanger (recuperator) ☐ Heat pipes ☐ Thermal wheel ☐ Run around coil ☐ Heat recovery seasonal efficiency: 50 / 60 / 65%				**Visual**, commissioning certificate, current test certificate, maintenance or operating manual, log book, schematic, building regulations approval, plans, specification, pre or post inspection questionnaire, cutting open (where): verbal (who): other: limitations **none** /
Mechanical supply ventilation	Any mechanical supply vents? Yes / **no** Supply specific fan power: ____ W/l/s Demand control ventilation Yes / **no** Activity requires high pressure drop air treatment: Yes / **no** Heat recovery present Yes / **no** Heat recovery seasonal efficiency ratio				**Visual**, commissioning certificate, current test certificate, maintenance or operating manual, log book, schematic, building regulations approval, plans, specification, pre or post inspection questionnaire, cutting open (where): verbal (who): other: limitations **none** /
Mechanical exhaust	Local mechanical exhaust systems **Yes** / no Extract rate: ____ l/s/m² Exhaust specific fan power: ____ W/l/s Scope of extract system: Extract fan is remote from the zone **Yes** / no Extract fan is within the zone **Yes** / no			*Manrose + Aidelle MEV's in toilets – internet search, but old + unable to locate any information*	**Visual**, commissioning certificate, current test certificate, maintenance or operating manual, log book, schematic, building regulations approval, plans, specification, pre or post inspection questionnaire, cutting open (where): verbal (who): other: limitations: none /

Continue on other sheet: Yes / **No**

Copyright Larry Russen June 2009

Commercial Energy Assessor's Handbook

RT2.D0671   COMMERCIAL EPC SITE / REFLECTION NOTES   File / page reference ___1954_/_28_

14 – Lighting systems and controls

Service	Type	Zone served	Other information, e.g. age, name, manufacturer, serial numbers etc	Evidence for identification, source, position seen, limitations and notes
Lighting – full survey	Full lighting design carried out: Yes /~~no~~ Total wattage:_____ (W) Total design illuminance:_____ (Lux)	/	/	Visual, commissioning certificate, current test certificate, maintenance or operating manual, log book, schematic, building regulations approval, plans, specification, pre or post inspection questionnaire, cutting open (where): verbal (who): other: limitations: none /
Lighting chosen, but calculation not carried out	Lumens / circuit wattage Lumens / circuit wattage Lumens / circuit wattage Lumens / circuit wattage Lumens / circuit wattage Lumens / circuit wattage Lumens / circuit wattage	C	*Philips TLD 58W/35 CE Made in France T8/1500 with starters see below*	**Visual**, commissioning certificate, current test certificate, maintenance or operating manual, log book, schematic, building regulations approval, plans, specification, pre or post inspection questionnaire, cutting open (where): verbal (who): other: limitations: **none** /
Lamp types	Tungsten lamp Tungsten halogen Metal halide Fluorescent (no details) Compact fluorescent T8 (25mm diameter) tri-phosphor coated fluorescent tube, high frequency ballast T8 (25mm diameter) halo-phosphate coated fluorescent tube, high frequency ballast T8 (25mm diameter) halo-phosphate coated fluorescent tube, standard ballast T12 (37mm diameter) halo-phosphate coated fluorescent tube T5 (16mm diameter) tri-phosphor coated fluorescent tube, high frequency ballast High pressure sodium High pressure mercury LEDs Air extracting luminaries fitted: Yes /~~no~~	A, E, K, I A, K B, G, J, L D, F D, H	*In zones A + D chose either most dominant, or inefficient, lamp* *Internet: lamps apparently halo-phosphor, old – assume 'standard' ballast. Lumens 4600. Action Energy Lighting Guide 007 suggests 12W for ballast.*	**Visual**, commissioning certificate, current test certificate, maintenance or operating manual, log book, schematic, building regulations approval, plans, specification, pre or post inspection questionnaire, cutting open (where): verbal (who): other: limitations: none / *Unable to see any lamp details due to poor access, other than in Zone 'C'. See sketches for more detail.*

Calculation lumens/circuit Watt for zone C
= 4600 ÷ [58W (lamp) + 12W (ballast)]
= 65.71, say 65 efficacy

Copyright Larry Russen June 2009

Appendix B

RT2.D0671 COMMERCIAL EPC SITE / REFLECTION NOTES   File / page reference __1954 / 29__

Display lighting	Display lighting present Yes /(no) Does display lighting use Yes / no energy efficient lamps: Lumens per circuit wattage/unit _____ W Time switching for Yes / no display lighting Type Hours off: _____ hrs Fraction off: ____ %			(Visual), commissioning certificate, current test certificate, maintenance or operating manual, log book, schematic, building regulations approval, plans, specification, pre or post inspection questionnaire, cutting open (where): verbal (who): other: limitations:(none)/
Lighting controls	**Plan** - Switches shown on plan (Yes)/ no **Type of switching** Local manual switching (Yes)/ no Photoelectric (Yes)/ no **Photoelectric options** (Yes)/ no Switching Yes / no Dimming Yes / no Different sensor to control Yes / no the back half of the zone **Type** Stand-alone sensors Addressable system Do you know the parasitic Yes / no power of the photoelectric device Parasitic power is ___ W/m² **Occupancy sensing** (None) Manual on/off and external Auto on dimmed Auto on/off Manual on dimmed Manual-on auto-off	*E, F, G, H, J, K, L* *All others*		(Visual), commissioning certificate, current test certificate, maintenance or operating manual, log book, schematic, building regulations approval, plans, specification, pre or post inspection questionnaire, cutting open (where): verbal (who): other: limitations:(none)/
Other				Visual, commissioning certificate, current test certificate, maintenance or operating manual, log book, schematic, building regulations approval, plans, specification, pre or post inspection questionnaire, cutting open (where): verbal (who): other: limitations: none /

Continue on other sheet: Yes /(No)

Copyright Larry Russen June 2009

Commercial Energy Assessor's Handbook

RT2.D0671   COMMERCIAL EPC SITE / REFLECTION NOTES   File / page reference 1954 / 30

15 – OVERALL / ZONE PLAN

External (ext) check measurements taken (Yes) / no

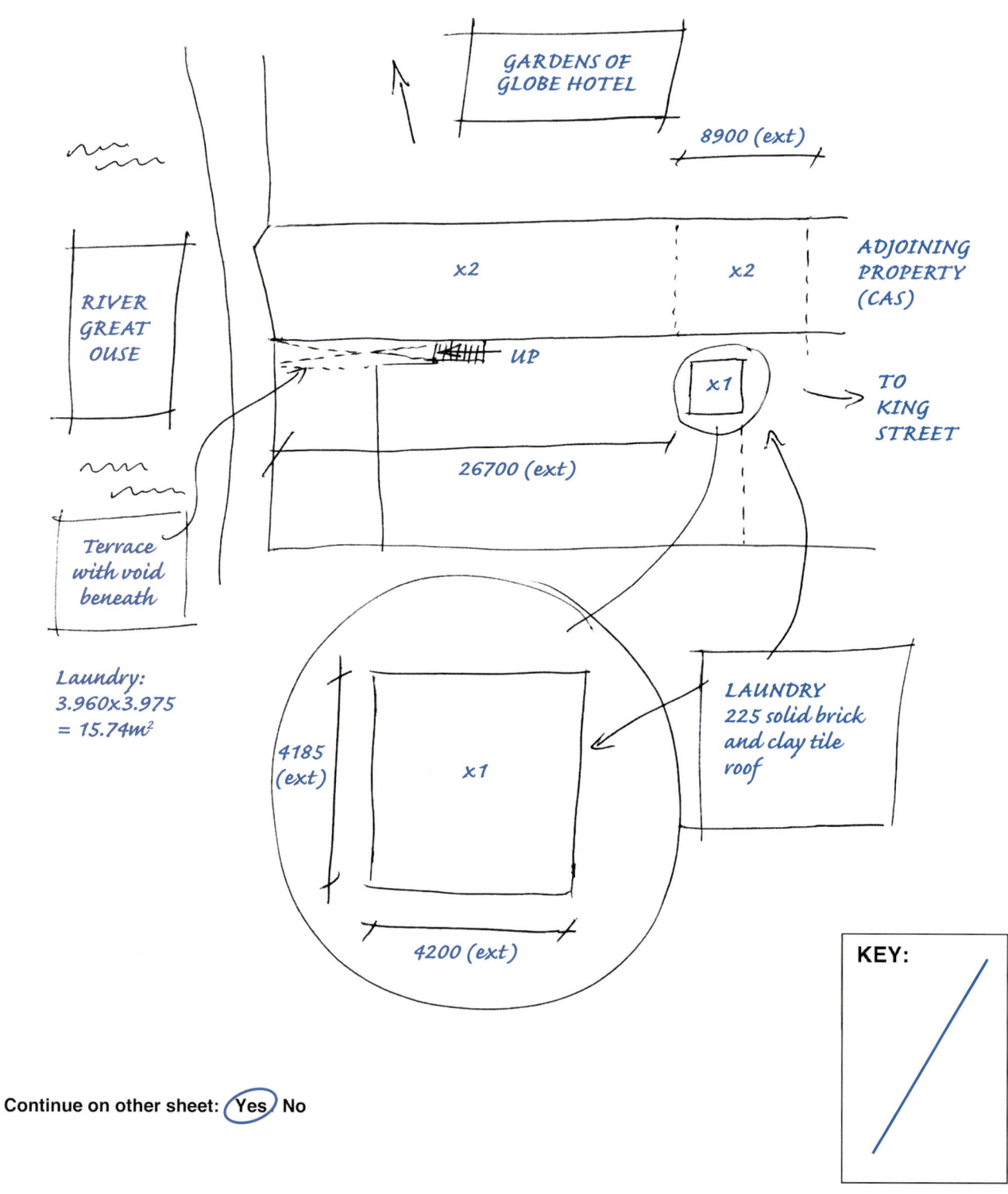

GARDENS OF GLOBE HOTEL

8900 (ext)

RIVER GREAT OUSE

x2   x2

ADJOINING PROPERTY (CAS)

UP   x1

TO KING STREET

26700 (ext)

Terrace with void beneath

Laundry: 3.960 x 3.975 = 15.74m²

4185 (ext)   x1

4200 (ext)

LAUNDRY 225 solid brick and clay tile roof

Continue on other sheet: (Yes) No

KEY:

Copyright Larry Russen June 2009

Appendix B

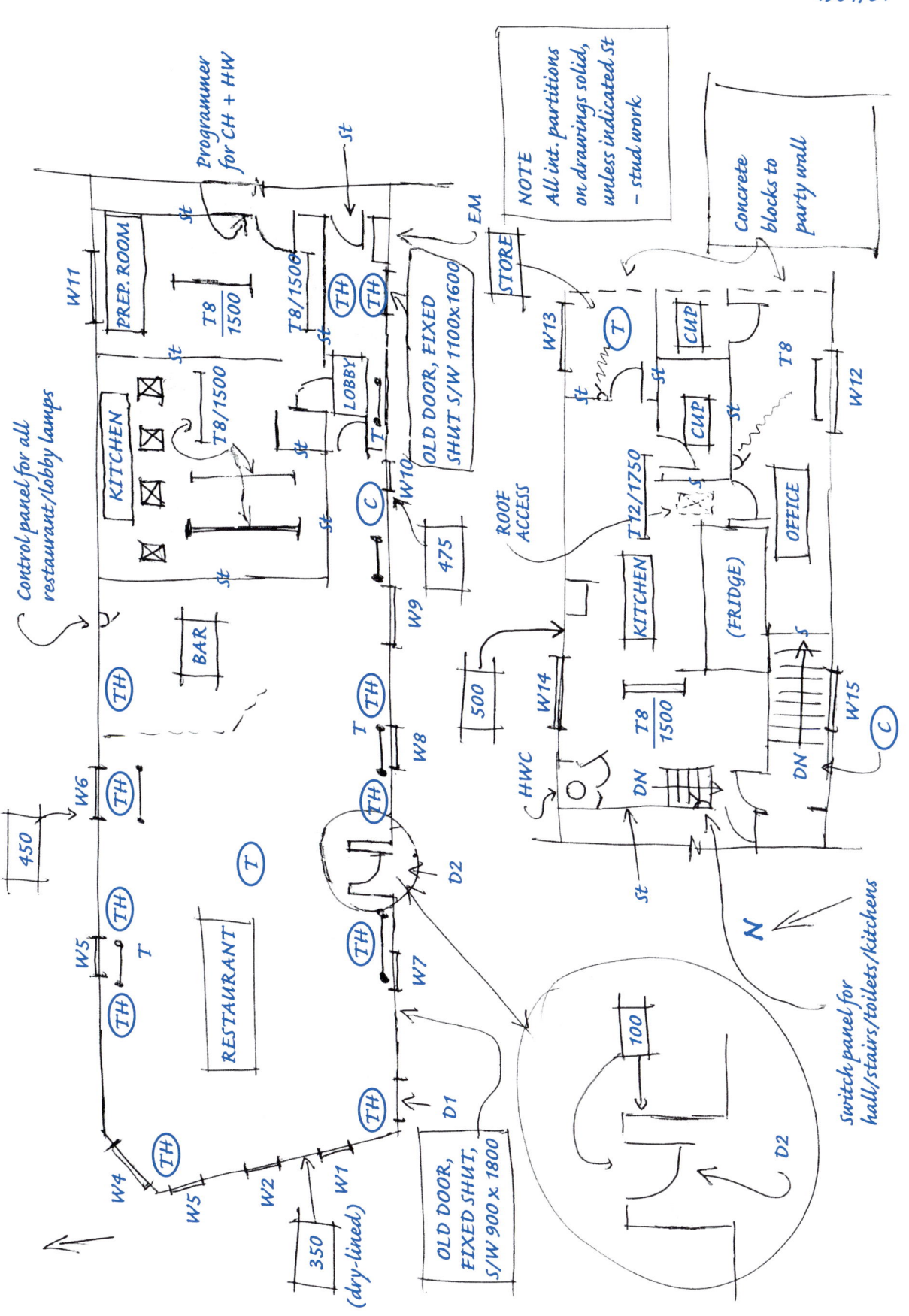

1954/31

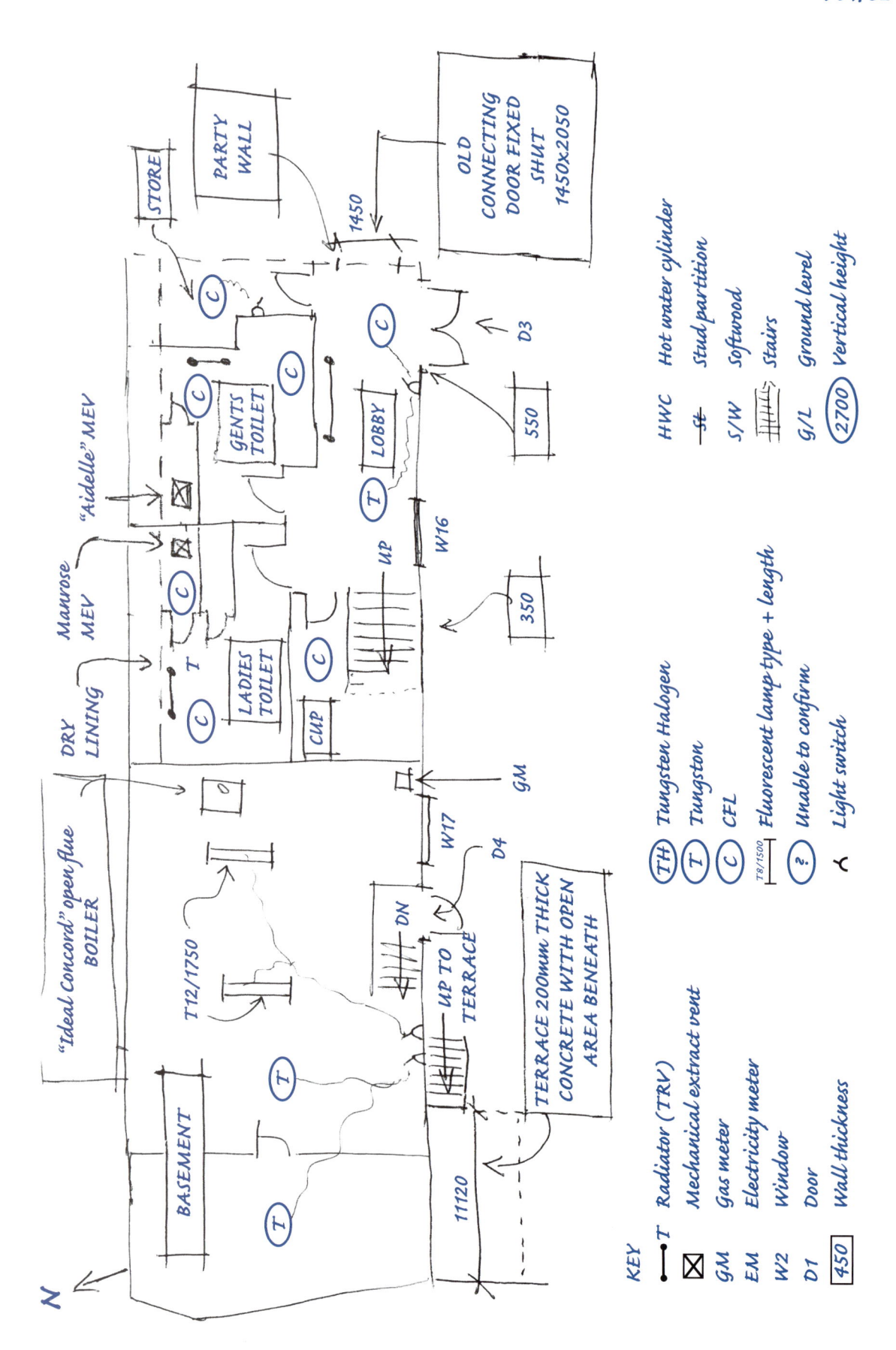

Commercial Energy Assessor's Handbook

1954/32

Appendix B

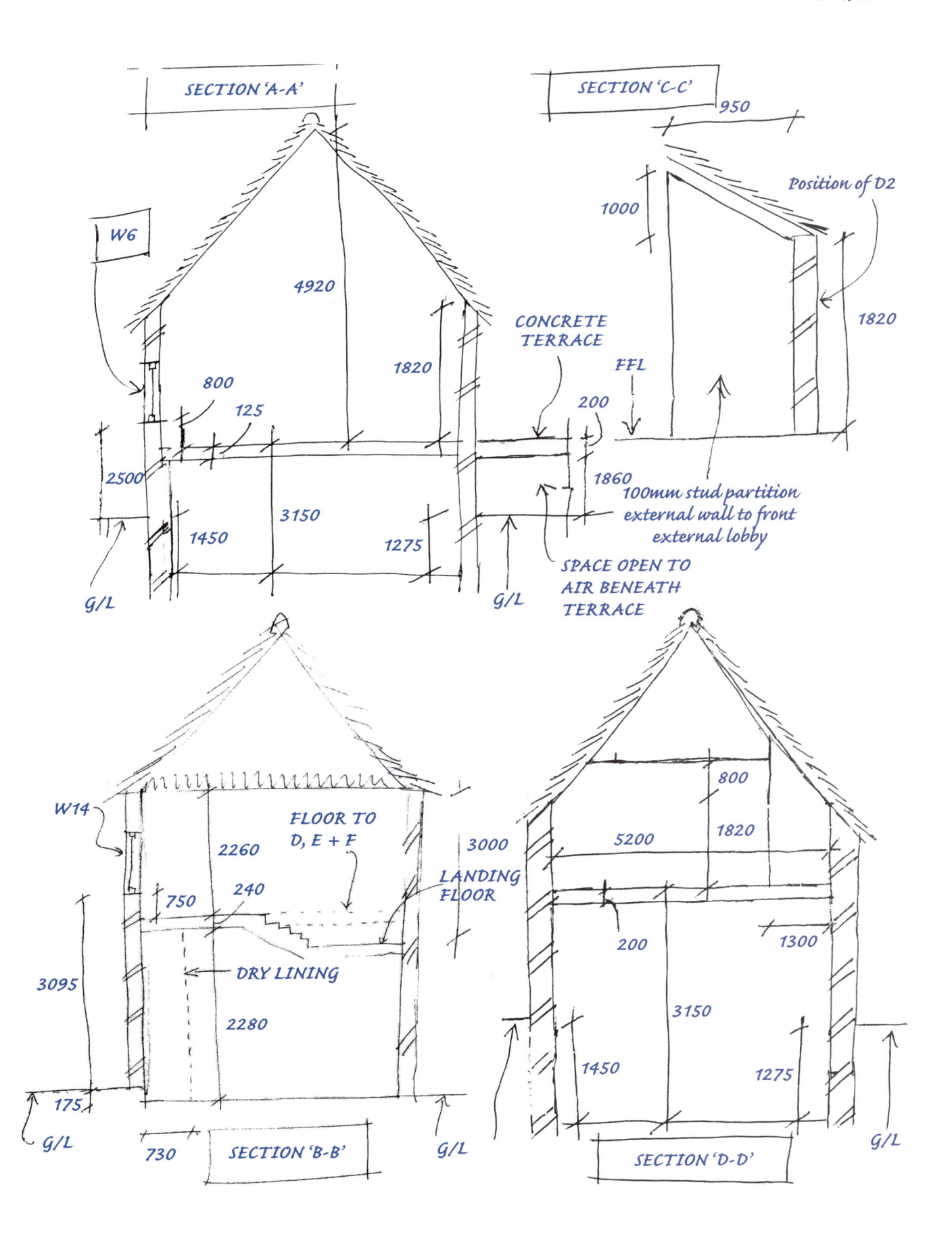

1954/33

229

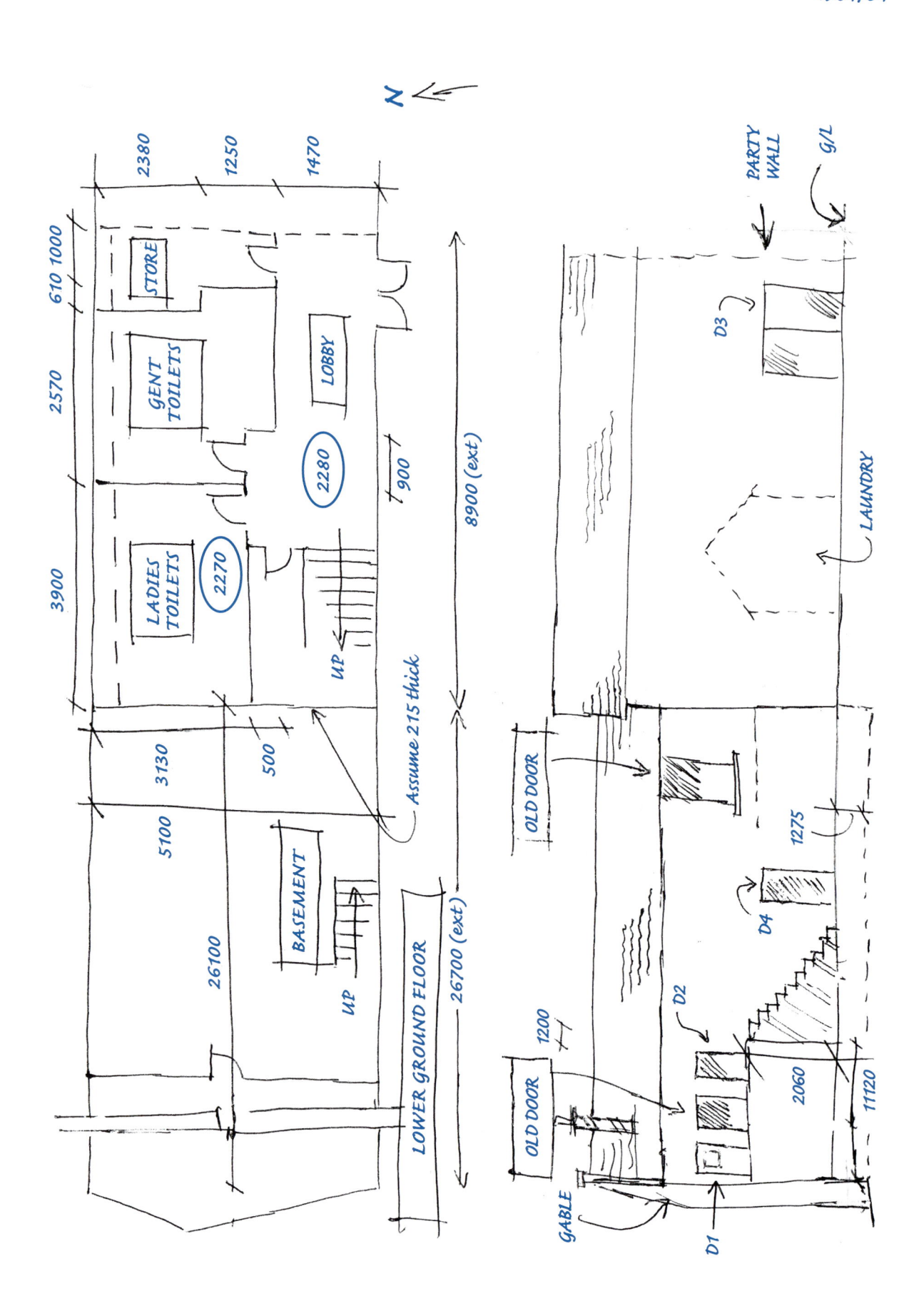

Commercial Energy Assessor's Handbook

1954/34

Appendix B

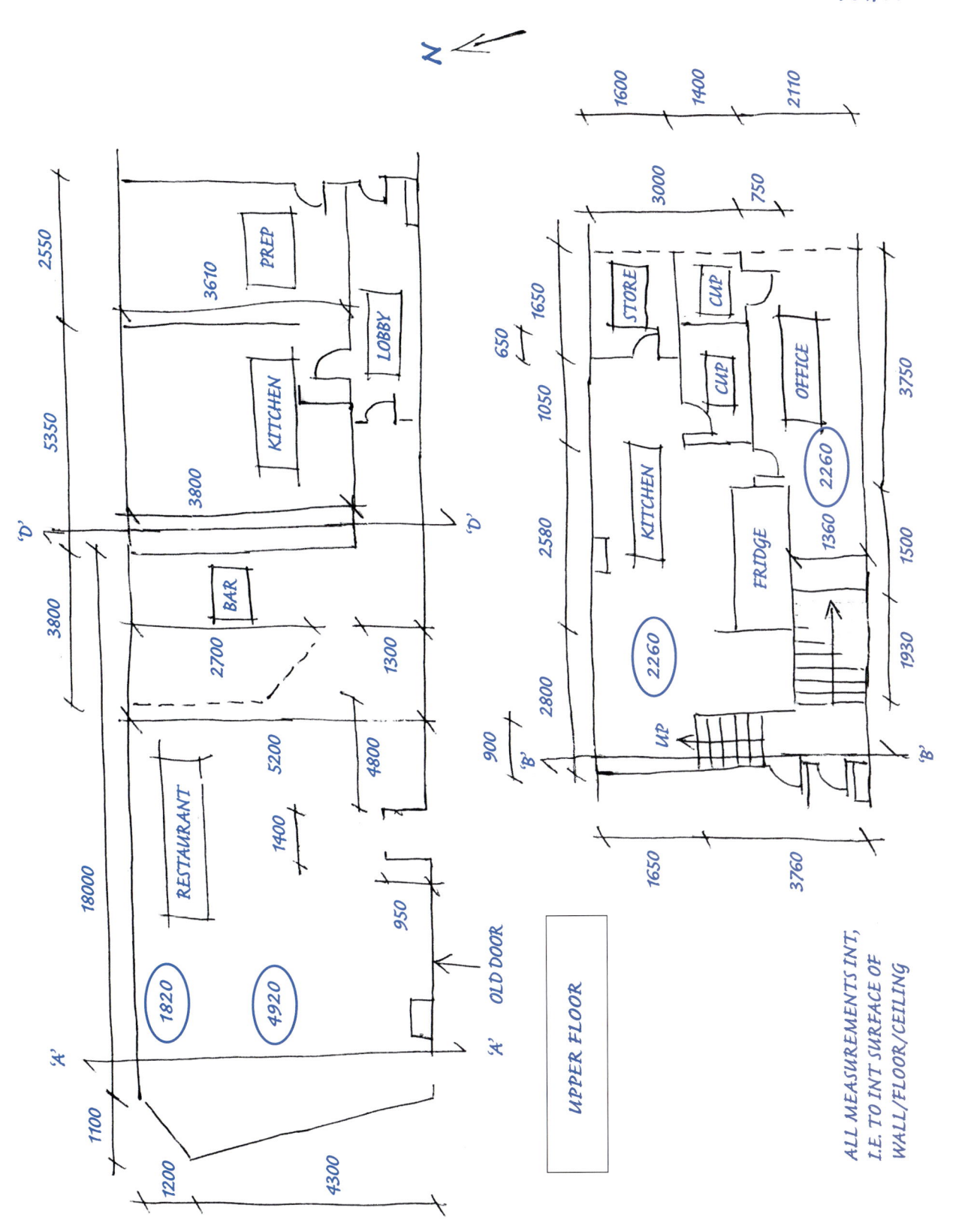

1954/35

UPPER FLOOR

ALL MEASUREMENTS INT,
I.E. TO INT SURFACE OF
WALL/FLOOR/CEILING

231

RT2.D0671   COMMERCIAL EPC SITE / REFLECTION NOTES   File / page reference_____/____

16 – Zone fabric and geometry

Zone name, activity and reference		Orient-ation	Adjoining condition type: E/SVS/UAS/CAS/U	Type (from construction schedule)	Length (m)	Height/ width (m)	Area m²	Door / Glazing ref / type %	Display Window ?
Zone height: Check: Activity HVAC Lighting system Daylight access	m ☐ ☐ ☐ ☐	S	E / SVS / UAS / CAS / U						
		SE	E / SVS / UAS / CAS / U						
		E	E / SVS / UAS / CAS / U						
		NE	E / SVS / UAS / CAS / U						
		N	E / SVS / UAS / CAS / U						
		NW	E / SVS / UAS / CAS / U						
		W	E / SVS / UAS / CAS / U						
		SW	E / SVS / UAS / CAS / U						
		Ceiling	E / SVS / UAS / CAS / U						
		Floor	E / SVS / UAS / CAS / U						
		Roof	E / SVS / UAS / CAS / U						
Zone height: Check: Activity HVAC Lighting system Daylight access	m ☐ ☐ ☐ ☐	S	E / SVS / UAS / CAS / U						
			E / SVS / UAS / CAS / U						
		E	E / SVS / UAS / CAS / U						
			E / SVS / UAS / CAS / U						
		N	E / SVS / UAS / CAS / U						
			E / SVS / UAS / CAS / U						
		W	E / SVS / UAS / CAS / U						
		Ceiling	E / SVS / UAS / CAS / U						
		Floor	E / SVS / UAS / CAS / U						
		Roof	E / SVS / UAS / CAS / U						
Zone height: Check: Activity HVAC Lighting system Daylight access	m ☐ ☐ ☐ ☐	S	E / SVS / UAS / CAS / U						
			E / SVS / UAS / CAS / U						
		E	E / SVS / UAS / CAS / U						
			E / SVS / UAS / CAS / U						
		N	E / SVS / UAS / CAS / U						
			E / SVS / UAS / CAS / U						
		W	E / SVS / UAS / CAS / U						
		Ceiling	E / SVS / UAS / CAS / U						
		Floor	E / SVS / UAS / CAS / U						
		Roof	E / SVS / UAS / CAS / U						
Zone height: Check: Activity HVAC Lighting system Daylight access	m ☐ ☐ ☐ ☐	S	E / SVS / UAS / CAS / U						
			E / SVS / UAS / CAS / U						
		E	E / SVS / UAS / CAS / U						
			E / SVS / UAS / CAS / U						
		N	E / SVS / UAS / CAS / U						
			E / SVS / UAS / CAS / U						
		W	E / SVS / UAS / CAS / U						
			E / SVS / UAS / CAS / U						
		Ceiling	E / SVS / UAS / CAS / U						
		Floor	E / SVS / UAS / CAS / U						
		Roof	E / SVS / UAS / CAS / U						
Total area									
Carry forward									

Copyright Larry Russen June 2009

At this stage of your learning, assuming you have read our book and the *BRE User Guide*, with the photographs and these notes together with plans; you should be in a position to enter data into your SBEM program and generate a result. Use the information we have given you; photocopy the blank 'Zone fabric and geometry' sheet we have provided and enter the envelope areas. Afterwards, compare your results with ours – go on, have a go, make some mistakes and start to learn how SBEM works, just like we had to! Don't be entirely surprised if your first attempt(s) at this assessment causes you some confusion, or even discouragement. We have deliberately chosen a difficult assessment as a learning exercise. It will help you develop your ability to visualise and consider a building in three dimensions.

COMPLETED ZONE PLANS, GEOMETRY DATA AND DATA REFLECTION NOTES

Figure B27 shows our plans coloured up to show the different zones, our geometry data and 'section 17 – final audit, ratings and results'. You may disagree with our zoning. That's okay; since there will always be variations in the way different CEAs approach zoning. We have omitted some of our calculation sheets for reasons of space – thus, we would normally record all of our calculations of envelope areas. In this case, there is some complicated geometry, e.g. in the restaurant, adjoining lobby and left-hand kitchen.

Commercial Energy Assessor's Handbook

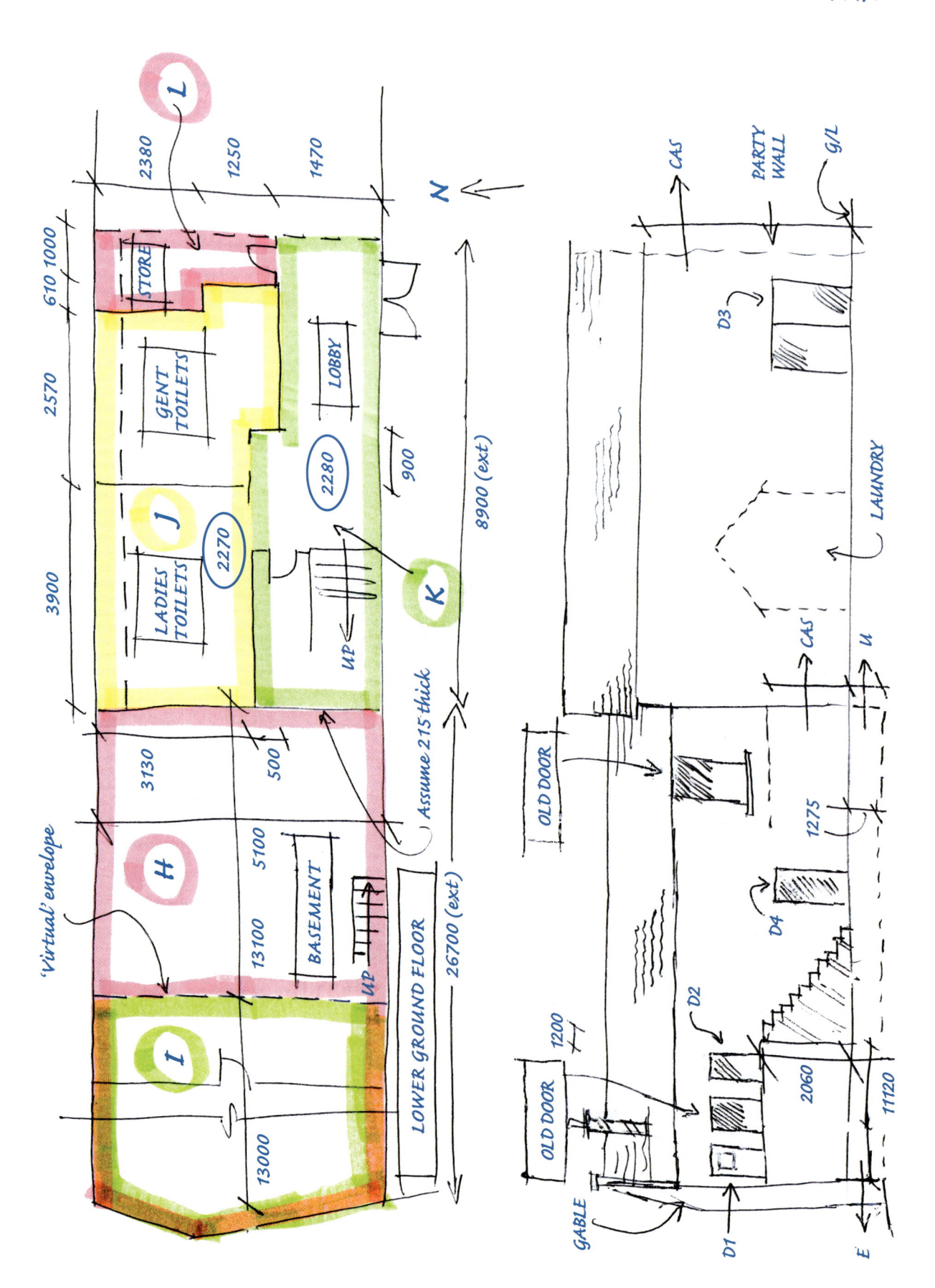

1954/36

Figure B27: Zone plans, geometry and reflection

Appendix B

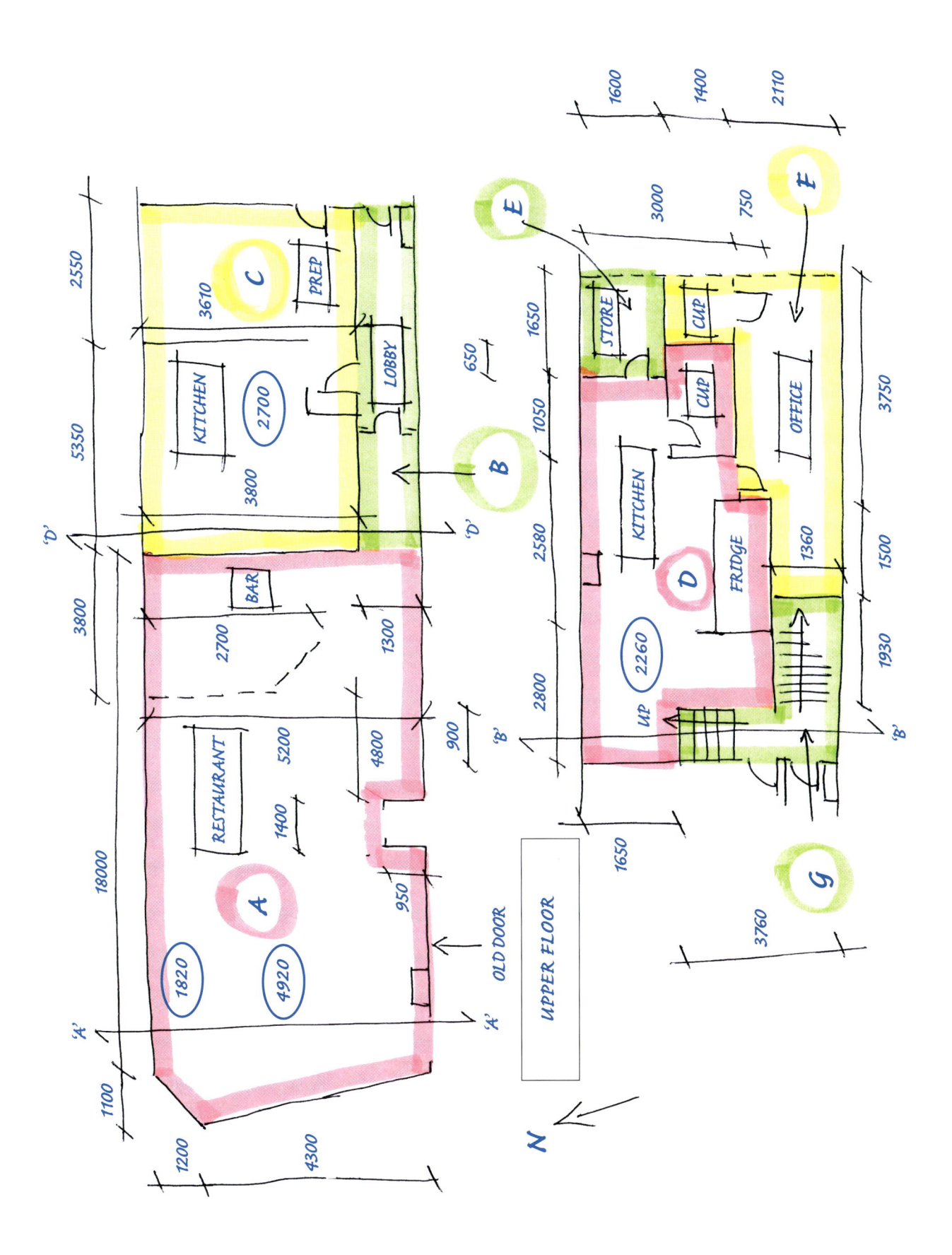

1954/37

Figure B27 (cont.)

Commercial Energy Assessor's Handbook

RT2.D0671 COMMERCIAL EPC SITE / REFLECTION NOTES File / page reference __1954__ / __38__

16 – Zone fabric and geometry

Zone name, activity and reference	Orient-ation	Adjoining condition type: E/SVS/UAS/CAS/U	Type (from construction schedule)	Length (m)	Height/width (m)	Area m² * *Excludes old door*	Door / Glazing ref / type %	Display Window?
'A' - Restaurant (Eating/drinking) Zone height: 1.82 m Check: Activity ✓ HVAC ✓ Lighting system ✓ Daylight access ✓	S	(E) SVS / UAS / CAS / U	W1	16.60	1.82	28.59*	W7-9+D1	No
	S̶E̶ S1	(E) SVS / UAS / CAS / U	W7	1.40	2.82	3.95	D2	
	E	E / SVS / UAS /(CAS)/ U	W6	–	–	9.56		
	N̶E̶ E1	(E) SVS / UAS / CAS / U	W6	–	–	3.85		
	N	(E) SVS / UAS / CAS / U	W1	18.00	1.82	32.76	W5+6	
	N̶W̶ E2	(E) SVS / UAS / CAS / U	W7	0.95	2.32	2.20		
	W	(E) SVS / UAS / CAS / U	W1	–	–	20.56	W1-4	
	S̶W̶ S2	(E) SVS / UAS / CAS / U	W2	0.90	1.80	1.62	Old door (W1)	
	Ceiling	(E) SVS / UAS / CAS / U	W7	0.95	2.32	2.20		
	Floor	E / SVS /(UAS)/ CAS / U	F3	–	–	97.10		
	Roof	(E) SVS / UAS / CAS / U	R1	–	–	147.18		
'B' – west lobby (circulation) Zone height: 2.59 m Check: Activity ✓ HVAC ✓ Lighting system ✓ Daylight access ✓	S	(E) SVS / UAS / CAS / U	W1	8.10	1.82	14.74	W10	No
		E / SVS / UAS / CAS / U						
	E	E / SVS / UAS /(CAS)/ U	W6	–	–	3.37		
		E / SVS / UAS / CAS / U						
	N	E / SVS / UAS /(CAS)/ U	W6	8.10	2.70	21.87		
	N1	(E) SVS / UAS / CAS / U	W7	8.10	0.65	5.27		
	W	E / SVS / UAS / CAS / U						
	Ceiling	E / SVS / UAS / CAS / U						
	Floor	E / SVS /(UAS)/ CAS / U	F3	1.35	8.10	10.94		
	Roof	(E) SVS / UAS / CAS / U	R1	8.10	2.05	16.61		
'C' kitchen (W) (food preparation) Zone height: 1.82 m Check: Activity ✓ HVAC ✓ Lighting system ✓ Daylight access ✓	S	E / SVS / UAS /(CAS)/ U	W6	8.10	2.70	21.87		No
		E / SVS / UAS / CAS / U						
	E	E / SVS / UAS /(CAS)/ U	W6	–	–	5.53		
	E1	E / SVS /(UAS)/ CAS / U	W6	1.70	2.26	3.84		
	N	(E) SVS / UAS / CAS / U	W1	8.10	1.82	14.74	W11	
		E / SVS / UAS / CAS / U						
	W	E / SVS / UAS /(CAS)/ U	W6	–	–	9.26		
	Ceiling	E / SVS / UAS / CAS / U						
	Floor	E / SVS /(UAS)/ CAS / U	F3	8.10	3.85	31.19		
	Roof	(E) SVS / UAS / CAS / U	R1	8.10	4.30	34.83		
'D' kitchen (E) (food preparation) Zone height: 2.26 m Check: Activity ✓ HVAC ✓ Lighting system ✓ Daylight access ✓	S	E / SVS / UAS /(CAS)/ U	W6	6.28	2.26	14.19		No
		E / SVS / UAS / CAS / U						
	E	E / SVS / UAS /(CAS)/ U	W6	2.15	2.26	4.86		
	E1	E / SVS /(UAS)/ CAS / U	W6	1.65	2.26	3.73		
	N	(E) SVS / UAS / CAS / U	W1	6.53	2.26	14.76	W14	
	N1	E / SVS /(UAS)/ CAS / U	W6	0.70	2.26	1.58		
	W	E / SVS / UAS / CAS / U						
		E / SVS / UAS / CAS / U						
	Ceiling	E / SVS / UAS / CAS / U						
	Floor	E / SVS / UAS /(CAS)/ U	F2	–	–	23.53		
	Roof	(E) SVS / UAS / CAS / U	R2	–	–	23.53		
Total area						162.76		
Carry forward						162.76		

(– in dimensions indicates calculated elsewhere on attached sheets)

Copyright Larry Russen June 2009

Appendix B

RT2.D0671   COMMERCIAL EPC SITE / REFLECTION NOTES   File / page reference ___1954___ / ___39___

Zone name, activity and reference	Orient-ation	Adjoining condition type: E/SVS/UAS/CAS/U	Type (from construction schedule)	Length (m)	Height/width (m)	Area m²	Door / Glazing ref / type %	Display Window ?
'E' – Store (storage)	S	E / SVS / UAS /(CAS)/ U	W6	1.05	2.26	2.37		No
	S1	E / SVS /(UAS)/ CAS / U	W6	0.65	2.26	1.47		
	E	E / SVS / UAS /(CAS)/ U	W4	1.65	2.26	3.73		
		E / SVS / UAS / CAS / U						
Zone height: 2.26 m	N	(E)/ SVS / UAS / CAS / U	W1	1.70	2.26	3.84	W13	
Check:		E / SVS / UAS / CAS / U						
Activity ✓	W	E / SVS /(UAS)/ CAS / U	W6	1.65	2.26	3.73		
HVAC ✓		E / SVS / UAS / CAS / U						
	Ceiling	E / SVS / UAS / CAS / U						
Lighting system ✓	Floor	E / SVS /(UAS)/ CAS / U	F3	1.70	1.65	2.81		
Daylight access ✓	Roof	(E)/ SVS / UAS / CAS / U	R2	1.70	1.65	2.81		
'F' – Office (cellular)	S	(E)/ SVS / UAS / CAS / U	W1	5.30	2.26	11.98	W12	No
		E / SVS / UAS / CAS / U						
	E	E / SVS / UAS /(CAS)/ U	W4	3.56	2.26	8.05		
Zone height: 2.26 m	Floor 1	E / SVS /(UAS)/ CAS / U	F3	2.04	1.10	2.24	(floor)	
Check:	N	E / SVS /(UAS)/ CAS / U	W6	5.30	2.26	11.98		
Activity ✓	W 1	E / SVS /(UAS)/ CAS / U	W6	2.16	2.26	4.88		
	W	E / SVS / UAS /(CAS)/ U	W6	1.40	2.26	3.16		
HVAC ✓	Ceiling	E / SVS / UAS / CAS / U						
Lighting system ✓	Floor	E / SVS / UAS /(CAS)/ U	F2	–	–	9.79		
Daylight access ✓	Roof	(E)/ SVS / UAS / CAS / U	R2	–	–	12.03		
'G' – Stairs (circulation)	S	(E)/ SVS / UAS / CAS / U	W1	2.88	3.00	8.64	W15	No
		E / SVS / UAS / CAS / U						
	E	E / SVS / UAS /(CAS)/ U	W6	1.41	3.00	4.23		
Zone height: 3.00 m		E / SVS / UAS / CAS / U						
Check:	N	E / SVS /(UAS)/ CAS / U	W6	2.88	3.00	8.64		
		E / SVS / UAS / CAS / U						
Activity ✓	W	E / SVS / UAS /(CAS)/ U	W6	3.76	3.00	11.28		
HVAC ✓	Ceiling	E / SVS / UAS / CAS / U						
Lighting system ✓	Floor	E / SVS / UAS /(CAS)/ U	F2	–	–	6.17		
Daylight access ✓	Roof	(E)/ SVS / UAS / CAS / U	R2	–	–	6.17		
'H' – Basement (E) (storage)	S	(E)/ SVS / UAS / CAS / U	W1	13.20	1.88	24.82	W17+D4	No
	S1	E / SVS / UAS / CAS /(U)	W3	13.20	1.28	16.90		
	E	E / SVS / UAS /(CAS)/ U	W5	5.10	1.88	9.59		
Zone height: 3.15 m	E1	E / SVS / UAS / CAS /(U)	W3	5.10	1.28	6.53		
Check:	N	(E)/ SVS / UAS / CAS / U	W1	13.20	1.70	22.44		
	N1	E / SVS / UAS / CAS /(U)	W3	13.20	1.45	19.14		
Activity ✓	W	E / SVS / UAS / CAS / U						
HVAC ✓	Ceiling	E / SVS / UAS /(CAS)/ U	F2	5.10	13.20	67.32		
Lighting system ✓	Floor	E / SVS / UAS / CAS /(U)	F1	5.10	13.20	67.32		
Daylight access ✓	~~Roof~~	E / SVS / UAS / CAS / U						
Total area						88.33		
Carry forward						251.09		

(Some small areas noted to be to different adjoining conditions – all very small and not likely to affect result)

Copyright Larry Russen June 2009

RT2.D0671   COMMERCIAL EPC SITE / REFLECTION NOTES   File / page reference ___1954___ / __40__

Zone name, activity and reference	Orient-ation	Adjoining condition type: E/SVS/UAS/CAS/U	Type (from construction schedule)	Length (m)	Height/width (m)	Area m²	Door / Glazing ref / type %	Display Window?
'I' = Basement (W) (storage) Zone height: 3.15 m Check: Activity ✓ HVAC ✓ Lighting system ✓ Daylight access ✓	S	(E) / SVS / UAS / CAS / U	W1	13.00	1.88	26.44		No
	S1	E / SVS / UAS / CAS / (U)	W3	13.00	1.28	16.64		
	E	E / SVS / UAS / CAS / U						
		E / SVS / UAS / CAS / U						
	N	(E) / SVS / UAS / CAS / U	W1	13.00	1.70	22.10		
	N1	E / SVS / UAS / CAS / (U)	W3	13.00	1.45	18.85		
	W	(E) / SVS / UAS / CAS / U	W1	6.10	3.15	19.22		
		E / SVS / UAS / (CAS) / U						
	Ceiling	E / SVS / UAS / (CAS) / U	F2	–	–	69.33		
	Floor	E / SVS / UAS / CAS / (U)	F1	–	–	69.33		
	Roof	E / SVS / UAS / CAS / U						
'J' – Toilets (toilet) Zone height: 2.52 m Check: Activity ✓ HVAC ✓ Lighting system ✓ Daylight access ✓	S	E / SVS / UAS / (CAS) / U	W5	7.34	2.52	18.50		No
		E / SVS / UAS / CAS / U						
	E	E / SVS / (UAS) / CAS / U	W5	3.68	2.52	9.27		
	E1	E / SVS / UAS / (CAS) / U	W5	3.18	2.52	8.01		
	N	(E) / SVS / UAS / CAS / U	W1	6.74	2.35	15.84		
	N1	E / SVS / UAS / CAS / (U)	W3	6.74	0.18	1.21		
	W	E / SVS / (UAS) / CAS / U	W5	3.18	2.52	8.01		
	Ceiling	E / SVS / (UAS) / CAS / U	F3	–	–	23.29		
	Floor	E / SVS / UAS / CAS / (U)	F1	–	–	23.29		
	Roof	E / SVS / UAS / CAS / U						
'K' – east lobby (circulation) Zone height: 2.52 m Check: Activity ✓ HVAC ✓ Lighting system ✓ Daylight access ✓	S	(E) / SVS / UAS / CAS / U	W1	8.39	2.52	21.14	W16+D3	No
		E / SVS / UAS / CAS / U						
	E	E / SVS / UAS / (CAS) / U	W4	1.52	2.52	3.83		
	E1	E / SVS / UAS / (CAS) / U	W5	0.55	2.52	1.39		
	N	E / SVS / UAS / (CAS) / U	W5	7.34	2.52	18.50		
	N1	E / SVS / (UAS) / CAS / U	W5	1.05	2.52	2.65		
	W	E / SVS / (UAS) / CAS / U	W5	2.02	2.52	5.09		
		E / SVS / UAS / CAS / U						
	Ceiling	E / SVS / UAS / (CAS) / U	F2	–	–	15.48		
	Floor	E / SVS / UAS / CAS / (U)	F1	–	–	15.48		
	Roof	E / SVS / UAS / CAS / U						
'L' – GF store (storage) Zone height: 2.52 m Check: Activity ✓ HVAC ✓ Lighting system ✓ Daylight access ✓	S	E / SVS / UAS / (CAS) / U	W5	1.66	2.52	4.18		No
		E / SVS / UAS / CAS / U						
	E	E / SVS / UAS / (CAS) / U	W4	3.78	2.52	9.53		
		E / SVS / UAS / CAS / U						
	N	(E) / SVS / UAS / CAS / U	W1	1.66	2.35	3.90		
	N1	E / SVS / UAS / CAS / (U)	W3	1.66	0.18	0.30		
	W	E / SVS / UAS / (CAS) / U	W5	3.78	2.52	9.53		
	Ceiling 1	E / SVS / UAS / (CAS) / U	F2	–	–	2.63		
	Ceiling 2	E / SVS / (UAS) / CAS / U	F3	1.66	1.70	2.82		
	Floor	E / SVS / UAS / CAS / (U)	F1	–	–	5.45		
	Roof	E / SVS / UAS / CAS / U						
Total area						113.55		
Carry forward						364.64		

*(Check measurement based on external measurements –
Ext length say 8.900 + 26.700 [allow for shape @ western end]
x average int. Depth 5.105 x 2 = 363.48m² – within <0.5% of actual)*

Copyright Larry Russen June 2009

Appendix B

RT2.D0671 COMMERCIAL EPC SITE / REFLECTION NOTES File / page reference 1954 / 41

17 – FINAL AUDIT, RATINGS AND RESULTS

Checks on data entry
Project database: Walls ✓ Roofs ✓ Floors ✓ Doors ✓ Windows ✓
Geometry: Building infiltration rate ✓ Zone area ✓ Zone height ✓ HVAC allocation ✓ Activity ✓ Doors / Windows added ✓
Building Services: ECA list ✓ M&T ✓ HVAC controls ✓ HWS storage ✓ Ventilation ✓ Exhaust ✓ Light controls ✓
Results page: Data Reflection – Actual Building ✓ Unassigned objects report ✓ Assigned objects report ✓

Results & benchmarks	Details / discussion	Notes
Initial EPC 'guess' before calculation	Rating: 105 – 110 Band: A / B / C / D / **E** / F / G	*Could be less, but boiler poor*
Actual rating 102	Rating: Band: A / B / C / **D** / **E** / F / G	—
BER and EPC seem 'about right' for building type / activity etc (**Yes**) / no	*Similar to other old buildings I have assessed.*	*Carefully checked lamps + HVAC as slightly lower than expected.*
Benchmark building ratings consistent with 'actual' rating? ['New' rating: 39] (**Yes**)/no ['Typical' rating: 64] (**Yes**)/no	*'New' is no surprise – insulation in existing is poor.* *'Typical' is as expected.*	—
'Condition' issues checked and recommendations added / (**deleted**) / (**edited**) (**Yes**)/no	—	*Nothing onerous noted.*
Recommendations		
Building recommendations viewed and / or edited ✓	*Removed 'cavity wall'*	*All walls solid*
HVAC recommendations viewed and / or edited ✓	*Removed 'time control'*	*System has programmer*
HWS recommendations viewed and / or edited ✓	—	—
Lighting recommendations viewed and / or edited ✓	*Removed 'replace GLS spotlights etc' – they are TH already*	
Own recommendations added ✓	*Added usual paragraph*	
Recalculate if recommendations altered or additional? (**Yes**)/no		
Final documents		
EPC checked ✓		
RR checked ✓	—	—
Sec RR checked ✓		
SBEM Output checked ✓		

Copyright Larry Russen June 2009

RT2.D0671	COMMERCIAL EPC SITE / REFLECTION NOTES	File / page reference 1954 / 42
EPC audit		
Construction	Wall, floor, roof constructions; window, rooflight, door specifications; all based on age, generic type	*Timber doors – adopted own values. Calculated 'U' value for roof 2.*
Geometry	Thermal bridges; air permeability; glazing shading	*Accepted air permeability default + thermal bridges*
HVAC and HWS	Heating and cooling system type; boiler efficiencies & chiller EERs based on age and ECA list; duct and AHU leakage, specific fan power, DHW system sizing, no metering, no renewables, high pressure drop air filtration from database	*Accepted defaults due to lack of any information on site or from internet*
Lighting	Lighting type unknown; local manual control; if automatic controls, default parasitic power	*Calculated efficacy in Zone 'C', otherwise default.*

Continue on other sheet: Yes (No)

Appendix B

ZONE PLAN DATA COLLECTION

Figure B28 is an example of the 'zone plan data collection' methodology discussed in Chapter 8, for zone 'F' only. You can consider and reflect on your type of approach.

COMMERCIAL EPC   **ZONE PLAN**   File / page reference _1954_ / _00_

Zone reference	Zone area	Ceiling height	Notes
'F'	*12.03*	*2260*	*For construction detail see project database*
Activity description	Zone height	North point	
'Cellular office'	*2260*		

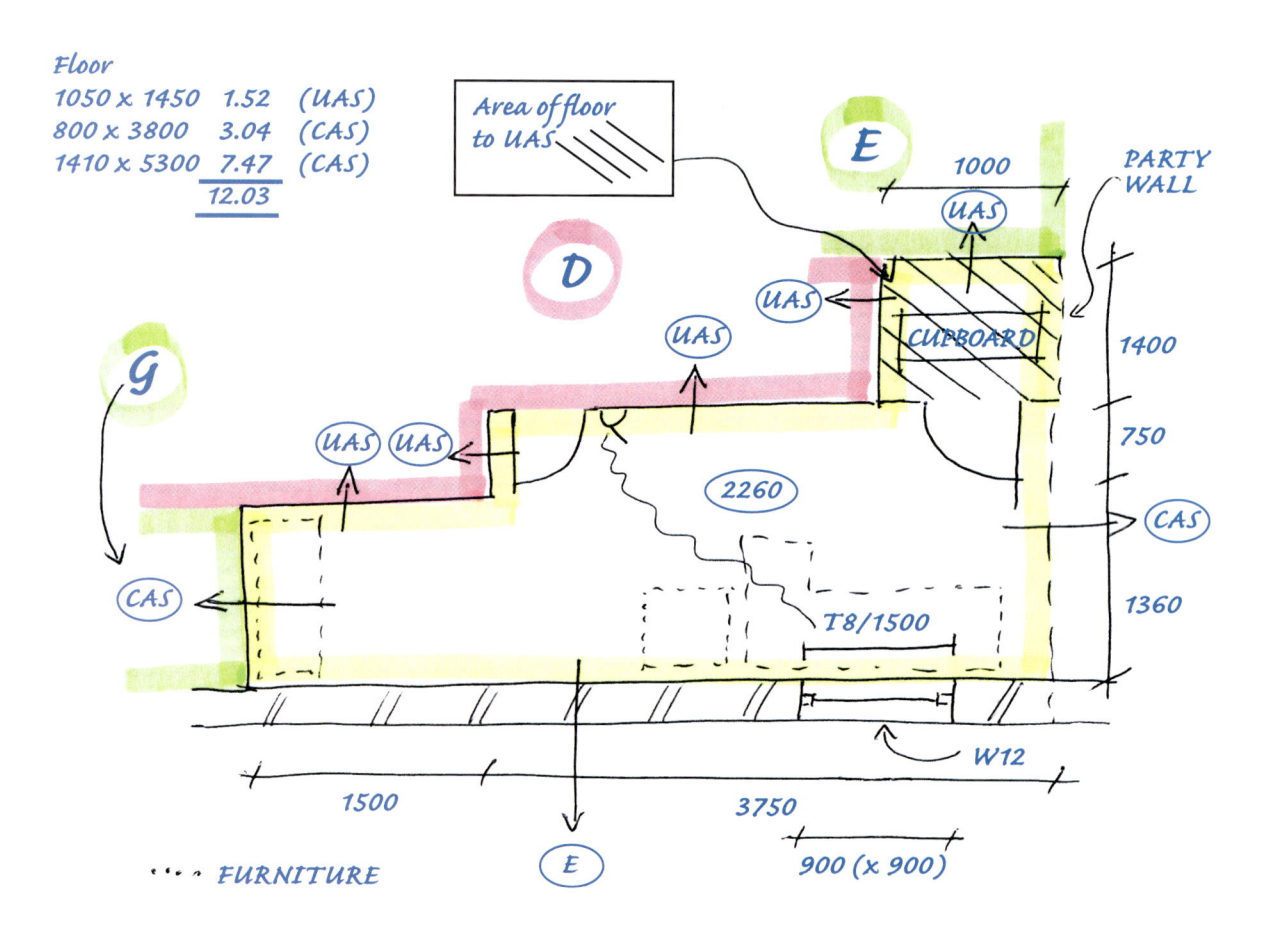

Floor
1050 x 1450 1.52 (UAS)
800 x 3800   3.04 (CAS)
1410 x 5300 7.47 (CAS)
*   12.03*

Area of floor to UAS

E   1000   PARTY WALL
UAS
D
UAS   CUPBOARD   1400
G   UAS   750
UAS   UAS   CAS
2260   1360
CAS   T8/1500
W12
1500   3750
···· *FURNITURE*   E   900 (x 900)

Construction type / detail		Services / detail			
Walls	*W1 – external* *W4 – party* *W6 – internal*	Heating / controls	*Electric resistance heating*	Supply ventilation	*None*
Roofs	*R2 to entirety*	heat source / fuel	*Portable electric heater*	Exhaust	*None*
Floors	*F2 – majority (to CAS)* *F3 – to UAS*	Cooling / controls	*None*	De-strat fans / Heat recovery	*None*
Doors	*None*	Cooling source / fuel	*None*	Lamps	*As shown, no details, starter, assume EM*
Windows / shading	*W12 – shaded, see detail pp 1954/14+16*	HWS / storage / circulation	*None*	Switching	*Local manual – <6m*

Figure B28: Alternative zone plan data methodology

Appendix B

DOCUMENTS GENERATED

Energy Performance Certificate

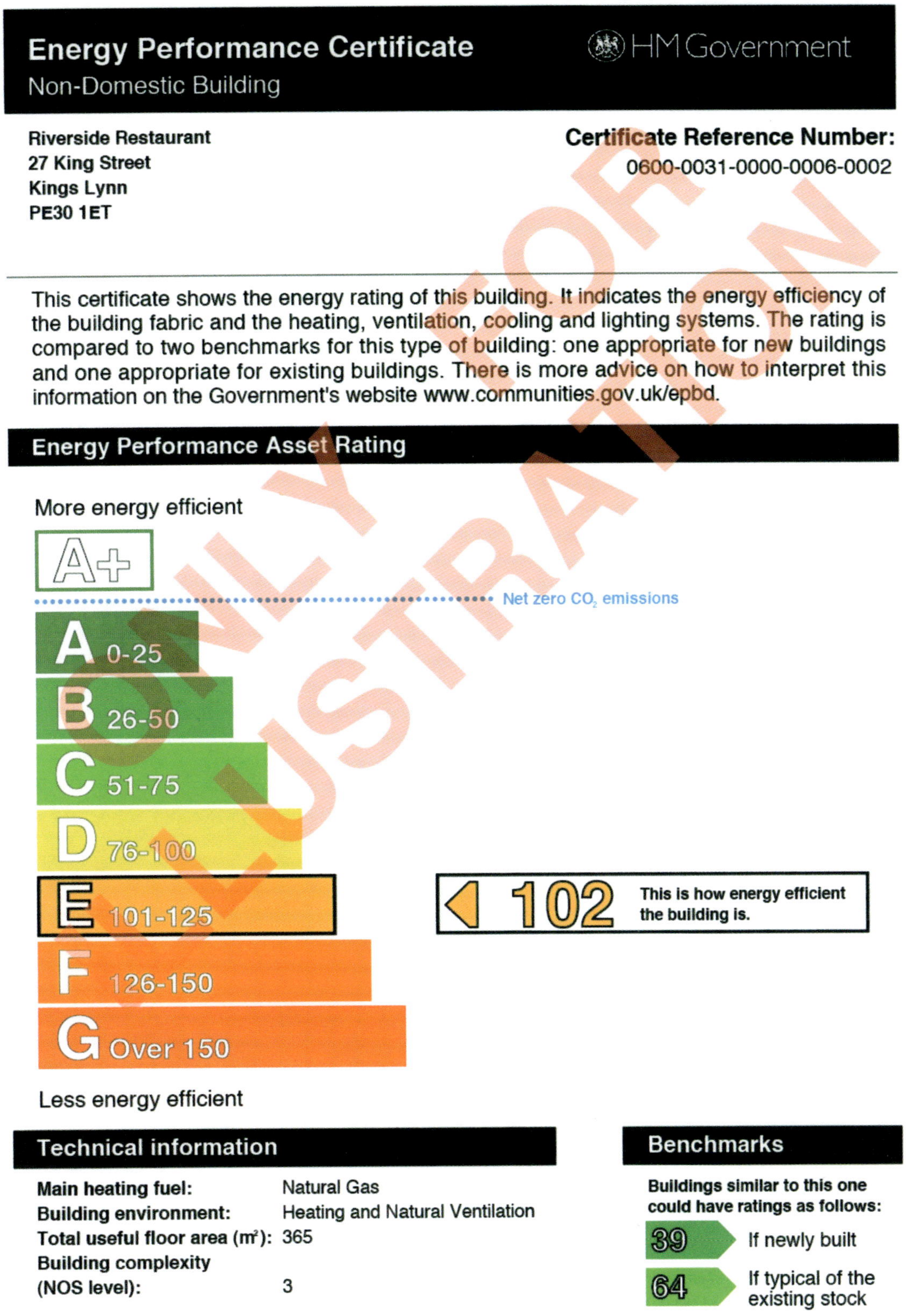

Energy Performance Certificate
Non-Domestic Building — HM Government

Riverside Restaurant
27 King Street
Kings Lynn
PE30 1ET

Certificate Reference Number:
0600-0031-0000-0006-0002

This certificate shows the energy rating of this building. It indicates the energy efficiency of the building fabric and the heating, ventilation, cooling and lighting systems. The rating is compared to two benchmarks for this type of building: one appropriate for new buildings and one appropriate for existing buildings. There is more advice on how to interpret this information on the Government's website www.communities.gov.uk/epbd.

Energy Performance Asset Rating

More energy efficient

- A+
- A 0-25
- B 26-50
- C 51-75
- D 76-100
- E 101-125
- F 126-150
- G Over 150

Net zero CO_2 emissions

◄ 102 This is how energy efficient the building is.

Less energy efficient

Technical information

Main heating fuel:	Natural Gas
Building environment:	Heating and Natural Ventilation
Total useful floor area (m²):	365
Building complexity (NOS level):	3

Benchmarks

Buildings similar to this one could have ratings as follows:

- 39 — If newly built
- 64 — If typical of the existing stock

Figure B29 – EPC

Administrative information

This is an Energy Performance Certificate as defined in SI2007:991 as amended

Assessment Software:	iSBEM v3.4.a using calculation engine SBEM v3.4.a
Property Reference:	000000000000
Assessor Name:	Mr Larry Russen
Assessor Number:	SAVA004349
Accreditation Scheme:	National Energy Services
Employer/Trading Name:	Allied Surveyors
Employer/Trading Address:	17 High Street, Kings Lynn PE30 1BP
Issue Date:	22 Aug 2009
Valid Until:	21 Aug 2019 (unless superseded by a later certificate)
Related Party Disclosure:	I know of no current conflicts of interest.

Recommendations for improving the property are contained in Report Reference Number: 0000-0060-0040-6000-0103

If you have a complaint or wish to confirm that the certificate is genuine

Details of the assessor and the relevant accreditation scheme are on the certificate. You can get contact details of the accreditation scheme from the Government's website at www.communities.gov.uk/epbd, together with details of the procedures for confirming authenticity of a certificate and for making a complaint.

CARBON TRUST

For advice on how to take action and to find out about technical and financial assistance schemes to help make buildings more energy efficient visit **www.carbontrust.co.uk** or call us on **0800 085 2005**

Recommendation Report

0000-0060-0040-6000-0103

Recommendation Report

HM Government

Report Reference Number: 0000-0060-0040-6000-0103

Riverside Restaurant
27 King Street
Kings Lynn
PE30 1ET

Building Type(s): Restaurant/public house

ADMINISTRATIVE INFORMATION	
Issue Date:	22 Aug 2009
Valid Until:	21 Aug 2019 (*)
Total Useful Floor Area (m^2):	365
Calculation Tool Used:	iSBEM v3.4.a using calculation engine SBEM v3.4.a
Property Reference:	000000000000
Energy Performance Certificate for the property is contained in Report Reference Number: 0600-0031-0000-0006-0002	

ENERGY ASSESSOR DETAILS	
Assessor Name:	Mr Larry Russen
Employer/Trading Name:	Allied Surveyors
Employer/Trading Address:	17 High Street, Kings Lynn PE30 1BP
Assessor Number:	SAVA004349
Accreditation scheme:	National Energy Services
Related Party Disclosure:	I know of no current conflicts of interest.

(*) Unless superseded by a later recommendation report

Page 1 of 8

Figure B30: RR

0000-0060-0040-6000-0103

Table of Contents

1. Background 3
2. Introduction 3
3. Recommendations 4
4. Next Steps 6
5. Glossary 8

Appendix B

0000-0060-0040-6000-0103

1. Background

Statutory Instrument 2007 No. 991, *The Energy Performance of Buildings (Certificates and Inspections) (England and Wales) Regulations 2007*, as amended, transposes the requirements of Articles 7.2 and 7.3 of the Energy Performance of Buildings Directive 2002/91/EC.

This report is a Recommendation Report as required under regulations 16(2)(a) and 19 of the Statutory Instrument SI 2007:991.

This section provides general information regarding the building:

Total Useful Floor Area (m^2):	365
Building Environment:	Heating and Natural Ventilation

2. Introduction

This Recommendation Report was produced in line with the Government's approved methodology and is based on calculation tool iSBEM v3.4.a using calculation engine SBEM v3.4.a .

In accordance with Government's current guidance, the Energy Assessor did undertake a walk around survey of the building prior to producing this Recommendation Report.

Commercial Energy Assessor's Handbook

0000-0060-0040-6000-0103

3. Recommendations

The following sections list recommendations selected by the energy assessor for the improvement of the energy performance of the building. The recommendations are listed under four headings: short payback, medium payback, long payback, and other measures.

a) Recommendations with a short payback

This section lists recommendations with a payback of less than 3 years:

Recommendation	Potential Impact
Replace 38mm diameter (T12) fluorescent tubes on failure with 26mm (T8) tubes.	LOW
Improve insulation on HWS storage.	LOW
Replace tungsten GLS lamps with CFLs: Payback period dependent on hours of use.	LOW
Some spaces have a significant risk of overheating. Consider solar control measures such as the application of reflective coating or shading devices to windows.	MEDIUM
Consider replacing T8 lamps with retrofit T5 conversion kit.	LOW
Consider replacing heating boiler plant with high efficiency type.	HIGH
Add optimum start/stop to the heating system.	MEDIUM

b) Recommendations with a medium payback

This section lists recommendations with a payback of between 3 and 7 years:

Recommendation	Potential Impact
The default heat generator efficiency is chosen. It is recommended that the heat generator system be investigated to gain an understanding of its efficiency and possible improvements.	HIGH
Introduce HF (high frequency) ballasts for fluorescent tubes: Reduced number of fittings required.	LOW
Some windows have high U-values - consider installing secondary glazing.	HIGH
Add local temperature control to the heating system.	MEDIUM

Appendix B

0000-0060-0040-6000-0103

Add weather compensation controls to heating system.	MEDIUM
Add local time control to heating system.	LOW

c) Recommendations with a long payback

This section lists recommendations with a payback of more than 7 years:

Recommendation	Potential Impact
Consider replacing heating boiler plant with a condensing type.	HIGH
Carry out a pressure test, identify and treat identified air leakage. Enter result in EPC calculation.	HIGH
Some glazing is poorly insulated. Replace/improve glazing and/or frames.	HIGH
Consider installing an air source heat pump.	MEDIUM
Consider installing a ground source heat pump.	MEDIUM

d) Other recommendations

This section lists other recommendations selected by the energy assessor, based on an understanding of the building, and / or based on a valid existing energy report.

Recommendation	Potential Impact
This Energy Performance Certificate (EPC) and Recommendation Report (RR) must be read and considered together with our accompanying Energy Report dated 6th June 2009. The Energy Report describes and explains the EPC and RR in much greater detail, in particular in relation to issues such as legal, planning and other matters.	LOW

Page 5 of 8

0000-0060-0040-6000-0103

4. Next steps

a) Your Recommendation Report

As the building occupier, regulation 10(1) of SI 2007:991 requires that an Energy Performance Certificate *"must be accompanied by a recommendation report"*.

You must be able to produce a copy of this Recommendation Report within seven days if requested by an Enforcement Authority under regulation 39 of SI 2007:991.

This Recommendation Report has also been lodged on the Government's central register. Access to the report, to the data used to compile the report, and to previous similar documents relating to the same building can be obtained by request through the Non-Dwellings Register (www.epcregister.com) using the report reference number of this document.

b) Implementing recommendations

The recommendations are provided as an indication of opportunities that appear to exist to improve the building's energy efficiency.

The calculation tool has automatically produced a set of recommendations, which the Energy Assessor has reviewed in the light of his / her knowledge of the building and its use. The Energy Assessor may have comments on the recommendations based on his / her knowledge of the building and its use. The Energy Assessor may have inserted additional measures in section 3d (Other Recommendations). He / she may have removed some automatically generated recommendations or added additional recommendations.

These recommendations do not include matters relating to operation and maintenance which cannot be identified from the calculation procedure.

Appendix B

0000-0060-0040-6000-0103

c) Legal disclaimer

The advice provided in this Recommendation Report is intended to be for information only. Recipients of this Recommendation Report are advised to seek further detailed professional advice before reaching any decision on how to improve the energy performance of the building.

d) Complaints

Details of the assessor and the relevant accreditation scheme are on this report and the energy performance certificate. You can get contact details of the accreditation scheme from our website at www.communities.gov.uk/epbd, together with details of their procedures for confirming authenticity of a certificate and for making a complaint.

0000-0060-0040-6000-0103

5. Glossary

a) Payback

The payback periods are based on data provided by Good Practice Guides and Carbon Trust energy survey reports and are average figures calculated using a simple payback method. It is assumed that the source data is correct and accurate using up to date information.

The figures have been calculated as an average across a range of buildings and may differ from the actual payback period for the building being assessed. Therefore, it is recommended that each suggested measure be further investigated before reaching any decision on how to improve the energy efficiency of the building.

b) Carbon impact

The High / Medium / Low carbon impact indicators against each recommendation are provided to distinguish, between the suggested recommendations, those that would have most impact on carbon emissions from the building. For automatically generated recommendations, the carbon impact indicators are determined by software, but may have been adjusted by the Energy Assessor based on his / her knowledge of the building. The impact of other recommendations are determined by the assessor.

c) Valid report

A valid report is a report that has been:
- Produced within the past 10 years
- Produced by an Energy Assessor who is accredited to produce Recommendation Reports through a Government Approved Accreditation Scheme
- Lodged on the Register operated by or on behalf of the Secretary of State.

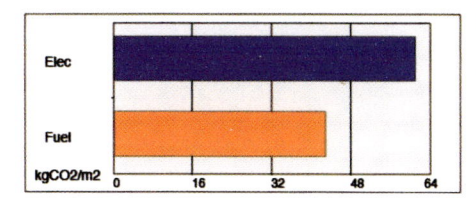

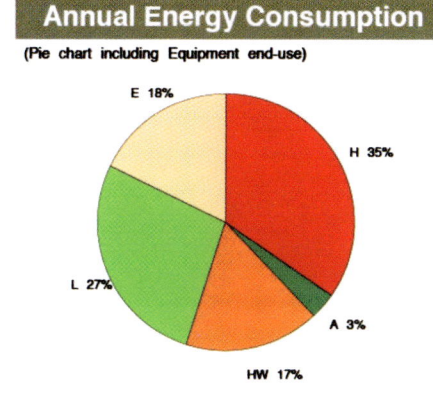

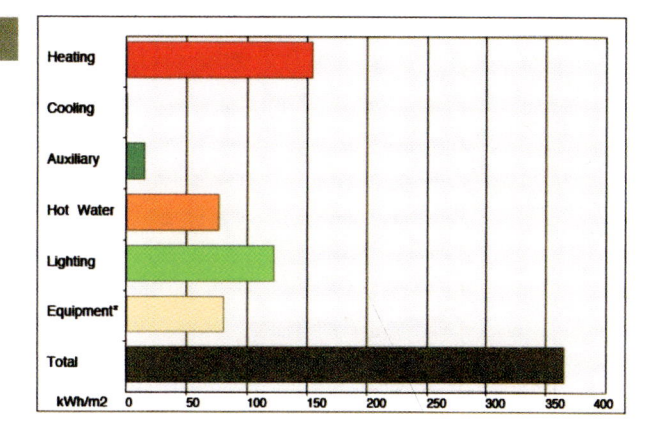

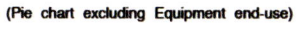

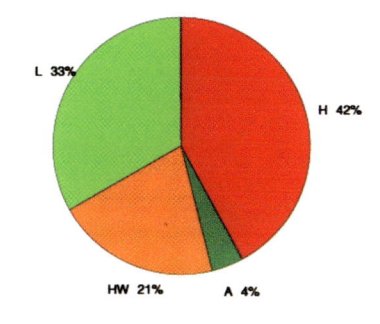

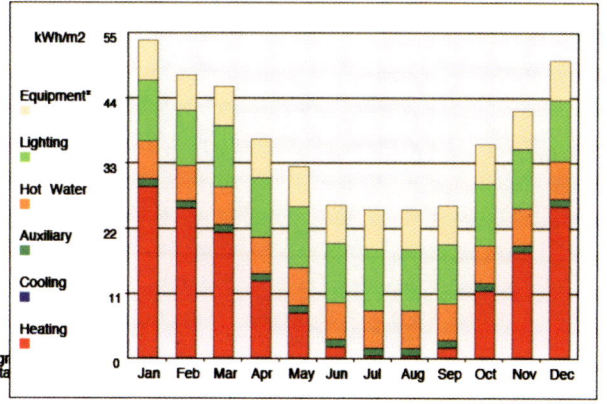

Main Calculation Output Document

Figure B31: Main output

Commercial Energy Assessor's Handbook

Secondary Recommendation Report

This includes all of the recommendations including those we excluded and those that do not have sufficient payback to appear in the main RR.

Secondary Recommendations Report
Not for Official Submission

Building name
Riverside Restaurant
Building type: Restaurant/public house   Date: Sat Aug 22 14:16:23 2009

This report lists recommendations for energy-efficiency improvements to the building.

Key to colour codes used in this report
Included by the calculation
Included by the user
Excluded by the user

Recommendations for HEATING

HEATING accounts for 45% of the CO2 emissions
(If hot water is provided by the HVAC system, then the % of CO2 emissions includes hot water provision)
The overall energy performance of HEATING provision is POOR
The overall CO2 performance of HEATING provision is POOR
The average energy efficiency of HEATING provision is POOR
The average CO2 efficiency of HEATING provision is POOR

This recommendation was excluded by the assessor.
Add time control to heating system.
Code:	EPC-H2
Energy Impact:	HIGH
CO2 Impact:	HIGH
CO2 Saved per £ Spent:	GOOD
Applicable to:	Whole building

Comments: No comments from assessor.

Add local time control to heating system.
Code:	EPC-H5
Energy Impact:	LOW
CO2 Impact:	LOW
CO2 Saved per £ Spent:	POOR
Applicable to:	Whole building

Comments:

Add local temperature control to the heating system.
Code:	EPC-H6
Energy Impact:	MEDIUM
CO2 Impact:	MEDIUM
CO2 Saved per £ Spent:	POOR
Applicable to:	Whole building

Comments:

Add optimum start/stop to the heating system.
Code:	EPC-H7

Figure B32: Secondary RR

Appendix B

Energy Impact: MEDIUM
CO2 Impact: MEDIUM
CO2 Saved per £ Spent: GOOD
Applicable to: Whole building

Comments:

Add weather compensation controls to heating system.
Code: EPC-H8
Energy Impact: MEDIUM
CO2 Impact: MEDIUM
CO2 Saved per £ Spent: POOR
Applicable to: Whole building

Comments:

Consider replacing heating boiler plant with high efficiency type.
Code: EPC-H1
Energy Impact: HIGH
CO2 Impact: HIGH
CO2 Saved per £ Spent: GOOD
Applicable to: Whole building

Comments:

Consider replacing heating boiler plant with a condensing type.
Code: EPC-H3
Energy Impact: HIGH
CO2 Impact: HIGH
CO2 Saved per £ Spent: POOR
Applicable to: Whole building

Comments:

The default heat generator efficiency is chosen. It is recommended that the heat generator system be investigated to gain an understanding of its efficiency and possible improvements.
Code: EPC-H4
Energy Impact: HIGH
CO2 Impact: HIGH
CO2 Saved per £ Spent: FAIR
Applicable to: Whole building

Comments:

The default heat generator efficiency is chosen. It is recommended that the heat generator system be investigated to gain an understanding of its efficiency and possible improvements.
Code: EPC-H4
Energy Impact: LOW
CO2 Impact: LOW
CO2 Saved per £ Spent: FAIR
Applicable to: Only Electric Heating

Comments:

Add time control to heating system.
Code: EPC-H2
Energy Impact: LOW
CO2 Impact: LOW
CO2 Saved per £ Spent: GOOD
Applicable to: Only Electric Heating

Comments:

Add local time control to heating system.
Code: EPC-H5
Energy Impact: LOW
CO2 Impact: LOW
CO2 Saved per £ Spent: POOR
Applicable to: Only Electric Heating

Comments:

Add local temperature control to the heating system.
Code: EPC-H6
Energy Impact: LOW
CO2 Impact: LOW
CO2 Saved per £ Spent: POOR
Applicable to: Only Electric Heating

Comments:

Add optimum start/stop to the heating system.
Code: EPC-H7
Energy Impact: LOW
CO2 Impact: LOW
CO2 Saved per £ Spent: FAIR
Applicable to: Only Electric Heating

Comments:

Add weather compensation controls to heating system.
Code: EPC-H8
Energy Impact: LOW
CO2 Impact: LOW
CO2 Saved per £ Spent: POOR
Applicable to: Only Electric Heating

Comments:

Consider replacing heating boiler plant with high efficiency type.
Code: EPC-H1
Energy Impact: HIGH
CO2 Impact: HIGH
CO2 Saved per £ Spent: GOOD
Applicable to: Heating 1

Comments:

Consider replacing heating boiler plant with a condensing type.
Code: EPC-H3
Energy Impact: HIGH
CO2 Impact: HIGH
CO2 Saved per £ Spent: POOR
Applicable to: Heating 1

Comments:

The default heat generator efficiency is chosen. It is recommended that the heat generator system be investigated to gain an understanding of its efficiency and possible improvements.
Code: EPC-H4
Energy Impact: HIGH
CO2 Impact: HIGH
CO2 Saved per £ Spent: FAIR
Applicable to: Heating 1

Comments:

Appendix B

Add local time control to heating system.
Code: EPC-H5
Energy Impact: LOW
CO2 Impact: LOW
CO2 Saved per £ Spent: POOR
Applicable to: Heating 1

Comments:

Add local temperature control to the heating system.
Code: EPC-H6
Energy Impact: MEDIUM
CO2 Impact: MEDIUM
CO2 Saved per £ Spent: POOR
Applicable to: Heating 1

Comments:

Add optimum start/stop to the heating system.
Code: EPC-H7
Energy Impact: MEDIUM
CO2 Impact: MEDIUM
CO2 Saved per £ Spent: GOOD
Applicable to: Heating 1

Comments:

Add weather compensation controls to heating system.
Code: EPC-H8
Energy Impact: LOW
CO2 Impact: MEDIUM
CO2 Saved per £ Spent: POOR
Applicable to: Heating 1

Comments:

Recommendations for COOLING

COOLING accounts for 0% of the CO2 emissions
The overall energy performance of COOLING provision is NOT APPLICABLE
The overall CO2 performance of COOLING provision is NOT APPLICABLE
The average energy efficiency of COOLING provision is NOT APPLICABLE
The average CO2 efficiency of COOLING provision is NOT APPLICABLE

There are no recommendations for COOLING

Recommendations for HOT-WATER

HOT-WATER accounts for 0% of the CO2 emissions
(If hot water is provided by the HVAC system, then hot water provision is included in the % of CO2 emissions due to HEATING)
The overall energy performance of HOT-WATER provision is GOOD
The overall CO2 performance of HOT-WATER provision is GOOD
The average energy efficiency of HOT-WATER provision is NOT APPLICABLE
The average CO2 efficiency of HOT-WATER provision is NOT APPLICABLE

Improve insulation on HWS storage.

Code: EPC-W3
Energy Impact: LOW
CO2 Impact: LOW
CO2 Saved per £ Spent: GOOD
Applicable to: Whole building

Comments:

Recommendations for LIGHTING

LIGHTING accounts for 49% of the CO2 emissions
The overall energy performance of LIGHTING provision is POOR
The overall CO2 performance of LIGHTING provision is POOR

Replace 38mm diameter (T12) fluorescent tubes on failure with 26mm (T8) tubes.

Code: EPC-L1
Energy Impact: LOW
CO2 Impact: LOW
CO2 Saved per £ Spent: GOOD
Applicable to: Whole building

Comments:

Replace tungsten GLS lamps with CFLs: Payback period dependent on hours of use.

Code: EPC-L2
Energy Impact: LOW
CO2 Impact: LOW
CO2 Saved per £ Spent: GOOD
Applicable to: Whole building

Comments:

This recommendation was excluded by the assessor.
Replace tungsten GLS spotlights with low-voltage tungsten halogen: Payback period dependent on hours of use.

Code: EPC-L4
Energy Impact: HIGH
CO2 Impact: HIGH
CO2 Saved per £ Spent: GOOD
Applicable to: Whole building

Comments: No comments from assessor

Consider replacing T8 lamps with retrofit T5 conversion kit.

Code: EPC-L5
Energy Impact: LOW
CO2 Impact: LOW
CO2 Saved per £ Spent: GOOD
Applicable to: Whole building

Comments:

Introduce HF (high frequency) ballasts for fluorescent tubes: Reduced number of fittings required.

Code: EPC-L7
Energy Impact: LOW
CO2 Impact: LOW
CO2 Saved per £ Spent: FAIR
Applicable to: Whole building

Comments:

Appendix B

Recommendations for RENEWABLES

Consider installing a ground source heat pump.

Code:	EPC-R1
Energy Impact:	MEDIUM
CO2 Impact:	MEDIUM
CO2 Saved per £ Spent:	POOR
Applicable to:	Whole building

Comments:

Consider installing building mounted wind turbine(s).

Code:	EPC-R2
Energy Impact:	LOW
CO2 Impact:	LOW
CO2 Saved per £ Spent:	POOR
Applicable to:	Whole building

Comments:

Consider installing solar water heating.

Code:	EPC-R3
Energy Impact:	LOW
CO2 Impact:	LOW
CO2 Saved per £ Spent:	POOR
Applicable to:	Whole building

Comments:

Consider installing PV.

Code:	EPC-R4
Energy Impact:	LOW
CO2 Impact:	LOW
CO2 Saved per £ Spent:	POOR
Applicable to:	Whole building

Comments:

Consider installing an air source heat pump.

Code:	EPC-R5
Energy Impact:	HIGH
CO2 Impact:	MEDIUM
CO2 Saved per £ Spent:	POOR
Applicable to:	Whole building

Comments:

Consider installing a ground source heat pump.

Code:	EPC-R1
Energy Impact:	MEDIUM
CO2 Impact:	MEDIUM
CO2 Saved per £ Spent:	POOR
Applicable to:	Only Electric Heating

Comments:

Consider installing an air source heat pump.

Code:	EPC-R5
Energy Impact:	MEDIUM
CO2 Impact:	MEDIUM

Commercial Energy Assessor's Handbook

CO2 Saved per £ Spent: POOR
Applicable to: Only Electric Heating

Comments:

Recommendations for OVERHEATING

The risk of some spaces in the building OVERHEATING is High risk

Some spaces have a significant risk of overheating. Consider solar control measures such as the application of reflective coating or shading devices to windows.
Code: EPC-V1
Energy Impact: MEDIUM
CO2 Impact: MEDIUM
CO2 Saved per £ Spent: POOR
Applicable to: Whole building

Comments:

Recommendations for ENVELOPE

This recommendation was excluded by the assessor.
Some walls have uninsulated cavities - introduce cavity wall insulation.
Code: EPC-E4
Energy Impact: HIGH
CO2 Impact: HIGH
CO2 Saved per £ Spent: GOOD
Applicable to: Whole building

Comments: There are no cavity walls.

Some windows have high U-values - consider installing secondary glazing.
Code: EPC-E5
Energy Impact: HIGH
CO2 Impact: HIGH
CO2 Saved per £ Spent: POOR
Applicable to: Whole building

Comments:

Carry out a pressure test, identify and treat identified air leakage. Enter result in EPC calculation.
Code: EPC-E7
Energy Impact: HIGH
CO2 Impact: HIGH
CO2 Saved per £ Spent: POOR
Applicable to: Whole building

Comments:

Some glazing is poorly insulated. Replace/improve glazing and/or frames.
Code: EPC-E8
Energy Impact: HIGH
CO2 Impact: HIGH
CO2 Saved per £ Spent: POOR
Applicable to: Whole building

Appendix B

Comments:

Recommendations for FUEL-SWITCHING

Consider switching from gas to biomass.
Code: EPC-F5
Energy Impact: LOW
CO2 Impact: HIGH
CO2 Saved per £ Spent: GOOD
Applicable to: Heating 1

Comments:

Recommendations for AUXILIARY

AUXILIARY accounts for 6% of the CO2 emissions
The overall energy performance of AUXILIARY provision is POOR
The overall CO2 performance of AUXILIARY provision is POOR

There are no recommendations for AUXILIARY

Recommendations for OTHER

This Energy Performance Certificate (EPC) and Recommendation Report (RR) must be read and considered together with our accompanying Energy Report dated 6th June 2009. The Energy Report describes and explains the EPC and RR in much greater detail, in particular in relation to issues such as legal, planning and other matters.
Code: USER
Energy Impact: LOW
CO2 Impact: LOW
CO2 Saved per £ Spent: POOR
Applicable to: Whole building

Comments: No comments from assessor

CLIENT REPORT

Here are some paragraphs from our report, with explanatory comments in *italics*.

This accompanying energy report forms an integral part of my advice and should be read with the EPC and RR. This report explains certain particular issues in the EPC and the RR in more detail. Please ensure that you pass a copy on to the purchaser or new tenant. I can provide further copies, at cost, if you need them.

My instructions were to produce an EPC, RR and Energy Report (this document) of the specified building. You confirmed my instructions and the limitations of my inspection in writing on 2 August 2009. I have attached a signed copy of the Conditions of Engagement to this report. Those Conditions of Engagement form an integral part of my report.

In very general terms, any owner who sells or rents a 'non-domestic' property is required to provide an EPC and RR (the 'required documents') at the earliest opportunity.

This report and the EPC and RR have been prepared by me, an accredited Energy Assessor, as required by UK legislation. I am also a Chartered Surveyor. However, this report is **not** a report on the condition of the property, services and/or associated matters. Further reports and tests will be required by appropriately qualified professionals if assurance regarding the condition of the property, or condition and capability of services is required.

The property was occupied at the time of my inspection. My inspection was restricted by the following matters:

- the property was almost entirely carpeted and furnished at the time of my inspection and this significantly restricted the inspection, in particular to floors;
- stored items and stock;
- business fixtures including desk, filing cabinets, kitchen and other cooking equipment and associated stored items.

I confirm my inspection was carried out on 18 August 2009.

The building I inspected for the purpose of preparing the EPC and RR comprises a two storey end-terrace restaurant with ancillary accommodation.

The property also includes a detached building used as a laundry. I have not included that building in my EPC and RR. In my opinion, that building falls outside the requirements of the relevant UK Government Regulations for non-domestic EPCs because it is a stand-alone building with a floor area of less than 50m² [Regulation 4(1) of *The Energy Performance of Buildings (Certificates and Inspections) (England and Wales) Regulations* 2007 (SI 2007/991)].

On the day of my visit I was able to inspect the entire premises, internally and externally.

I have attached the following documents:

- Energy Performance Certificate (EPC);
- SBEM Recommendations Report (RR);
- SBEM Main Calculation Output Document.

The **EPC** is intended to provide information about the energy performance of a building, so energy efficiency can be considered as part of the investment and business decisions relating to that premises. It is a legal document that is required for buildings when they are constructed, modified, sold, or let. An EPC prepared for the sale or letting of premises is valid for a period of 10 years, or until a more recent EPC is prepared; whichever is the sooner.

The results of the energy performance survey are graded in bands of A to G, with A being the most efficient. The energy performance 'asset rating' for this property is 102, which means the property has a Band E rating. You will note that the EPC confirms that buildings similar to this building could have ratings as follows:

- if newly built and complying with current Regulations – 39;
- if typical of similar existing stock – 64. However, this description is not entirely accurate, as the definition of a 'typical' building for non-domestic energy purposes is one constructed in 1995 and complying with the Regulations at that date; whereas, your building is considerably older.

The adequacy of the property construction, fabric and design so far as energy-related issues are concerned is considered to be fair. However, this is to be expected in our view, due to the age and character of the building.

The adequacy of the property services so far as energy-related issues are concerned is considered to be poor so far as the boiler and hot water systems are concerned and fair for the lighting. However, this is to be expected for similar reasons.

The **Recommendations Report** provides recommendations on how the energy performance of the building can be enhanced, together with an indication of the pay back period.

I have only included improvement recommendations for your property that are appropriate bearing in mind the age and planning status of the property. I make the following particular points about some of the recommendations.

The recommendations are provided in four categories:

- short term pay back – less than three years;
- medium term pay back – between three and seven years;
- long term pay back – greater than seven years;
- other recommendations based upon the Energy Assessor's inspection and knowledge.

Table B1: Recommendation report

Recommendations with payback estimated at less than 3 years			
Item	Recommendation	Comment	Benefits
1.	Replace 38mm diameter (T12) fluorescent tubes on failure with 26mm (T8) tubes.	Relatively simple improvement when the old tubes fail.	Savings in operational costs.
2.	Improve insulation on HWS.	This is the hot water cylinder in the cupboard in the east kitchen.	Savings in cost of storing and re-heating water, although in practice there is little space in the cupboard to effect this improvement.
3.	Replace tungsten GLS lamps with CFLs.	'GLS lamps' are ordinary light bulbs. 'CFLs' are compact fluorescent lights	Reduced energy bills, and (sometimes) longer-lasting bulbs.
4.	Some spaces have a significant risk of overheating. Consider solar control measures such as the application of reflective coating or shading devices to windows.	You could provide blinds to the windows on the south and western (riverside) elevations.	Would help reduce temperatures in the summer; and also any glare in the building.
5.	Consider replacing T8 lamps with retrofit T5 conversion kit.	'T5' lamps are slimmer fluorescent tube lights.	See items 1 and 3 above.
6.	Consider replacing heater boiler plant with high efficiency type.	The boiler is old, inefficient and does not comply with modern standards.	Reduced energy costs. A condensing boiler would be considerably more energy efficient.
7.	Add optimum start/stop to the heating system.	This is a system that improves control of the heating system by reacting to external temperatures and the building's fabric.	If operating efficiently would reduce the heating bills.

(You could include similar tables for other recommendations. Note how we explain for our client some of the terms SBEM automatically generates – e.g. 'Heat generator', 'HWS', 'lamps' and 'CFLs' are not expressions in common use.)

In addition to the above recommendations, you should also consider:

- I have assumed there is no insulation above the sloping ceilings in the restaurant and in the roofs over the main (western) kitchen, preparation room and lobby. As you know, there is no access into those areas. Provision of insulation here would make this part of the building more energy efficient. However, placing insulation above these areas would be difficult;
- you should lay further insulation over the ceiling at the eastern end of the building as the existing insulation is inadequate;
- you should install thermostatic radiator valves to all radiators – only four radiators have such control. This would enable you to control the heating better.

(You could include these recommendations in your RR.)

I believe you could carry out the works at items 1, 3 and 6 reasonably easily. However, providing external shading devices to windows could be more difficult because of the fact your property is listed and in the conservation area. In addition, you will need to obtain the landlord's consent.

Provision of other 'renewable' energy sources such as wind turbines, photovoltaic and solar energy arrays on the roof and ground and air source heat pumps could also contribute to reducing operating costs.

Installation of the improvements in the RR and our other suggested improvements are likely to result in an improved energy rating.

The assessment has been based on data collated during my recent inspection, and has been prepared using government approved iSBEM (v3.4a) software. The data has been lodged with Landmark, an arm of the Land Registry office. The full set of documentation can be accessed online, for file download and/or printing, at the following site:

www.ndepcregister.com

You should enter the following report reference: 0000-0060-0040-6000-0103.

I would stress an occupier is under no legal obligation to act on the recommendations in that Report.

The **SBEM Main Calculation Output Document** shows a diagrammatic representation of the energy consumption and CO_2 emissions from the building. You will note the amount of energy used by the heating system in your building. However, you might also like to take particular note of the amount of energy, and therefore cost, that is attributable to lighting and hot water.

(Some CEAs do not include this document as the message contained in the watermark that sometimes appears can confuse clients)

I have inspected the lighting in so far as this was possible. However, I could not fully analyse many of the lighting units and I have entered data into the software program using a simple approach based on the types of lamp.

There is an alternative approach that involves entering data confirming the exact amount of power consumed by the lamps. In this particular case this was not possible (other than in one room) due to limited access to the lamps.

It is possible that a better energy rating could be obtained by carrying out such an exercise. I will be happy to provide further advice if required.

In the attached Recommendation Report I have recommended that any occupier considers carrying out some improvements to various parts of the fabric and services of the property. An occupier must not carry out any of those improvements before they take advice from appropriate professionals such as their solicitor, surveyor, building services engineer, accountant or other appropriately qualified person. In particular, they must take care in the following instances:

- if the property is in a sensitive area such as a Conservation Area, or is 'listed' as being a building of special and/or architectural historical interest, most external and some internal alterations will require planning or similar permissions from the local authority – the property is currently listed and in a conservation area;
- some energy improvement works can attract 'enhanced capital allowances', i.e. there are tax advantages if you carry out certain types of energy-efficient works, e.g. installation of certain types of boiler – talk to me and/or discuss the list of recommendations with your tax adviser/accountant;
- some alterations might affect the property's structure, health and safety, conservation of fuel and power (energy efficiency) or some other relevant matter requiring Building Regulations approval; e.g. installation of a new boiler, new wall insulation or similar;
- if the occupier or owner is a tenant leasing, or landlord letting, a property; most alterations or improvements could significantly affect the legal relationship, i.e. the terms of the lease, between the landlord and tenant. This can sometimes result in major financial claims for damages or dilapidations, or affect the amount of rent that is payable, or cause other disagreements between the two parties. In most instances a tenant must obtain written approval from the landlord for any work – seek advice from your solicitor or other legal adviser;
- some improvements or alterations could affect the buildings or contents insurance policy, in the event of a claim in some instances possibly affecting the amount of any financial settlement – seek advice from an insurance broker and/or insurance company;
- if any of the works that I have recommended could affect the party wall, the occupier/owner might need to serve notice on adjoining owners under the Party Wall etc Act 1996;
- some work requires that you must gain access onto an adjoining owners' land. There is generally no automatic right of access in such cases, and it is sometimes necessary to negotiate an access;
- any contractor who carries out work on the premises may need to comply with the Health and Safety at Work Act and associated Regulations. However, some works require application of special health and safety rules, under the Construction, Design and Management Regulations.

In all instances my advice would be to discuss the matter in full with the appropriate person and ideally in writing to record such discussions and advice for the future in the event of any dispute arising.

In addition, you should note that my advice is current. The EPC and RR are intended to be valid for 10 years. The energy improvement market is in a state of constant change. Before any improvements are carried out, please ask my advice to ensure that the proposals are the best 'fit' bearing in mind the current technology.

In particular, you should note that provision of external attachments to the property such as wind turbines, solar energy systems and photovoltaic arrays might not receive planning approval, without overcoming local objections and/or possible local authority refusal of such permissions.

(This ends our client report)

DISCUSSION

We could not discover much about the levels of insulation or thermal capacities. We therefore generally chose from the SBEM 'inference procedures'. We chose 'activities' from the menu, based on our observations at the time of our inspection.

Section 1 – the asbestos register helps to confirm a reduced health and safety risk when compared with many old buildings. It is important to note the restrictions to your inspection, particularly the staff working on the day of your inspection. Your client has worked in the building for years – a helpful source of information. The building's situation in the conservation area, and listed status, has implications for recommendations. You will need a boat to view the west elevation, at high water! Note the extensive checklists to ensure we do not forget to check all relevant issues.

Section 2 – there are a number of wall types, as follows:

- solid external wall, identified from the thickness, age and various bonds, predominantly 'English garden wall' – with 'exterior' adjoining condition;
- old timber doors on the front wall. We saw no evidence of insulation and chose to use a description from the library with a 'U' value and K_m value that seemed 'about right';

- the same wall as 1, but with an adjoining condition of 'underground';
- another solid wall, but this is the party wall at the eastern end, with an assumed 'conditioned adjoining space';
- modern concrete blocks forming some internal partitions, with 'conditioned adjoining space' adjoining;
- timber stud work, with 'conditioned adjoining space';
- timber stud work, with 'external' adjoining condition.

Section 3 and 4 – we found insulation in the right hand roof. However, we cannot confirm any insulation in the sloping ceilings, without good documentary evidence and/or opening up. Although there is a reasonable possibility this roof may be insulated from the 1980s refurbishment, this is impractical in this case. Thus, we chose two roof types, with 'external' conditions, as follows:

- uninsulated pitched roof above the restaurant and central area (left-hand kitchen/lobby); and
- insulated area over Zones D, E, F and G, 'U' value from a calculation program and 'K$_m$' taken from a similar roof using inference procedures – this seems reasonable. The calculation is 0.433 – we adopt 0.45.

Section 5 – floors are reasonably visible and accessible (but note we make careful note of restrictions) and comprise:

- old (in the cellar), and possibly modern solid floors in the lower ground floor, both to 'underground'. The lower ground floor is unlikely to have been insulated in the refurbishment in the 1980s as the Building Regulations' requirement for floor insulation dates from the 1990s;
- suspended timber floor over the lower ground floor, no evidence of insulation – 'conditioned adjoining space';
- suspended timber floor over the basement. That area is unheated and unlikely to be heated, although there is some residual heat from the boiler. We chose 'unheated adjoining space'.

Section 6 – the doors are all doors, none of them are 'windows', although we note some glazing. All are solid, uninsulated, timber. We chose just one door type, with a 'U' value of 3 and K$_m$ value of 11.25, both figures apparently reasonable.

Section 7 – windows are all similar, i.e. single glazed, softwood frames, clear with no coating and no insulation. On the south and north elevations we note some possible 'overhangs' providing some shading. We consult the May 2009 *BRE User Guide*, p. 102, and enter the appropriate transmission factors.

Section 8 – we confirm our 'U value' calculation in Figure B33.

Section 9 – we have insufficient information to calculate electric power factor so accept the default, and we find no air-conditioning.

Sections 10 and 11 – we take particular note of the heat source and fuel in the basement, recording the relevant details on the manufacturer's label. We check whether the heat source is on the ETL (extremely unlikely); it isn't. We research the internet, but the boiler is too old for current information to be available. We search the SEDBUK database to help confirm the age – without success (thereby probably confirming it

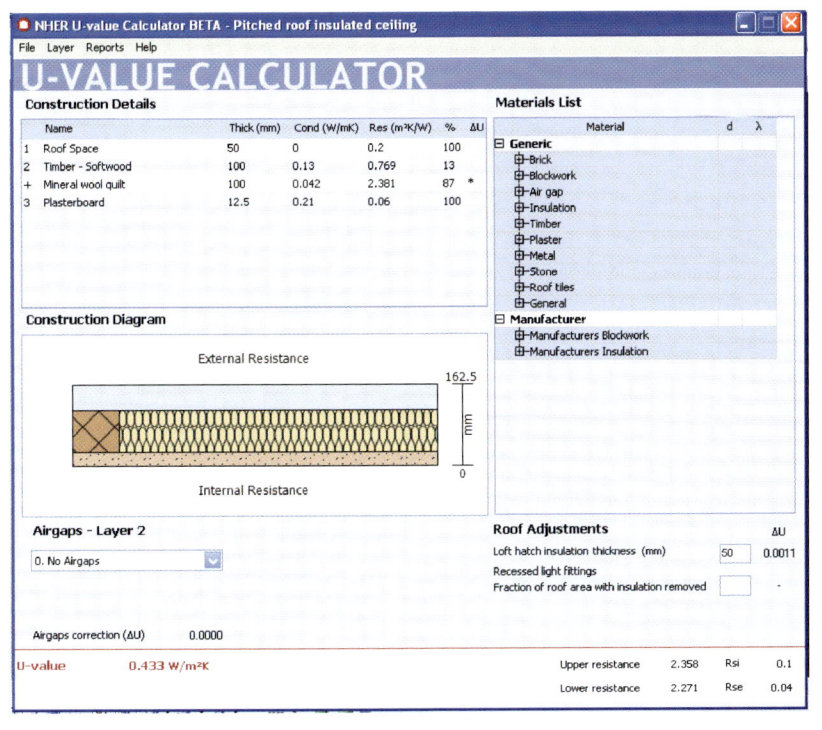

Figure B33: 'U' value calculation from program

is old). In view of the apparent age of the installation as confirmed by the owner on the questionnaire ('around 1980'), and with no other documentary evidence, we conclude the boiler is older than 1989. With no other information available from the manufacturer we must accept the default figure of 0.65 for the 'effective heat generating seasonal efficiency'.

We note the programmer, and enter this as 'Central Time Control'. We see four thermostatic radiator valves ('local temperature control', i.e. 'room by room') in the restaurant (two), west lobby and ladies toilet only. We cannot reasonably apply such controls to the system when entering data with so few TRVs in place compared with the total number of radiators – an issue to note for the RR and the client report. We cannot therefore consider altering the seasonal efficiency of the heat generator in accordance with Section 2.6, Table 7 of the CLG *Non-Domestic Heating, Cooling and Ventilation Guide*.

We conclude some zones are directly conditioned by the radiators – A, B, J and K.

We conclude zones C and G are indirectly conditioned because there is likely to be a 'high level of interaction' between the conditioned zones and those areas. This is a matter for judgment, and frankly, disagreement – your own conclusions will depend on the evidence on the day. This is one reason why it is vital to record all of your decisions and take good photographs to support those decisions.

Zone F has a portable electric heater, in use on the day – we note this. We conclude this zone is likely to have fixed heating installed and allocate electric resistance heating – we remember the spirit of the EPBD is to capture the likely CO_2 emissions. Zones D, E, H, L and I are unheated. We do not believe they are either indirectly conditioned or likely to have fixed heating provided. We allocate 'zones without HVAC system' in the program.

Section 11 – hot water is primarily heated by the main boiler, with storage in the hot water cylinder in the east kitchen. The cylinder has an electric immersion heater element, but SBEM does not recognise that type of system and thus 'generator type' is 'same as HVAC'. We estimate the 'storage volume' of the cylinder at 90 litres.

We look at the hot water distribution pipes and turn the taps on at all of the hot water outlets. Hot water takes some time to arrive and we conclude there is no 'secondary circulation' system. We are unlikely to encounter such a system in a building of this type and age anyway.

We note hot water dead-legs in zones C and J and enter the lengths.

Section 12 – we find no renewable energy systems. However, note the fact we have carefully recorded the nature of our observation – 'visual' and limitations 'none', to help confirm, if challenged later, the fact we conducted a careful inspection. You will be amazed at how easy it is to 'miss' a solar collector on a roof.

Section 13 – all ventilation is 'natural', i.e. through windows. We see three mechanical extract units. We note as much information as is available on the units and try to locate information regarding efficiencies, but to no avail. We must therefore accept the default values for 'ventilation flow' and the 'exhaust specific fan powers'. We see no evidence of any heat recovery systems.

Section 14 – we record all of the available information regarding the lamps, but lack of access to luminaires means we cannot confirm the efficacies of the lamps except in one zone where we conduct a 'lumens/circuit watt' exercise.

In the cellar we have two lamp types. We decide to create two zones, with a 'virtual' boundary between them, assigning 'T12s' to the eastern zone and 'tungsten' to the western zone. We do this as the program is particularly sensitive to tungsten lamps because of their poor efficacy.

Section 15 – an overall plan, with north point, ensures we can demonstrate we have inspected the entire property and confirm measurements of the building we do not include in the assessment – the laundry.

Section 16 – the difficult geometry of this building means we must spend some time entering data if we want to ensure the envelope areas are correct. Note the fact we took external check measurements as an added 'insurance' for the geometry entry. In particular, we must take note of:

Zone A – we must carefully measure and enter the difficult shape at the west (left-hand) end of the zone. In addition, the shape of the roof is an issue we need to consider with care – note how much larger the roof is when compared with the floor. At the eastern end the geometry is particularly difficult because of the sloping roof. In addition, note the fact that in this zone the envelope height extends up to the ridge of the roof, whereas in the adjoining zone to the right (the kitchen, or zone 'C'), the ceiling is lower. Thus, we have a conditioned space (zone 'A') with an eastern envelope that in part adjoins an unheated space (the roof space over the kitchen). We adopt a cautious approach and enter the adjoining condition as 'exterior'. Correct entry of the window data is time-consuming.

Zone B – we must address a similar issue of an envelope (to the north) onto the unheated roof space above the kitchen. The old door next to the electric meters requires separate data entry as 'wall type 2'.

Zone C – we find this area difficult to measure due to kitchen equipment and people working. The roof envelope shape is as in the previous zones except for the fact the ceiling does not extend up to the ridge.

Zone D – although this zone is an awkward shape, it is simpler because of the flat ceiling. There are two lamp types. We adopt T8s as there are more of them, but there is an argument for two zones.

Zone E – we could have decided to include this store in the right-hand kitchen, with some justification. Remember to consider whether the adjoining condition to the east, in the adjoining building, is indeed 'conditioned'. Without firm evidence to the contrary, we believe you should normally assume such is the case.

Zone F – a difficult zone for geometry.

Zone G – challenging to 'visualise' and measure because of the two flights of stairs.

Zone H – a zone with a number of issues. We need to take care with the adjoining conditions – parts of the south and north envelopes are to 'underground'. Part of the east envelope is to 'underground', where our careful measurement of vertical heights should have revealed the floor level in the cellar is lower than in the adjoining lower ground floor – consider 'section A-A'.

Zone I – adjoining conditions are an issue here as for the preceding zone. The west envelope is entirely to 'exterior'.

Zone J – a relatively simple shape, but with a short vertical section of the north envelope to 'underground' – look at 'section B-B'. We note a large space to the rear behind dry-lining (also in zone L). We decide to include that space in our data entry to capture the full envelope sizes. We include the two toilets as one zone to reflect the zoning 'rules'. However, the solid block partition between the gents and ladies toilets is 'heavy construction'. We must enter the wall into the program so SBEM is aware of the thermal capacity and can 'determine how the building retains and emits heat' – *BRE User Guide*, section 3.3. We must remember to enter 'unheated adjoining space' for the west and east envelopes.

Zone K – a relatively simple data entry exercise for this zone. We decide to ignore minor differences in envelope areas beneath the landing floor due to their small sizes.

Zone L – the last zone, and an easy one – you deserve it!

Section 17 – an opportunity for reflection and careful checking to ensure our assessment is correct, or at least 'about right'.

SUMMARY

Note the careful way we consider and record our observations in each section and the source of our evidence. We note where our inspection is obscured. We 'strike through' with a bold pen-stroke those sections that do not apply, to indicate we have considered those possible aspects of the data entry – a legitimate argument is that SBEM data collection (and entry) is for 'plodders'; **or** professionals who take care at every stage of their job.

We hope you appreciate we have approached the various questions that arose during our reflection about the data entry in a logical manner; and by looking at the site notes how we have provided a thorough audit trail. We adopted the same approach for the RR and the client report.

Appendix C – Case Study 3 – Level 4 HVAC Assessment of the PACE building, De Montfort University, Leicester LE1 9BH

INTRODUCTION

As case study 2 provided a complete inspection with site notes and EPC, we have not provided similar material here but focus entirely on assessment of the HVAC systems following the same steps as in case study 2.

DESCRIPTION OF THE BUILDING

This building is used for performance arts and includes:

- studio theatre;
- dance studios;
- general offices;
- technicians offices;
- meeting rooms;
- toilets and changing rooms.

BUILDING INSPECTION

The following describes a process we find useful for this building type. We record all necessary information in our site notes and take appropriate photographs.

External inspection

You can learn many things walking around the whole building:

- the front elevation is heavily glazed with opening windows (Figure C1) – some rooms must have natural ventilation;
- one facade has glazed areas without windows (Figure C2) – some rooms are mechanically ventilated or air conditioned but we see no equipment;
- one facade has small windows with a number of extract fan outlets adjacent (Figure C3) – some rooms have local extract;
- near the louvre doors we see two fan-assisted balanced flue outlets (Figure C4) – there is combustion equipment to check.

It is worth noting some things we did not find – fuel storage facilities, condenser units or ventilation openings (other than the extractor fan outlets). This suggests the anticipated air conditioning equipment may be on the roof.

Figure C1: Front elevation showing opening windows and main entrance

Commercial Energy Assessor's Handbook

Figure C2: One elevation has some large fixed glazing

Figure C3: One elevation shows a number of small extract fan outlets

Figure C4: One pair of louvre doors and two flue outlets suggest this is a good place to start inspecting the building's HVAC systems

Boiler room inspection

We choose to investigate what looks like a plant room and find the following equipment:

- two wall-hung boilers (Figure C5);
- gas meter;
- water tank;
- motor control panel;
- electrical switchgear panel;

- sets of variable speed pumps and LTHW pipes (Figure C6).

We can immediately ignore the tank and switchgear panel as these have nothing to do with energy assessment. Closer inspection of the boiler and pipe connections show they are gas-fired condensing LTHW boilers (not combi type). We record information from the boiler labels (Figure C7).

The control panel refers to pumps but also has a BMS display (Figure C8). This suggests comprehensive control features and we should not need to make recommendations about controls in the RR.

Some notable absences in the boiler room are hot water and chilled water systems.

Figure C5: Main boiler and gas meter

Figure C6: Variable speed LTHW pumps

Figure C7: Boiler control panel and label

Figure C8: Control panel includes a screen with information from the BMS

Lobby inspection

We enter the building and inspect the lobby and associated corridor. The first item we find is an overdoor heater (Figure C9). The lobby is the glazed space we earlier noted as being naturally ventilated (Figure C10). We cannot see any radiators or convectors – perhaps there is underfloor heating? There are no grilles so there can be no warm air heating. Closer inspection of the ceiling reveals long panels alongside regular ceiling tiles – a radiant heating system (Figure C11). The *iSBEM User Guide* advises we should represent radiant panels as radiators. We find similar panels on the level above (Figure C12). We find one room on the ground floor with a conventional radiator (Figure C13).

Figure C9: Overdoor heater at main entrance

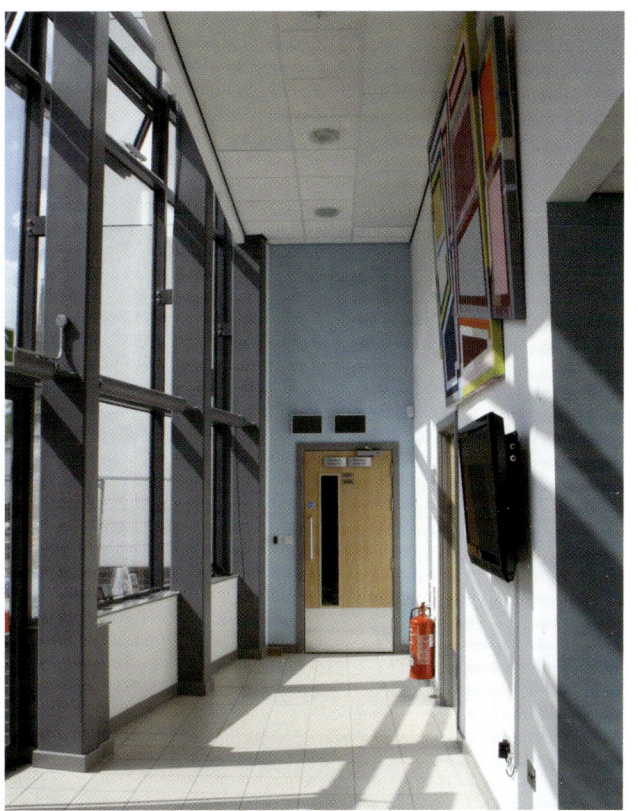

Figure C10: Opening windows in lobby

Figure C11: Lobby and corridor LTHW radiant panels

Figure C12: Radiant panels on level 2

Figure C13: Office with conventional radiator

Theatre rehearsal room

The largest space on the ground floor is a theatre rehearsal room. There is no suspended ceiling and we inspect the building services – even if everything is painted black! We see large ductwork with conventional diffusers (Figure C14 and C15). We cannot see any terminals such as VAV boxes, reheating coils, associated pipes or other types of heat emitter. We reasonably conclude this is a type of CAV system.

Figure C14: General view of theatre rehearsal room

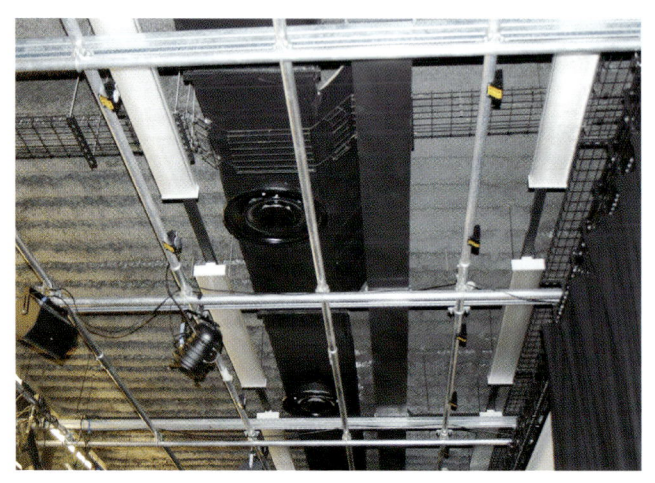

Figure C15: Large ducts and diffusers visible in rehearsal room

Toilets and changing rooms

There are male and female toilets and changing rooms off the corridor on the ground floor and two levels above. These have the small domestic pattern extract fans we saw outside (Figure C16). The slightly unusual heat emitters are radiators (Figure C17). There are basins in the room – inspection of the pipes confirms no secondary circulation system pipework (Figure C18).

Figure C16: The toilets contain a number of small extract fans

Figure C17: Radiators in the changing rooms are enclosed and operate with lower surface temperatures than conventional radiators

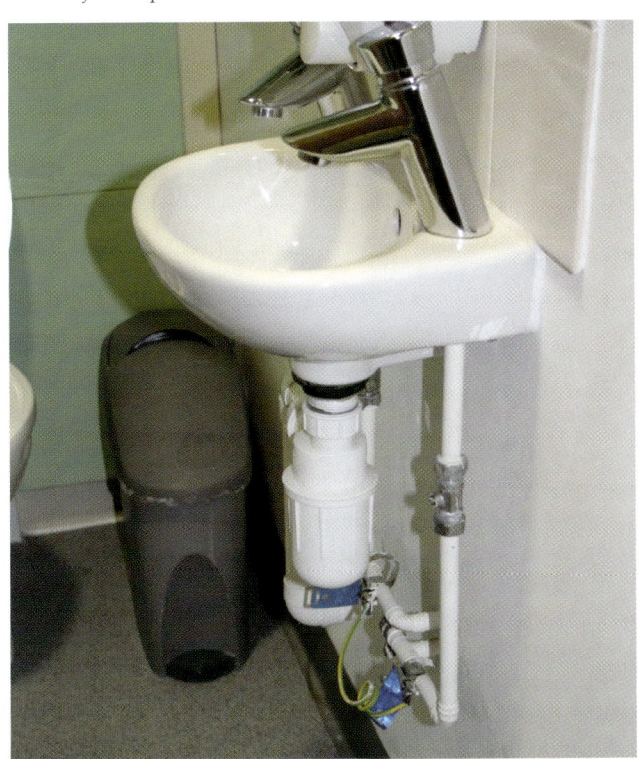

Figure C18: Pipes are visible below wash basins – no evidence of secondary circulation

We find a HWS storage cylinder tucked away in a services cupboard (Figure 5.28.2). This has only electrical connections, not LTHW; so we note the HWS fuel type as electricity. We record the storage capacity from the label. The orange cable tucked under the pipe insulation and controller confirms a trace heating system (Figure 5.26). This system is clearly supplying hot water to the changing rooms. We find a kitchen on this floor and checking under the sink reveals an instantaneous electric water heater (Figure C19).

Figure C19: Kitchen uses a small instantaneous HWS

Main rehearsal room

This is on the second level with a tall suspended ceiling incorporating a number of supply and extract grilles (Figure C20). This could indicate a variety of air conditioning systems. We find a controller on the wall for what looks like a split system (Figure C21). If there is a split system the indoor units must be above the ceiling, connected by ductwork to the diffusers. There might also be a zone supply/extract fresh air ventilation system.

Offices

There are a number of small technician and meeting rooms without external windows. These have cassette units in the suspended ceiling (Figure C22) and we easily find the associated controllers on the wall (Figure C23). We record controller details and brand name. The single ceiling grille suggests a fresh air supply working alongside the refrigeration system, i.e. zone mechanical supply/extract ventilation.

We inspect and record a third level with similar office and meeting rooms.

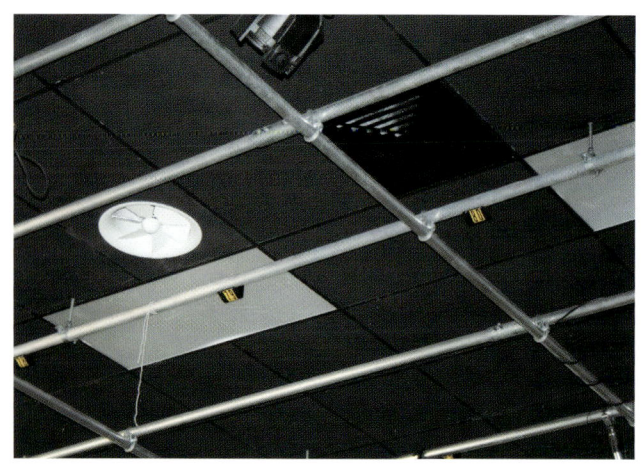

Figure C20: One of the dance rehearsal rooms has a suspended ceiling with combination of round 'swirl' and square diffusers

Figure C21: Rehearsal room has two digital controllers marked 'Daikin' – one probably for time control and the other for temperature

Figure C22: Technicians rooms have cassette units and ventilation grilles

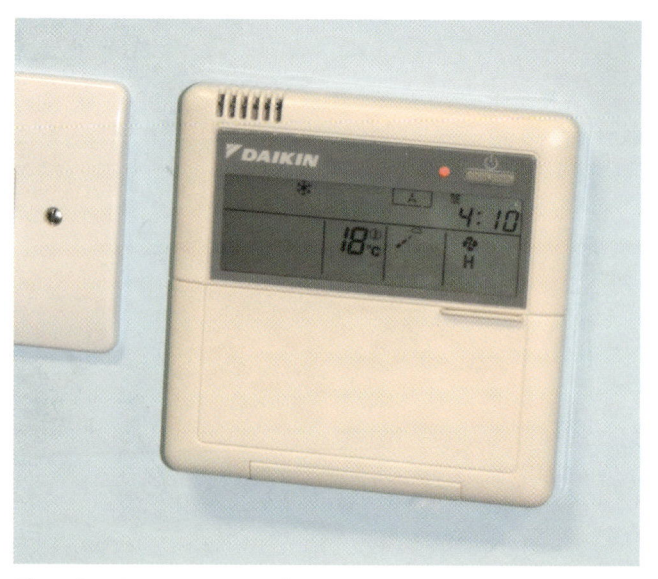

Figure C23: Cassette unit controllers are slightly different to those in the rehearsal rooms

Figure C24: One dance rehearsal room has a suspended ceiling and three cassette units

Dance rehearsal rooms

Two large dance rehearsal rooms take up most of the fourth level. One room has a tall suspended ceiling with three cassette units and some supply diffusers (Figure C24). We find a controller on the wall similar to that in the main rehearsal room.

The second dance rehearsal room has no suspended ceiling. There are three cassette units suspended from the roof (Figure C25) and much ductwork, some passing through without connecting to anything. There are three small ducts directing air to the back of the cassette units (Figure C26).

Roof level inspection

Part of the roof is flat, with several pieces of HVAC equipment including two AHUs and a number of condenser units along with a control panel. There are four large condenser units with the same branding as the cassette units inside (Figure C27). Fortunately somebody has labelled each unit with the associated room numbers. We can see that each condenser unit is coupled to one rehearsal room, i.e. each is connected to three cassette units. We record notes from the labels and will check if these also provide heating.

The fourth condenser unit has three connecting refrigerant pipes and is labelled as 'VRV' (Figure C28).

Appendix C

Figure C25: Cassette units in this rehearsal room suspended from the roof

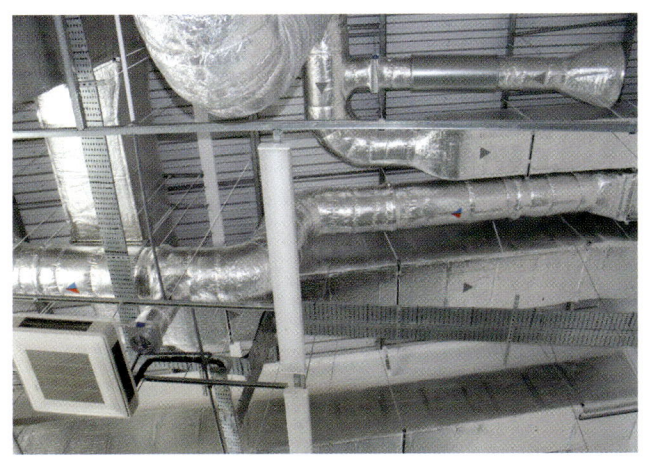

Figure C26: Each cassette has a supply duct above it. Triangular ID stickers on the ducts give a clue to their function

Figure C27: Four large condenser units on flat roof

Figure C28: One condenser unit marked as a VRV (VRF) system

Figure C29: Two AHUs of very similar appearance. Gas pipe connections can be readily identified

We record information from the labels to help in later research. The labels show this is connected to all other cassette units in the small rooms. This system should provide heating (this means selecting the heating source as 'heat pump (electric)' and a split system as the cooling system).

There are two double-deck AHUs with intake and exhaust louvres, and ducts passing into the building (Figure C29). We already identified a CAV system in the theatre rehearsal room. The only other ducted system we have seen is a fresh air supply to the rooms with cassette units. This explains the functions of these AHUs and we can see how the ducts on the roof correspond to those we saw on level 3 inside.

275

The AHUs have telltale yellow gas pipe connections that indicate the heating sections and at the back of the AHUs we see corresponding flues (Figure C30). We conclude both CAV systems have indirect gas fired heaters. We cannot find dampers or mixing box so we assume they are 'fixed fresh air' CAV systems. Condensate drains on one AHU confirms a cooling coil. There are refrigeration pipe connections with the control valves showing where they connect to the AHU for the theatre rehearsal room (Figure C32). We trace the refrigerant pipes to large condenser units (Figure C34). The other AHU does not have cooling.

SBEM does not require much information about the primary air system for the VRV and multi-split

Figure C30: AHUs have gas fired heaters with flues at the rear

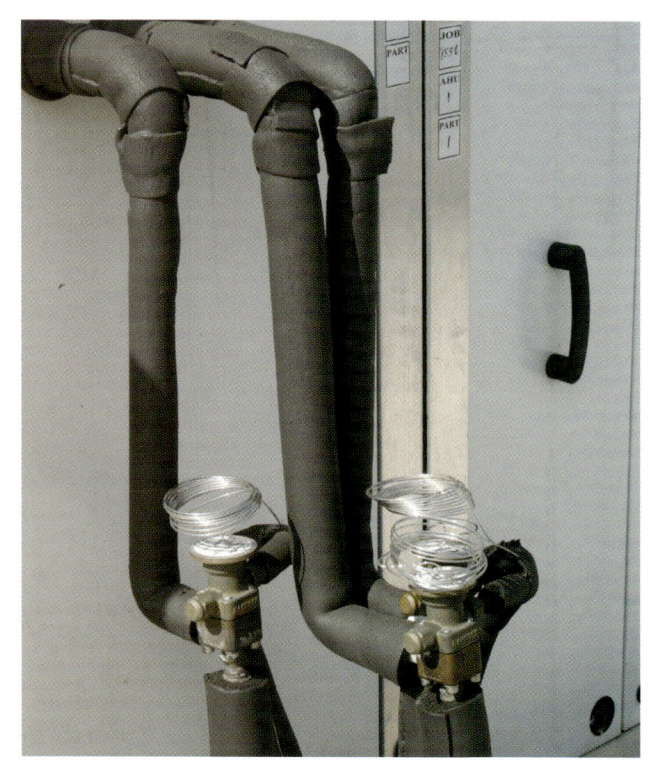

Figure C31: Refrigerant pipes and condensate drain indicate a DX cooling coil and grid electricity as the fuel type

Figure C32: The DX cooling coil is coupled to three condenser units

Figure C33: Control panel switches confirm thermal wheel heat recovery

systems – not even the fuel type. This ventilation system is represented in the calculations by selecting mechanical supply/extract ventilation for the zones with cassette systems. We need to know whether this AHU has a heat recovery device. The control panel shows both AHUs have a thermal wheel (Figure C33).

Note that it has not been necessary to stop any equipment or open any AHU to discover the system types or fuels.

REFLECTION

We now know the heating and cooling fuels for both AHUs – gas and grid electricity. It is not clear how we represent the gas-fired heating and DX cooling in SBEM. A 'work-around' is needed to deal with the gas heater in order to get the correct emission rate. For a given HVAC system the energy delivered is related to the carbon emissions by:

$$\text{EnergyDemand} \times \frac{\text{EmmisionFactor}_{Fuel}}{\text{SEff}} = CO_2 \text{EmissionRate}$$

Energy demand is calculated by SBEM and the fuel emission factor applied depending on which of the available fuels we select. What we can change is the seasonal efficiency. Our aim is to arrive at the correct ratio of fuel emission factor to seasonal efficiency. We decide to select grid electricity as the fuel. We then need to find an effective efficiency so that:

$$\frac{\text{EmmisionFactor}_{Gas}}{\text{SEff}_{GasHeater}} = \frac{\text{EmissionFactor}_{Electricity}}{\text{SEff}_{effective}}$$

The heater in the AHU is very similar to that in a flued forced convection heater and we assume the same level of seasonal efficiency applies – the default is 0.65. The emission factors for electricity and gas are 0.422 KgCO$_2$/kWh and 0.194 KgCO$_2$/kWh respectively. The calculation to find the effective efficiency is therefore:

$$\text{SEff}_{effective} = \frac{\text{EmissionFactor}_{Electricity}}{\text{EmmisionFactor}_{Gas}} \times \text{SEff}_{GasHeater}$$

$$\text{SEff}_{effective} = \frac{0.422}{0.194} \times 0.65 = 1.41$$

The seasonal efficiency is artificially high because electricity has a much higher emission factor than natural gas.

The DX cooling coil and local condenser isn't specifically listed as a cooling source in SBEM but this is physically similar to an air source heat pump and so we can expect similar efficiency. We consequently choose 'heat pump (electric)' as the cooling source for the CAV system. We don't have any specific information about the condenser equipment and so use the default SEER for the heat pump.

We still need to check the energy technology list to see whether any of our systems attract enhanced capital allowances – a task to complete in the office.

We can summarise our findings about the building's HVAC systems as follows:

- natural gas fired LTHW boilers;
- radiators and radiant panels for space heating;
- electric hot water generation with trace heating;
- local extract in toilets;
- fixed fresh air CAV serving the theatre rehearsal room with a DX cooling coil and gas-fired heating;
- primary fresh air for the VAV and multi-split system provided by a fixed fresh air AHU with gas heater;
- mechanical ventilation systems have heat recovery;
- multi-split or VRV in other spaces – with zone mechanical ventilation.

Appendix D – Pipe and Duct Identification

You can often identify piped building services by reference to regularly spaced colour bands and labels applied around the pipe or insulation. Such labels may not be common in smaller premises with heating systems on a domestic scale but labelling of pipe services has been common practice in non-domestic HVAC installations for a number of decades. An example is shown in Figure D1. Pipe paint colour and identification band colour coding is defined in British Standard 1710 *Standard for Specification of Pipelines and Services*. A table illustrating the bands on common piped services is shown in Figure D2. This is not an exhaustive listing. You will often see the bands applied along with arrows showing direction of flow and sometimes text with an abbreviation of the service name (e.g. LTHW – see Appendix E, Table E5).

Figure D1: An example of pipe services identified using BS1710 colour coding

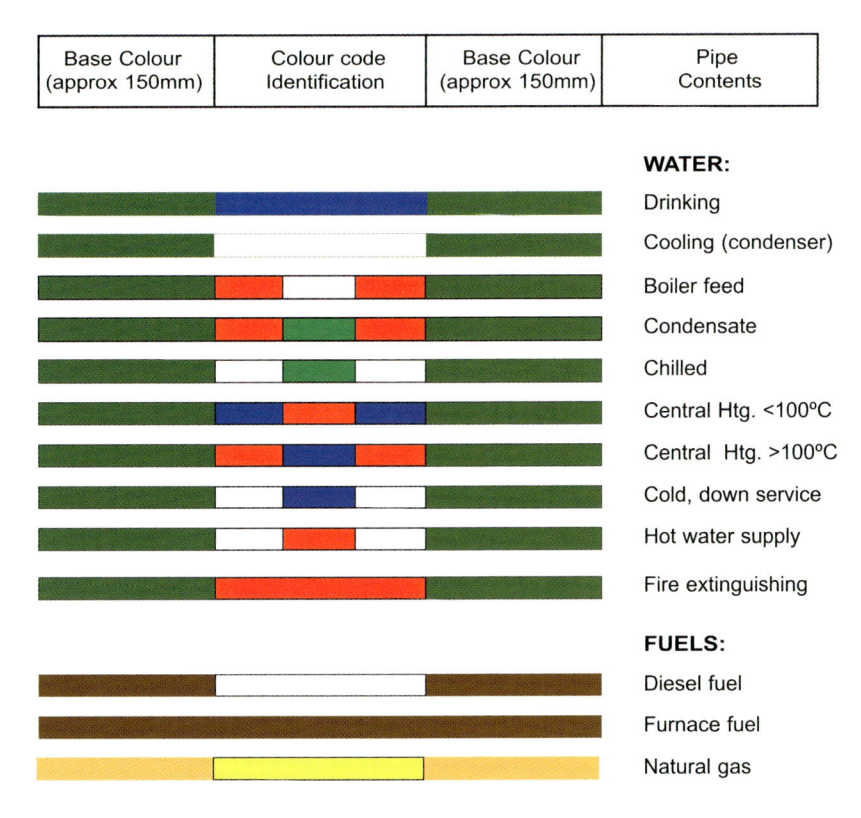

Base Colour (approx 150mm)	Colour code Identification	Base Colour (approx 150mm)	Pipe Contents

WATER:

- Drinking
- Cooling (condenser)
- Boiler feed
- Condensate
- Chilled
- Central Htg. <100°C
- Central Htg. >100°C
- Cold, down service
- Hot water supply
- Fire extinguishing

FUELS:

- Diesel fuel
- Furnace fuel
- Natural gas

Figure D2: Examples of British Standard colour identification bands for pipe services

Coloured triangles are sometimes used to identify ducts and their purpose. Standard practice is set out by the Heating and Ventilating Contractors Association document 'Specification for Sheet Metal Ductwork' DW/144: appendix B. Some common symbols are shown in Figure D3.

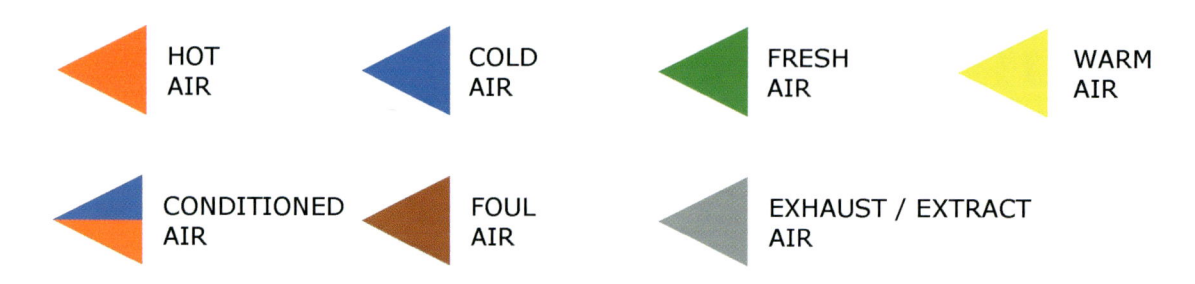

◄ HOT AIR   ◄ COLD AIR   ◄ FRESH AIR   ◄ WARM AIR

◄ CONDITIONED AIR   ◄ FOUL AIR   ◄ EXHAUST / EXTRACT AIR

Figure D3: Examples of identification symbols used on ductwork

Appendix E – Drawing Symbols and Abbreviations

One of the most useful sources of information for assessment of HVAC systems are schematics of heating and cooling systems. You can find schematics displayed in boiler and plant rooms and can learn much from them (Figure E1). You should find information about which systems are used and which zones they serve. Like other sources of information schematics are not 100% reliable and may be out of date by virtue of refurbishments and alterations. You need to become familiar with the symbols and abbreviations commonly used to understand and make the most of the information – this appendix will help.

The drawing symbols are grouped into four tables. Symbols for pipe system schematics included in BS 1553-1:1977 are shown in Table E1. You will also find many drawings that do not strictly comply with this standard (e.g. you may find valve symbols shaded black). Other symbols are in common use and we include some of these in Table E2.

Symbols commonly used in simplified schematic representations of AHUs and air distribution systems are shown in Table E3. We have used these symbols in a number of the schematics in Chapter 6. Symbols used to draw air handling unit components and show overall configurations vary considerably. We have included some in Table E4 that are commonly used. You will find these symbols in schematics but they are often used for diagrams showing the physical configuration of equipment. We have used these symbols to show AHU configurations in Chapter 6.

Abbreviations are used to denote particular services (e.g. LTHW) but also particular components. Common abbreviations are listed in Table E5 and E6.

Figure E1: An example of a schematic provided on a plant room wall

Table E1: Standard symbols for pipe schematics (selected from British Standard 1553-1:1977)

Description	Symbol	Description	Symbol
Pipeline (major flow line)		Radiator	
Convector, fan		Convector, natural	
Junction (connected)		Indication of flow direction	
Valve: In-line (any type or pattern)		Valve: Angle (simple screw-down)	
Valve: Relief (in-line)		Radiant panel, ceiling mounted	
Valve: Relief (angle, pressure)		Pipe coils, plain and finned	
Valve: Check (non-return or reflux)		Valve: Flanged (alternative)	
Valve: 3-way		Valve: Butterfly	
Valve: Globe		Radiant strip: in front of, or above, section	
Valve: Ball		Towel rail	
Open tank (basic symbol)		Closed tank (basic symbol)	

Table E1 (cont.)

Description	Symbol	Description	Symbol
Fired heater/boiler (basic symbol)		Heater/cooler unit, floor mounted type Heater: H or + Cooler: C or −	
Air filter		Heater/cooler unit, horizontal type Heater: H or + Cooler: C or −	
Heat exchanger (basic symbols) Alternative:		Heater/cooler unit, downward type Heater: H or + Cooler: C or −	
Heating/cooling coil (basic symbol)		Air-blown cooler	
Rotary pump, fan or simple compressor (basic symbol)		Centrifugal pump or centrifugal fan	
Cooling tower: forced/induced draft (Fans included as appropriate)		Evaporative condenser	
Water chilling evaporator		Water-cooled condenser	
Air-cooled condenser, natural/forced draft		Air cooling evaporator	

Table E2: Other common symbols used in HVAC schematics

Description	Symbol	Description	Symbol
Meter, e.g. gas or heat meter		Temperature gauge	
Pressure gauge, e.g. at pump outlet		Temperature sensor, e.g. for BMS control system	
Pump (common symbol). Triangle may not be filled		Pump – twin set	
Test Point		Non-return valve/check valve (alternative symbol)	
Flexible connector		Three way valve with actuator	
Orifice measuring device		Commissioning station valve	
Strainer			

Appendix E

Table E3: Air system component symbols used in single line schematics

Description	Symbol	Description	Symbol
Heating coil	▯ with +	Cooling coil	▯ with −
Heating coil (alternative 1)	▯ with ⊕	Cooling coil (alternative 1)	▯ with ⊖
Heating coil (alternative 2)	▯ with ⊕ and diagonal	Cooling coil (alternative 2)	▯ with ⊖ and diagonal
Centrifugal fan	(circle with line)	Axial fan	(hourglass shape)
Filter (panel type)	(hatched rectangle)	Louvre	(angled lines)
Motorised damper	(damper with motor box)	damper	(diagonal line with dot)

285

Table E4: Air handling unit diagrammatic symbols (flow right to left)

Description	Symbol	Description	Symbol
Centrifugal fan		Plenum (plug) fan	
Access section		Panel Filter	
Cooling coil section		Bag Filter	
Heating coil section			
Thermal wheel		Plate heat exchanger	
Sound attenuator		Sound attenuator (alternative)	
Evaporative cooler section		Steam Humidifier section	

Table E5: Abbreviations for piped services

Piped service	Abbreviation
Chilled water	CHW or CHWS
Cold water main	MWS or MCWS
Cold water down service (from tank)	CWS
Tank cold water service (as above)	TCWS
Cold water drinking	DWS
Condensate	C
Compressed air	CA
Cooling water	CLW
Liquefied petroleum gas	LPG
Low temperature (or pressure) hot water heating	LTHW (or LPHW)
Medium temperature (or pressure) hot water heating	MTHW (or MPHW)
High temperature (or pressure) hot water heating	HTHW (or HPHW)
Steam	S

Table E6: Abbreviations for pipe system components

Pipework components	Abbreviation
Cold water storage cistern	CWSC
Hot water storage cistern	HWSC
Strainer	ST
Expansion vessel	ExVl
Air release valve	ARV
Air vent	AV
Automatic air vent	AAV
Check valve or non-return valve	CV or NRV
Commissioning station	CS
Double check valve	DCV
Drain tap or drain cock	DT or DC
Gate valve	GV
Isolating valve	IV
Lockshield valve	LSV
Stop cock or valve	SC or SV
Pressure reducing valve	PRV
Temperature and pressure relief valve	TPRV
Thermostatic radiator valve	TRV
Gas solenoid valve	GSV
Gas cock	GC

Appendix F: Surface Area Ratio (as cited in Chapter 9, External windows and doors)

The surface area ratio of a window or rooflight is specified in SBEM as the ratio of the developed area to the projected area. The developed area is the area of the glass and the frame. The projected area is the area of the opening in the wall or roof in which the window or rooflight is fitted. The default value for a window is 1 and for a rooflight it is 1.3. These values may need to be adjusted, particularly for rooflights where the area of the glazing and frame is often bigger than the opening. The default value cannot be less than 1.

Figure F1 shows a traditional rooflight assembly. Its developed area is 3.664m² (i.e. total area of glazing) and its projected area is 1.440m² (i.e. the opening 1.6m × 0.9m). The surface area ratio of these two figures is 2.54. The same approach would be taken with domed and conical rooflights.

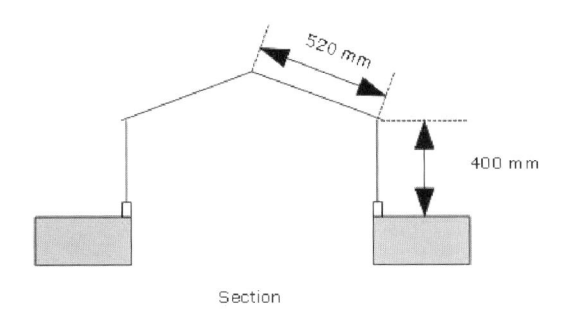

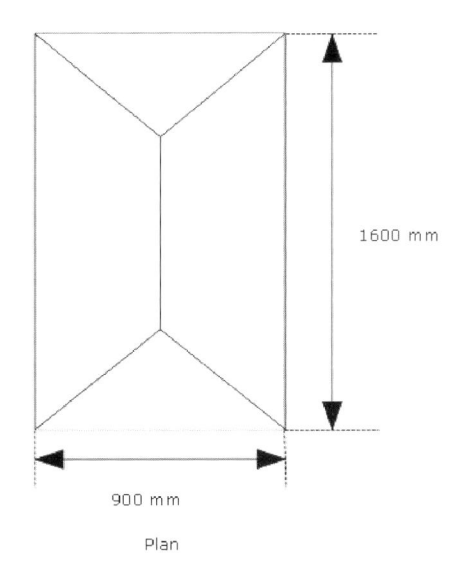

Figure F1

Appendix G – Area Ratio Covered (applied to rooflights only)

Area ratio covered is the ratio of the roof area covered by an array of rooflights to the total area of the rooflight glazing. It is used by SBEM for automatic daylight zoning but the program user must confirm the default value or adjust it. Figures G1 and G2 show the data to be considered in an assessment and how the ratio is calculated.

Figure G1: An array of rooflights

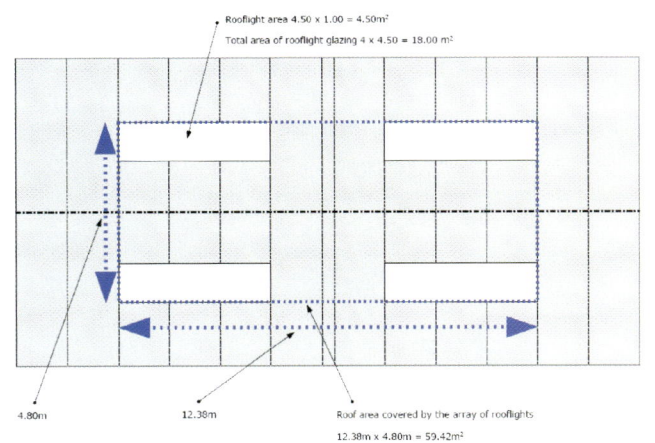

Figure G2: Plan of rooflight array

In this example we established the roof is less than 10° by trigonometry from the dimensions of its height at the eaves and the ridge, which we recorded using a laser measuring device. We estimated the dimensions for the rooflights and the area covered by the array from the spacing of the purlins and the width of the roof cladding.

The *area ratio covered* is the ratio of the roof area covered by the array (59.42m^2) to the total area of the rooflight glazing (18.00m^2). The ratio in this case is 59.42/18.00 = 3.30. SBEM currently limits the data entry for the *area ratio covered* to a range between 1 and 4. We would enter a ratio of 4, if the calculated ratio had exceeded 4.

Appendix H: Transmission Factor Correction (as cited in Chapter 9, Fins and overhangs)

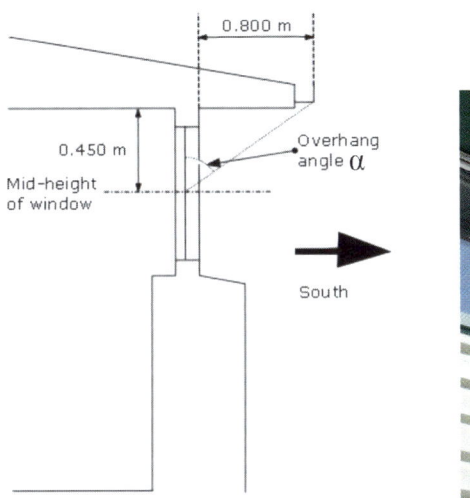

The building in this example is located in Littlehampton, which is at Latitude 50.82° N (decimal degrees). You can get the latitude of specific locations from Ordnance Survey maps and some internet sites, such as www.getlatlon.com. At the end of this example we have stated the latitude of the locations that are currently included in the SBEM database and a few more.

The dimensions shown in the diagram were measured on site. Luckily the windows opened and the depth of the overhang could be measured safely using a steel tape, without leaning out of the window. (Check first that you will not injure anyone or damage any property, such as parked cars, should you accidently drop your tape!) We would have estimated the depth if we could not get access from the windows for any reason. We have also included the width of the gutter in the overhang depth particularly because of its width and the low position in which it is fixed. In the photograph you can see the gutter contributes to the overall shading from its reflection in the window.

The calculation for correcting transmission values through windows given in the *iSBEM User Guide* is:

$$TS = F_o F_f$$

Where F_o is the partial shading correction factor for overhangs, and F_f is the partial shading correction factor for fins.

Our example considers the correction factor for overhangs only, but the whole calculation is important.

Values for F_o and F_f are given in two tables in the *iSBEM User Guide*. The table for F_o is reproduced below and marked to identify the data that are relevant to our example. We have calculated the overhang angle using trigonometry and the dimensions shown on the diagram.

Tan α = opposite/adjacent

Tan α = 0.80/0.45 = 1.78

α = Tan^{-1} 1.78 = 60°

An alternative to a trigonometry calculation might be to prepare a sketch showing a section through the window and overhang and measuring the angle with a protractor.

Commercial Energy Assessor's Handbook

50.82° is between the two given latitudes

Overhang angle	45° N lat.			55° N lat.			65° N lat.		
	S	E/W	N	S	E/W	N	S	E/W	N
0°	1.00	1.00	1.00	1.00	1.00	1.00	1.00	1.00	1.00
30°	0.90	0.89	0.91	0.93	0.91	0.91	0.95	0.92	0.90
45°	0.74	0.76	0.80	0.80	0.79	0.80	0.85	0.81	0.80
60°	0.50	0.58	0.66	0.60	0.61	0.65	0.66	0.65	0.66

Table 10: Partial shading correction factor for overhang, Fo

The window in the example faces south.

The table from the *iSBEM User Guide* only shows three latitude angles and three overhang angles. The latitude for the location of the building in our example is not shown on the table so a value for *Fo* must be interpolated from the values given. We have completed the interpolation for the partial shading correction factors of both overhangs and fins in the six tables at the end of this appendix.

Using the table for south facing windows, an overhang angle of 60° and rounding up the actual latitude from 50.82° to 51° you should find a value for *Fo* of 0.56.

Correction factor

You will note from the tables that where there is no shading (0°) the correction factor is 1. The same is true for shading from fins, so the value for *Ff* in our example is 1. We can now calculate the correction factor: *TS = FoFf* = 0.56 × 1 = 0.56. This value is entered into the SBEM program for the windows affected by the shading.

Latitudes

Location	Latitude	Location	Latitude
Aberdeen	57.15° N	Newcastle	54.98° N
Belfast	54.60° N	Norwich	52.63° N
Birmingham	52.48° N	Nottingham	52.96° N
Cardiff	51.48° N	Penzance	50.12° N
Edinburgh	55.95° N	Plymouth	50.37° N
Inverness	57.48° N	Saint Helier	49.19° N
Leeds	53.80° N	Southampton	50.90° N
London	51.50° N	Swindon	51.56° N
Manchester	53.48° N	Thurso	58.59° N

Note: the latitudes in this table are shown in decimal degrees for ease of calculation. Latitude is more commonly stated in degrees, minutes and seconds.

For example, the latitude for London would be 51° 30' 00"N because 0.5 of a degree is 30 minutes. Latitude converters, such as http://convertalot.com/latitude_longtitude_converter.html, are available on the internet.

South facing window partial shading correction factors for overhangs, Fo

Latitude / Overhang Angle	45	46	47	48	49	50	51	52	53	54	55	56	57	58	59	60	61	62	63	64	65
0	1.00	1.00	1.00	1.00	1.00	1.00	1.00	1.00	1.00	1.00	1.00	1.00	1.00	1.00	1.00	1.00	1.00	1.00	1.00	1.00	1.00
1	1.00	1.00	1.00	1.00	1.00	1.00	1.00	1.00	1.00	1.00	1.00	1.00	1.00	1.00	1.00	1.00	1.00	1.00	1.00	1.00	1.00
2	0.99	0.99	0.99	0.99	0.99	0.99	0.99	0.99	0.99	0.99	0.99	0.99	0.99	0.99	0.99	0.99	0.99	0.99	0.99	0.99	0.99
3	0.99	0.99	0.99	0.99	0.99	0.99	0.99	0.99	0.99	0.99	0.99	0.99	0.99	0.99	0.99	0.99	0.99	0.99	0.99	0.99	0.99
4	0.98	0.99	0.99	0.99	0.99	0.99	0.99	0.99	0.99	0.99	0.99	0.99	0.99	0.99	0.99	0.99	0.99	0.99	0.99	0.99	0.99
5	0.98	0.98	0.98	0.98	0.98	0.99	0.99	0.99	0.99	0.99	0.99	0.99	0.99	0.99	0.99	0.99	0.99	0.99	0.99	0.99	0.99
6	0.98	0.98	0.98	0.98	0.98	0.98	0.98	0.98	0.98	0.98	0.98	0.98	0.98	0.98	0.98	0.98	0.98	0.98	0.98	0.98	0.98
7	0.97	0.98	0.98	0.98	0.98	0.98	0.98	0.98	0.98	0.98	0.98	0.98	0.98	0.98	0.98	0.98	0.98	0.98	0.98	0.98	0.98
8	0.97	0.97	0.97	0.97	0.97	0.97	0.98	0.98	0.98	0.98	0.98	0.98	0.98	0.98	0.98	0.98	0.98	0.98	0.98	0.98	0.98
9	0.97	0.97	0.97	0.97	0.97	0.97	0.97	0.97	0.97	0.97	0.97	0.97	0.98	0.98	0.98	0.98	0.98	0.98	0.98	0.98	0.98
10	0.96	0.96	0.97	0.97	0.97	0.97	0.97	0.97	0.97	0.97	0.97	0.97	0.97	0.97	0.97	0.97	0.98	0.98	0.98	0.98	0.98
11	0.96	0.96	0.96	0.96	0.96	0.97	0.97	0.97	0.97	0.97	0.97	0.97	0.97	0.97	0.97	0.97	0.97	0.97	0.97	0.97	0.97
12	0.96	0.96	0.96	0.96	0.96	0.96	0.96	0.96	0.97	0.97	0.97	0.97	0.97	0.97	0.97	0.97	0.97	0.97	0.97	0.97	0.97
13	0.96	0.95	0.96	0.96	0.96	0.96	0.96	0.96	0.96	0.96	0.96	0.96	0.97	0.97	0.97	0.97	0.97	0.97	0.97	0.97	0.97
14	0.95	0.95	0.95	0.95	0.96	0.96	0.96	0.96	0.96	0.96	0.96	0.96	0.96	0.96	0.96	0.96	0.96	0.96	0.96	0.96	0.96
15	0.95	0.95	0.95	0.95	0.95	0.95	0.96	0.96	0.96	0.96	0.96	0.96	0.96	0.96	0.96	0.96	0.96	0.96	0.96	0.96	0.96
16	0.95	0.95	0.95	0.95	0.95	0.95	0.95	0.95	0.95	0.95	0.96	0.96	0.96	0.96	0.96	0.96	0.96	0.96	0.96	0.96	0.96
17	0.94	0.94	0.95	0.95	0.95	0.95	0.95	0.95	0.95	0.95	0.95	0.95	0.95	0.95	0.96	0.96	0.96	0.96	0.96	0.96	0.96
18	0.94	0.94	0.94	0.94	0.94	0.94	0.94	0.94	0.95	0.95	0.95	0.95	0.95	0.95	0.95	0.95	0.95	0.95	0.95	0.95	0.95
19	0.94	0.94	0.94	0.94	0.94	0.94	0.94	0.94	0.94	0.94	0.94	0.94	0.95	0.95	0.95	0.95	0.95	0.95	0.95	0.95	0.95
20	0.93	0.93	0.93	0.93	0.94	0.94	0.94	0.94	0.94	0.94	0.94	0.94	0.94	0.94	0.94	0.94	0.94	0.94	0.94	0.94	0.94
21	0.93	0.93	0.93	0.93	0.93	0.93	0.93	0.94	0.94	0.94	0.94	0.94	0.94	0.94	0.94	0.94	0.94	0.94	0.94	0.94	0.94
22	0.92	0.93	0.93	0.93	0.93	0.93	0.93	0.93	0.93	0.93	0.93	0.93	0.93	0.93	0.94	0.94	0.94	0.94	0.94	0.94	0.94
23	0.92	0.92	0.92	0.93	0.93	0.93	0.93	0.93	0.93	0.93	0.93	0.93	0.93	0.93	0.93	0.93	0.93	0.93	0.93	0.93	0.93
24	0.92	0.92	0.92	0.92	0.92	0.92	0.93	0.93	0.93	0.93	0.93	0.93	0.93	0.93	0.93	0.93	0.93	0.93	0.93	0.93	0.93
25	0.92	0.92	0.92	0.92	0.92	0.92	0.92	0.92	0.92	0.92	0.92	0.92	0.92	0.92	0.93	0.93	0.93	0.93	0.93	0.93	0.93
26	0.91	0.91	0.92	0.92	0.92	0.92	0.92	0.92	0.92	0.92	0.92	0.92	0.92	0.92	0.92	0.92	0.92	0.92	0.92	0.92	0.92
27	0.91	0.91	0.91	0.91	0.91	0.92	0.92	0.92	0.92	0.92	0.92	0.92	0.92	0.92	0.92	0.92	0.92	0.92	0.92	0.92	0.92
28	0.91	0.91	0.91	0.91	0.91	0.91	0.91	0.91	0.92	0.92	0.92	0.92	0.92	0.92	0.92	0.92	0.92	0.92	0.92	0.92	0.92
29	0.91	0.91	0.91	0.91	0.91	0.91	0.91	0.91	0.91	0.91	0.91	0.91	0.91	0.91	0.91	0.91	0.91	0.91	0.92	0.92	0.92
30	0.90	0.90	0.91	0.91	0.91	0.91	0.91	0.91	0.91	0.91	0.91	0.91	0.91	0.91	0.91	0.91	0.91	0.91	0.91	0.91	0.91
31	0.89	0.89	0.90	0.90	0.90	0.90	0.90	0.90	0.91	0.91	0.91	0.91	0.91	0.91	0.91	0.91	0.91	0.91	0.91	0.91	0.91
32	0.88	0.88	0.89	0.89	0.89	0.89	0.89	0.90	0.90	0.90	0.90	0.90	0.90	0.90	0.90	0.90	0.90	0.90	0.90	0.90	0.90
33	0.87	0.87	0.88	0.88	0.88	0.89	0.89	0.89	0.89	0.89	0.89	0.89	0.90	0.90	0.90	0.90	0.90	0.90	0.90	0.90	0.90
34	0.86	0.86	0.87	0.87	0.88	0.88	0.88	0.88	0.88	0.88	0.89	0.89	0.89	0.89	0.89	0.89	0.89	0.89	0.89	0.89	0.89
35	0.85	0.85	0.86	0.86	0.87	0.87	0.87	0.87	0.87	0.88	0.88	0.88	0.88	0.88	0.88	0.88	0.88	0.88	0.88	0.88	0.88
36	0.84	0.84	0.85	0.85	0.85	0.86	0.86	0.86	0.86	0.87	0.87	0.87	0.87	0.87	0.87	0.88	0.88	0.88	0.88	0.88	0.88
37	0.83	0.83	0.83	0.84	0.84	0.85	0.85	0.85	0.86	0.86	0.86	0.86	0.86	0.87	0.87	0.87	0.87	0.87	0.87	0.87	0.87
38	0.81	0.82	0.82	0.83	0.83	0.84	0.84	0.84	0.85	0.85	0.85	0.86	0.86	0.86	0.86	0.86	0.86	0.86	0.87	0.87	0.87
39	0.80	0.81	0.81	0.82	0.82	0.83	0.83	0.83	0.84	0.84	0.85	0.85	0.85	0.85	0.85	0.85	0.86	0.86	0.86	0.86	0.86
40	0.79	0.80	0.80	0.81	0.81	0.82	0.82	0.83	0.83	0.83	0.84	0.84	0.84	0.84	0.85	0.85	0.85	0.85	0.85	0.85	0.85
41	0.78	0.79	0.79	0.80	0.80	0.81	0.81	0.82	0.82	0.82	0.83	0.83	0.83	0.83	0.84	0.84	0.84	0.84	0.85	0.85	0.85
42	0.77	0.78	0.78	0.79	0.79	0.80	0.80	0.81	0.81	0.82	0.82	0.82	0.83	0.83	0.83	0.83	0.84	0.84	0.84	0.84	0.84
43	0.76	0.77	0.77	0.78	0.78	0.79	0.79	0.80	0.80	0.81	0.81	0.82	0.82	0.82	0.83	0.83	0.83	0.83	0.83	0.83	0.84
44	0.75	0.76	0.76	0.77	0.77	0.78	0.78	0.79	0.79	0.80	0.80	0.81	0.81	0.81	0.82	0.82	0.82	0.82	0.83	0.83	0.83
45	0.74	0.75	0.75	0.76	0.76	0.77	0.77	0.78	0.79	0.79	0.80	0.80	0.80	0.80	0.81	0.81	0.82	0.82	0.82	0.82	0.82
46	0.72	0.73	0.74	0.74	0.75	0.76	0.76	0.77	0.77	0.78	0.79	0.79	0.80	0.80	0.79	0.81	0.80	0.81	0.81	0.82	0.82
47	0.71	0.71	0.72	0.73	0.73	0.74	0.75	0.75	0.76	0.77	0.77	0.78	0.78	0.79	0.79	0.80	0.80	0.80	0.81	0.81	0.81
48	0.69	0.70	0.71	0.71	0.72	0.73	0.73	0.74	0.75	0.75	0.76	0.77	0.77	0.78	0.78	0.79	0.79	0.80	0.80	0.81	0.81
49	0.68	0.68	0.69	0.70	0.70	0.71	0.72	0.73	0.73	0.74	0.75	0.75	0.76	0.76	0.77	0.77	0.78	0.78	0.79	0.79	0.80
50	0.66	0.67	0.67	0.68	0.69	0.70	0.70	0.71	0.72	0.73	0.73	0.74	0.74	0.75	0.75	0.76	0.77	0.77	0.78	0.78	0.79
51	0.64	0.65	0.66	0.67	0.67	0.68	0.69	0.70	0.70	0.71	0.72	0.73	0.73	0.74	0.74	0.75	0.75	0.76	0.76	0.77	0.77
52	0.63	0.64	0.64	0.65	0.66	0.67	0.68	0.68	0.69	0.70	0.71	0.71	0.72	0.72	0.73	0.73	0.74	0.74	0.75	0.76	0.76
53	0.61	0.62	0.63	0.64	0.64	0.65	0.66	0.67	0.68	0.69	0.69	0.70	0.71	0.71	0.72	0.72	0.73	0.73	0.74	0.74	0.75
54	0.60	0.60	0.61	0.62	0.63	0.64	0.65	0.65	0.66	0.67	0.68	0.69	0.69	0.70	0.70	0.71	0.71	0.72	0.72	0.73	0.74
55	0.58	0.59	0.60	0.60	0.61	0.62	0.63	0.64	0.65	0.66	0.67	0.67	0.68	0.68	0.69	0.69	0.70	0.71	0.71	0.72	0.72
56	0.56	0.57	0.58	0.59	0.60	0.61	0.62	0.63	0.64	0.64	0.65	0.66	0.66	0.67	0.68	0.68	0.69	0.69	0.70	0.70	0.71
57	0.55	0.56	0.57	0.58	0.58	0.59	0.60	0.61	0.62	0.63	0.64	0.65	0.65	0.66	0.66	0.67	0.67	0.68	0.69	0.69	0.70
58	0.53	0.54	0.55	0.56	0.57	0.58	0.59	0.60	0.61	0.62	0.63	0.63	0.64	0.64	0.65	0.66	0.66	0.67	0.67	0.68	0.69
59	0.52	0.53	0.54	0.55	0.55	0.56	0.57	0.58	0.59	0.60	0.61	0.62	0.63	0.63	0.64	0.64	0.65	0.65	0.66	0.67	0.67
60	0.50	0.51	0.52	0.53	0.54	0.55	0.56	0.57	0.58	0.59	0.60	0.61	0.61	0.62	0.62	0.63	0.64	0.64	0.65	0.65	0.66

East or West facing window partial shading correction factors for overhangs, Fo

Latitude Overhang Angle	45	46	47	48	49	50	51	52	53	54	55	56	57	58	59	60	61	62	63	64	65
0	1.00	1.00	1.00	1.00	1.00	1.00	1.00	1.00	1.00	1.00	1.00	1.00	1.00	1.00	1.00	1.00	1.00	1.00	1.00	1.00	1.00
1	1.00	1.00	1.00	1.00	1.00	1.00	1.00	1.00	1.00	1.00	1.00	1.00	1.00	1.00	1.00	1.00	1.00	1.00	1.00	1.00	1.00
2	0.99	0.99	0.99	0.99	0.99	0.99	0.99	0.99	0.99	0.99	0.99	0.99	0.99	0.99	0.99	0.99	0.99	0.99	0.99	0.99	0.99
3	0.99	0.99	0.99	0.99	0.99	0.99	0.99	0.99	0.99	0.99	0.99	0.99	0.99	0.99	0.99	0.99	0.99	0.99	0.99	0.99	0.99
4	0.99	0.99	0.99	0.99	0.99	0.99	0.99	0.99	0.99	0.99	0.99	0.99	0.99	0.99	0.99	0.99	0.99	0.99	0.99	0.99	0.99
5	0.98	0.98	0.98	0.98	0.98	0.98	0.98	0.98	0.98	0.98	0.98	0.98	0.98	0.98	0.98	0.98	0.98	0.98	0.98	0.98	0.98
6	0.98	0.98	0.98	0.98	0.98	0.98	0.98	0.98	0.98	0.98	0.98	0.98	0.98	0.98	0.98	0.98	0.98	0.98	0.98	0.98	0.98
7	0.97	0.97	0.97	0.97	0.97	0.97	0.97	0.97	0.97	0.97	0.97	0.97	0.97	0.97	0.97	0.97	0.97	0.97	0.97	0.97	0.98
8	0.97	0.97	0.97	0.97	0.97	0.97	0.97	0.97	0.97	0.97	0.97	0.97	0.97	0.97	0.97	0.97	0.97	0.97	0.97	0.97	0.97
9	0.97	0.97	0.97	0.97	0.97	0.97	0.97	0.97	0.97	0.97	0.97	0.97	0.97	0.97	0.97	0.97	0.97	0.97	0.97	0.97	0.97
10	0.96	0.96	0.96	0.96	0.96	0.96	0.96	0.96	0.96	0.96	0.96	0.96	0.96	0.96	0.96	0.96	0.96	0.96	0.96	0.96	0.96
11	0.96	0.96	0.96	0.96	0.96	0.96	0.96	0.96	0.96	0.96	0.96	0.96	0.96	0.96	0.96	0.96	0.96	0.96	0.96	0.96	0.96
12	0.96	0.96	0.96	0.96	0.96	0.96	0.96	0.96	0.96	0.96	0.96	0.96	0.96	0.96	0.96	0.96	0.96	0.96	0.96	0.96	0.96
13	0.95	0.95	0.95	0.95	0.95	0.95	0.95	0.95	0.95	0.95	0.95	0.95	0.95	0.95	0.95	0.95	0.95	0.95	0.95	0.95	0.95
14	0.95	0.95	0.95	0.95	0.95	0.95	0.95	0.95	0.95	0.95	0.95	0.95	0.95	0.95	0.95	0.95	0.95	0.95	0.95	0.95	0.95
15	0.94	0.94	0.94	0.94	0.94	0.94	0.94	0.94	0.94	0.94	0.94	0.94	0.94	0.94	0.94	0.94	0.94	0.94	0.94	0.94	0.94
16	0.94	0.94	0.94	0.94	0.94	0.94	0.94	0.94	0.94	0.94	0.94	0.94	0.94	0.94	0.94	0.94	0.94	0.94	0.94	0.94	0.94
17	0.94	0.94	0.94	0.94	0.94	0.94	0.94	0.94	0.94	0.94	0.94	0.94	0.94	0.94	0.94	0.94	0.94	0.94	0.94	0.94	0.94
18	0.93	0.93	0.93	0.93	0.93	0.93	0.93	0.93	0.93	0.93	0.93	0.93	0.93	0.93	0.93	0.93	0.93	0.93	0.93	0.93	0.93
19	0.93	0.93	0.93	0.93	0.93	0.93	0.93	0.93	0.93	0.93	0.93	0.93	0.93	0.93	0.93	0.93	0.93	0.93	0.93	0.93	0.93
20	0.93	0.92	0.92	0.93	0.93	0.92	0.93	0.93	0.93	0.93	0.93	0.93	0.93	0.93	0.93	0.93	0.93	0.93	0.93	0.93	0.93
21	0.92	0.92	0.92	0.92	0.92	0.92	0.92	0.92	0.92	0.92	0.92	0.92	0.92	0.92	0.92	0.92	0.92	0.92	0.92	0.92	0.92
22	0.92	0.92	0.92	0.92	0.92	0.92	0.92	0.92	0.92	0.92	0.92	0.92	0.92	0.92	0.92	0.92	0.92	0.92	0.92	0.92	0.93
23	0.92	0.92	0.92	0.92	0.92	0.92	0.92	0.92	0.92	0.92	0.92	0.92	0.92	0.92	0.92	0.92	0.92	0.92	0.92	0.92	0.93
24	0.91	0.91	0.91	0.91	0.91	0.91	0.91	0.91	0.91	0.91	0.91	0.91	0.91	0.91	0.92	0.92	0.92	0.92	0.92	0.92	0.92
25	0.91	0.91	0.91	0.91	0.91	0.91	0.91	0.91	0.91	0.91	0.91	0.91	0.91	0.91	0.91	0.92	0.92	0.92	0.92	0.92	0.92
26	0.90	0.91	0.91	0.91	0.91	0.91	0.91	0.91	0.91	0.91	0.91	0.91	0.91	0.91	0.91	0.91	0.92	0.92	0.92	0.92	0.92
27	0.90	0.90	0.90	0.90	0.90	0.91	0.91	0.91	0.91	0.91	0.91	0.91	0.91	0.91	0.91	0.91	0.91	0.91	0.92	0.92	0.92
28	0.90	0.90	0.90	0.90	0.90	0.90	0.90	0.90	0.90	0.91	0.91	0.91	0.91	0.91	0.91	0.91	0.91	0.91	0.91	0.92	0.92
29	0.89	0.90	0.90	0.90	0.90	0.90	0.90	0.90	0.90	0.90	0.90	0.91	0.91	0.91	0.91	0.91	0.91	0.91	0.91	0.91	0.91
30	0.89	0.89	0.89	0.89	0.89	0.90	0.90	0.90	0.90	0.90	0.90	0.90	0.90	0.90	0.91	0.91	0.91	0.91	0.91	0.91	0.91
31	0.88	0.89	0.89	0.89	0.89	0.89	0.89	0.89	0.90	0.90	0.90	0.90	0.90	0.90	0.90	0.90	0.90	0.90	0.90	0.90	0.90
32	0.87	0.88	0.88	0.88	0.88	0.88	0.89	0.89	0.89	0.89	0.89	0.89	0.90	0.89	0.89	0.89	0.89	0.89	0.89	0.89	0.89
33	0.86	0.87	0.87	0.87	0.87	0.87	0.88	0.88	0.88	0.88	0.89	0.89	0.89	0.89	0.89	0.89	0.89	0.89	0.90	0.90	0.90
34	0.86	0.86	0.86	0.86	0.86	0.87	0.87	0.87	0.87	0.87	0.88	0.88	0.88	0.88	0.88	0.88	0.89	0.89	0.89	0.89	0.89
35	0.85	0.85	0.85	0.85	0.86	0.86	0.86	0.86	0.87	0.87	0.87	0.87	0.87	0.87	0.87	0.88	0.88	0.88	0.88	0.88	0.88
36	0.84	0.84	0.85	0.85	0.85	0.85	0.85	0.85	0.86	0.86	0.86	0.86	0.86	0.87	0.87	0.87	0.87	0.87	0.87	0.87	0.87
37	0.83	0.83	0.83	0.84	0.84	0.84	0.84	0.85	0.85	0.85	0.85	0.86	0.86	0.86	0.86	0.86	0.86	0.86	0.86	0.86	0.87
38	0.82	0.82	0.83	0.83	0.83	0.83	0.83	0.84	0.84	0.84	0.84	0.85	0.85	0.85	0.85	0.85	0.85	0.85	0.86	0.86	0.86
39	0.81	0.81	0.82	0.82	0.82	0.82	0.82	0.83	0.83	0.83	0.84	0.84	0.84	0.84	0.84	0.85	0.85	0.85	0.85	0.85	0.85
40	0.80	0.81	0.81	0.81	0.81	0.82	0.82	0.82	0.82	0.83	0.83	0.83	0.83	0.84	0.84	0.84	0.84	0.84	0.84	0.85	0.85
41	0.79	0.80	0.80	0.80	0.81	0.81	0.81	0.81	0.82	0.82	0.82	0.82	0.83	0.83	0.83	0.83	0.83	0.83	0.84	0.84	0.84
42	0.79	0.79	0.79	0.80	0.80	0.80	0.80	0.81	0.81	0.81	0.81	0.82	0.82	0.82	0.82	0.82	0.82	0.83	0.83	0.83	0.83
43	0.78	0.78	0.78	0.79	0.79	0.79	0.79	0.80	0.80	0.80	0.81	0.81	0.81	0.81	0.81	0.82	0.82	0.82	0.82	0.82	0.82
44	0.77	0.77	0.77	0.78	0.78	0.78	0.79	0.79	0.79	0.80	0.80	0.80	0.80	0.80	0.81	0.81	0.81	0.81	0.81	0.81	0.81
45	0.76	0.76	0.77	0.77	0.77	0.78	0.78	0.78	0.78	0.79	0.79	0.79	0.79	0.80	0.80	0.80	0.80	0.80	0.81	0.81	0.81
46	0.75	0.75	0.75	0.76	0.76	0.76	0.77	0.77	0.77	0.78	0.78	0.78	0.78	0.78	0.79	0.79	0.79	0.79	0.80	0.80	0.80
47	0.74	0.74	0.74	0.75	0.75	0.75	0.75	0.76	0.76	0.76	0.77	0.77	0.77	0.77	0.78	0.78	0.78	0.78	0.78	0.79	0.79
48	0.72	0.73	0.73	0.73	0.74	0.74	0.74	0.75	0.75	0.75	0.75	0.76	0.76	0.76	0.76	0.77	0.77	0.77	0.77	0.78	0.78
49	0.71	0.72	0.72	0.72	0.73	0.73	0.73	0.74	0.74	0.74	0.74	0.74	0.75	0.75	0.75	0.75	0.76	0.76	0.76	0.76	0.77
50	0.70	0.70	0.71	0.71	0.71	0.72	0.72	0.72	0.73	0.73	0.73	0.73	0.74	0.74	0.74	0.74	0.75	0.75	0.75	0.75	0.76
51	0.69	0.69	0.69	0.70	0.70	0.70	0.71	0.71	0.71	0.72	0.72	0.72	0.72	0.73	0.73	0.73	0.73	0.74	0.74	0.74	0.75
52	0.68	0.68	0.68	0.69	0.69	0.69	0.70	0.70	0.70	0.70	0.71	0.71	0.71	0.71	0.72	0.72	0.72	0.72	0.73	0.73	0.74
53	0.66	0.67	0.67	0.67	0.68	0.68	0.69	0.69	0.69	0.69	0.69	0.70	0.70	0.70	0.71	0.71	0.71	0.71	0.72	0.72	0.72
54	0.65	0.66	0.66	0.66	0.66	0.67	0.67	0.67	0.68	0.68	0.68	0.69	0.69	0.69	0.69	0.70	0.70	0.70	0.70	0.71	0.71
55	0.64	0.64	0.65	0.65	0.65	0.66	0.66	0.66	0.67	0.67	0.67	0.67	0.68	0.68	0.68	0.69	0.69	0.69	0.69	0.70	0.70
56	0.63	0.63	0.63	0.64	0.64	0.64	0.65	0.65	0.65	0.66	0.66	0.66	0.67	0.67	0.67	0.68	0.68	0.68	0.69	0.69	0.69
57	0.62	0.62	0.62	0.63	0.63	0.63	0.64	0.64	0.65	0.65	0.65	0.65	0.66	0.66	0.66	0.66	0.67	0.67	0.67	0.68	0.68
58	0.60	0.61	0.61	0.61	0.62	0.62	0.62	0.63	0.63	0.63	0.63	0.64	0.64	0.65	0.65	0.65	0.66	0.66	0.66	0.67	0.67
59	0.59	0.60	0.60	0.60	0.60	0.61	0.61	0.61	0.62	0.62	0.62	0.63	0.63	0.63	0.64	0.64	0.65	0.65	0.65	0.66	0.66
60	0.58	0.58	0.59	0.59	0.59	0.60	0.60	0.60	0.60	0.61	0.61	0.61	0.62	0.62	0.63	0.63	0.63	0.64	0.64	0.65	0.65

Appendix H

North facing window partial shading correction factors for overhangs, Fo

Latitude Overhang Angle	45	46	47	48	49	50	51	52	53	54	55	56	57	58	59	60	61	62	63	64	65
0	1.00	1.00	1.00	1.00	1.00	1.00	1.00	1.00	1.00	1.00	**1.00**	1.00	1.00	1.00	1.00	1.00	1.00	1.00	1.00	1.00	**1.00**
1	1.00	1.00	1.00	1.00	1.00	1.00	1.00	1.00	1.00	1.00	1.00	1.00	1.00	1.00	1.00	1.00	1.00	1.00	1.00	1.00	1.00
2	0.99	0.99	0.99	0.99	0.99	0.99	0.99	0.99	0.99	0.99	0.99	0.99	0.99	0.99	0.99	0.99	0.99	0.99	0.99	0.99	0.99
3	0.99	0.99	0.99	0.99	0.99	0.99	0.99	0.99	0.99	0.99	0.99	0.99	0.99	0.99	0.99	0.99	0.99	0.99	0.99	0.99	0.99
4	0.98	0.98	0.98	0.98	0.98	0.98	0.98	0.98	0.98	0.98	0.98	0.98	0.98	0.98	0.98	0.98	0.98	0.98	0.98	0.98	0.98
5	0.98	0.98	0.98	0.98	0.98	0.98	0.98	0.98	0.98	0.98	0.98	0.98	0.98	0.98	0.98	0.98	0.98	0.98	0.98	0.98	0.98
6	0.98	0.98	0.98	0.98	0.98	0.98	0.98	0.98	0.98	0.98	0.98	0.98	0.98	0.98	0.98	0.98	0.98	0.98	0.98	0.98	0.98
7	0.98	0.98	0.98	0.98	0.98	0.98	0.98	0.98	0.98	0.98	0.98	0.98	0.98	0.98	0.98	0.98	0.98	0.98	0.98	0.98	0.98
8	0.97	0.97	0.97	0.97	0.97	0.97	0.97	0.97	0.97	0.97	0.97	0.97	0.97	0.97	0.97	0.97	0.97	0.97	0.97	0.97	0.97
9	0.97	0.97	0.97	0.97	0.97	0.97	0.97	0.97	0.97	0.97	0.97	0.97	0.97	0.97	0.97	0.97	0.97	0.97	0.97	0.97	0.97
10	0.97	0.97	0.97	0.97	0.97	0.97	0.97	0.97	0.97	0.97	0.97	0.97	0.97	0.97	0.97	0.97	0.97	0.97	0.97	0.97	0.97
11	0.96	0.96	0.96	0.96	0.96	0.96	0.96	0.96	0.96	0.96	0.96	0.96	0.96	0.96	0.96	0.96	0.96	0.96	0.96	0.96	0.96
12	0.96	0.96	0.96	0.96	0.96	0.96	0.96	0.96	0.96	0.96	0.96	0.96	0.96	0.96	0.96	0.96	0.96	0.96	0.96	0.96	0.96
13	0.96	0.96	0.96	0.96	0.96	0.96	0.96	0.96	0.96	0.96	0.96	0.96	0.96	0.96	0.96	0.96	0.96	0.96	0.96	0.96	0.96
14	0.95	0.95	0.95	0.95	0.95	0.95	0.95	0.95	0.95	0.95	0.95	0.95	0.95	0.95	0.95	0.95	0.95	0.95	0.95	0.95	0.95
15	0.95	0.95	0.95	0.95	0.95	0.95	0.95	0.95	0.95	0.95	0.95	0.95	0.95	0.95	0.95	0.95	0.95	0.95	0.95	0.95	0.95
16	0.95	0.95	0.95	0.95	0.95	0.95	0.95	0.95	0.95	0.95	0.95	0.95	0.95	0.95	0.95	0.95	0.95	0.95	0.95	0.95	0.95
17	0.94	0.94	0.94	0.94	0.94	0.94	0.94	0.94	0.94	0.94	0.94	0.94	0.94	0.94	0.94	0.94	0.94	0.94	0.94	0.94	0.94
18	0.94	0.94	0.94	0.94	0.94	0.94	0.94	0.94	0.94	0.94	0.94	0.94	0.94	0.94	0.94	0.94	0.94	0.94	0.94	0.94	0.94
19	0.94	0.94	0.94	0.94	0.94	0.94	0.94	0.94	0.94	0.94	0.94	0.94	0.94	0.94	0.94	0.94	0.94	0.94	0.94	0.94	0.94
20	0.93	0.93	0.93	0.93	0.93	0.93	0.93	0.93	0.93	0.93	0.93	0.93	0.93	0.93	0.93	0.93	0.93	0.93	0.93	0.93	0.93
21	0.93	0.93	0.93	0.93	0.93	0.93	0.93	0.93	0.93	0.93	0.93	0.93	0.93	0.93	0.93	0.93	0.93	0.93	0.93	0.93	0.93
22	0.93	0.93	0.93	0.93	0.93	0.93	0.93	0.93	0.93	0.93	0.93	0.93	0.93	0.93	0.93	0.93	0.93	0.93	0.93	0.93	0.93
23	0.92	0.92	0.92	0.92	0.92	0.92	0.92	0.92	0.92	0.92	0.92	0.92	0.92	0.92	0.92	0.92	0.92	0.92	0.92	0.92	0.92
24	0.92	0.92	0.92	0.92	0.92	0.92	0.92	0.92	0.92	0.92	0.92	0.92	0.92	0.92	0.92	0.92	0.92	0.92	0.92	0.92	0.92
25	0.92	0.92	0.92	0.92	0.92	0.92	0.92	0.92	0.92	0.92	0.92	0.92	0.92	0.92	0.92	0.92	0.92	0.92	0.92	0.92	0.92
26	0.92	0.92	0.92	0.92	0.92	0.92	0.92	0.92	0.92	0.92	0.92	0.92	0.92	0.92	0.92	0.92	0.92	0.92	0.92	0.92	0.92
27	0.91	0.91	0.91	0.91	0.91	0.91	0.91	0.91	0.91	0.91	0.91	0.91	0.91	0.91	0.91	0.91	0.91	0.91	0.91	0.91	0.91
28	0.91	0.91	0.91	0.91	0.91	0.91	0.91	0.91	0.91	0.91	0.91	0.91	0.91	0.91	0.91	0.91	0.91	0.91	0.91	0.91	0.91
29	0.91	0.91	0.91	0.91	0.91	0.91	0.91	0.91	0.91	0.91	0.91	0.91	0.91	0.91	0.91	0.91	0.91	0.91	0.91	0.90	0.90
30	**0.91**	0.91	0.91	0.91	0.91	0.91	0.91	0.91	0.91	0.91	**0.91**	0.91	0.91	0.91	0.91	0.91	0.91	0.90	0.90	0.90	**0.90**
31	0.90	0.90	0.90	0.90	0.90	0.90	0.90	0.90	0.90	0.90	0.90	0.90	0.90	0.90	0.90	0.90	0.90	0.90	0.89	0.89	0.89
32	0.90	0.90	0.90	0.90	0.90	0.90	0.90	0.90	0.90	0.90	0.90	0.89	0.89	0.89	0.89	0.89	0.89	0.89	0.89	0.89	0.89
33	0.89	0.89	0.89	0.89	0.89	0.89	0.89	0.89	0.89	0.89	0.89	0.89	0.89	0.88	0.88	0.88	0.88	0.88	0.88	0.88	0.88
34	0.88	0.88	0.88	0.88	0.88	0.88	0.88	0.88	0.88	0.88	0.88	0.88	0.88	0.88	0.88	0.88	0.88	0.88	0.87	0.87	0.87
35	0.87	0.87	0.87	0.87	0.87	0.87	0.87	0.87	0.87	0.87	0.87	0.87	0.87	0.87	0.87	0.87	0.87	0.87	0.87	0.87	0.86
36	0.87	0.87	0.87	0.87	0.87	0.87	0.87	0.87	0.87	0.87	0.87	0.87	0.87	0.86	0.86	0.86	0.86	0.86	0.86	0.86	0.86
37	0.86	0.86	0.86	0.86	0.86	0.86	0.86	0.86	0.86	0.86	0.86	0.86	0.86	0.86	0.86	0.86	0.86	0.86	0.86	0.85	0.85
38	0.85	0.85	0.85	0.85	0.85	0.85	0.85	0.85	0.85	0.85	0.85	0.85	0.85	0.85	0.85	0.85	0.85	0.85	0.85	0.85	0.85
39	0.84	0.84	0.84	0.84	0.84	0.84	0.84	0.84	0.84	0.84	0.84	0.84	0.84	0.84	0.84	0.84	0.84	0.84	0.84	0.84	0.84
40	0.84	0.84	0.84	0.84	0.84	0.84	0.84	0.84	0.84	0.84	0.84	0.84	0.84	0.84	0.84	0.84	0.83	0.83	0.83	0.83	0.83
41	0.83	0.83	0.83	0.83	0.83	0.83	0.83	0.83	0.83	0.83	0.83	0.83	0.83	0.83	0.83	0.83	0.83	0.83	0.83	0.83	0.83
42	0.82	0.82	0.82	0.82	0.82	0.82	0.82	0.82	0.82	0.82	0.82	0.82	0.82	0.82	0.82	0.82	0.82	0.82	0.82	0.82	0.82
43	0.82	0.82	0.82	0.82	0.82	0.82	0.82	0.82	0.82	0.82	0.82	0.82	0.81	0.81	0.81	0.81	0.81	0.81	0.81	0.81	0.81
44	0.81	0.81	0.81	0.81	0.81	0.81	0.81	0.81	0.81	0.81	0.81	0.81	0.81	0.81	0.81	0.81	0.81	0.81	0.81	0.81	0.81
45	**0.80**	0.80	0.80	0.80	0.80	0.80	0.80	0.80	0.80	0.80	**0.80**	0.80	0.80	0.80	0.80	0.80	0.80	0.80	0.80	0.80	**0.80**
46	0.79	0.79	0.79	0.79	0.79	0.79	0.79	0.79	0.79	0.79	0.79	0.79	0.79	0.79	0.79	0.79	0.79	0.79	0.79	0.79	0.79
47	0.78	0.78	0.78	0.78	0.78	0.78	0.78	0.78	0.78	0.78	0.78	0.78	0.78	0.78	0.78	0.78	0.78	0.78	0.78	0.78	0.78
48	0.77	0.77	0.77	0.77	0.77	0.77	0.77	0.77	0.77	0.77	0.77	0.77	0.77	0.77	0.77	0.77	0.77	0.77	0.77	0.77	0.77
49	0.76	0.76	0.76	0.76	0.76	0.76	0.76	0.76	0.76	0.76	0.76	0.76	0.76	0.76	0.76	0.76	0.76	0.76	0.76	0.76	0.76
50	0.75	0.75	0.75	0.75	0.75	0.75	0.75	0.75	0.75	0.75	0.75	0.75	0.75	0.75	0.75	0.75	0.75	0.75	0.75	0.75	0.75
51	0.74	0.74	0.74	0.74	0.74	0.74	0.74	0.74	0.74	0.74	0.74	0.74	0.74	0.74	0.74	0.74	0.74	0.74	0.74	0.74	0.74
52	0.73	0.73	0.73	0.73	0.73	0.73	0.73	0.73	0.73	0.73	0.73	0.73	0.73	0.73	0.73	0.73	0.73	0.73	0.73	0.73	0.73
53	0.72	0.72	0.72	0.72	0.72	0.72	0.72	0.72	0.72	0.72	0.72	0.72	0.72	0.72	0.72	0.72	0.72	0.72	0.72	0.72	0.72
54	0.72	0.72	0.71	0.71	0.71	0.71	0.71	0.71	0.71	0.71	0.71	0.71	0.71	0.71	0.71	0.71	0.71	0.71	0.71	0.71	0.71
55	0.71	0.71	0.71	0.70	0.70	0.70	0.70	0.70	0.70	0.70	0.70	0.70	0.70	0.70	0.70	0.70	0.70	0.70	0.70	0.70	0.70
56	0.70	0.70	0.70	0.70	0.69	0.69	0.69	0.69	0.69	0.69	0.69	0.69	0.69	0.69	0.69	0.69	0.69	0.69	0.69	0.69	0.69
57	0.69	0.69	0.69	0.69	0.69	0.68	0.68	0.68	0.68	0.68	0.68	0.68	0.68	0.68	0.68	0.68	0.68	0.68	0.68	0.68	0.68
58	0.68	0.68	0.68	0.68	0.68	0.68	0.67	0.67	0.67	0.67	0.67	0.67	0.67	0.67	0.67	0.67	0.67	0.67	0.67	0.67	0.67
59	0.67	0.67	0.67	0.67	0.67	0.67	0.66	0.66	0.66	0.66	0.66	0.66	0.66	0.66	0.66	0.66	0.66	0.66	0.66	0.66	0.67
60	**0.66**	0.66	0.66	0.66	0.66	0.66	0.65	0.65	0.65	0.65	**0.65**	0.66	0.65	0.65	0.65	0.66	0.66	0.66	0.66	0.66	**0.66**

South facing window partial shading correction factors for fins, Ff

Latitude / Fin Angle	45	46	47	48	49	50	51	52	53	54	55	56	57	58	59	60	61	62	63	64	65
0	1.00	1.00	1.00	1.00	1.00	1.00	1.00	1.00	1.00	1.00	1.00	1.00	1.00	1.00	1.00	1.00	1.00	1.00	1.00	1.00	1.00
1	1.00	1.00	1.00	1.00	1.00	1.00	1.00	1.00	1.00	1.00	1.00	1.00	1.00	1.00	1.00	1.00	1.00	1.00	1.00	1.00	1.00
2	1.00	1.00	1.00	1.00	1.00	1.00	1.00	1.00	1.00	1.00	1.00	1.00	1.00	1.00	1.00	1.00	1.00	1.00	1.00	1.00	1.00
3	1.00	1.00	1.00	0.99	0.99	0.99	0.99	0.99	0.99	0.99	0.99	0.99	0.99	0.99	0.99	0.99	0.99	0.99	0.99	0.99	0.99
4	0.99	0.99	0.99	0.99	0.99	0.99	0.99	0.99	0.99	0.99	0.99	0.99	0.99	0.99	0.99	0.99	0.99	0.99	0.99	0.99	0.99
5	0.99	0.99	0.99	0.99	0.99	0.99	0.99	0.99	0.99	0.99	0.99	0.99	0.99	0.99	0.99	0.99	0.99	0.99	0.99	0.99	0.99
6	0.99	0.99	0.99	0.99	0.99	0.99	0.99	0.99	0.99	0.99	0.99	0.99	0.99	0.99	0.99	0.99	0.99	0.99	0.99	0.99	0.99
7	0.99	0.99	0.99	0.99	0.99	0.99	0.99	0.99	0.99	0.99	0.99	0.99	0.99	0.99	0.99	0.99	0.99	0.99	0.99	0.99	0.99
8	0.98	0.98	0.98	0.98	0.98	0.98	0.98	0.98	0.98	0.98	0.98	0.98	0.98	0.98	0.98	0.98	0.98	0.98	0.98	0.98	0.98
9	0.98	0.98	0.98	0.98	0.98	0.98	0.98	0.98	0.98	0.98	0.98	0.98	0.98	0.98	0.98	0.98	0.98	0.98	0.98	0.98	0.98
10	0.98	0.98	0.98	0.98	0.98	0.98	0.98	0.98	0.98	0.98	0.98	0.98	0.98	0.98	0.98	0.98	0.98	0.98	0.98	0.98	0.98
11	0.98	0.98	0.98	0.98	0.98	0.98	0.98	0.98	0.98	0.98	0.98	0.98	0.98	0.98	0.98	0.98	0.98	0.98	0.98	0.98	0.98
12	0.98	0.98	0.98	0.98	0.98	0.98	0.98	0.98	0.98	0.98	0.98	0.98	0.98	0.98	0.98	0.98	0.98	0.98	0.98	0.98	0.98
13	0.97	0.97	0.97	0.97	0.97	0.97	0.97	0.97	0.97	0.97	0.97	0.97	0.97	0.97	0.97	0.97	0.97	0.97	0.97	0.97	0.97
14	0.97	0.97	0.97	0.97	0.97	0.97	0.97	0.97	0.97	0.97	0.97	0.97	0.97	0.97	0.97	0.97	0.97	0.97	0.97	0.97	0.97
15	0.97	0.97	0.97	0.97	0.97	0.97	0.97	0.97	0.97	0.97	0.97	0.97	0.97	0.97	0.97	0.97	0.97	0.97	0.97	0.97	0.97
16	0.97	0.97	0.97	0.97	0.97	0.97	0.97	0.97	0.97	0.97	0.97	0.97	0.97	0.97	0.97	0.97	0.97	0.97	0.97	0.97	0.97
17	0.97	0.97	0.97	0.97	0.97	0.97	0.97	0.97	0.97	0.97	0.97	0.97	0.97	0.97	0.97	0.97	0.97	0.97	0.97	0.97	0.97
18	0.96	0.96	0.96	0.96	0.96	0.96	0.96	0.96	0.96	0.96	0.96	0.96	0.96	0.96	0.96	0.96	0.96	0.96	0.96	0.96	0.96
19	0.96	0.96	0.96	0.96	0.96	0.96	0.96	0.96	0.96	0.96	0.96	0.96	0.96	0.96	0.96	0.96	0.96	0.96	0.96	0.96	0.96
20	0.96	0.96	0.96	0.96	0.96	0.96	0.96	0.96	0.96	0.96	0.96	0.96	0.96	0.96	0.96	0.96	0.96	0.96	0.96	0.96	0.96
21	0.96	0.96	0.96	0.96	0.96	0.96	0.96	0.96	0.96	0.96	0.96	0.96	0.96	0.96	0.96	0.96	0.96	0.96	0.96	0.96	0.96
22	0.96	0.96	0.96	0.96	0.96	0.96	0.96	0.96	0.96	0.96	0.96	0.96	0.96	0.96	0.96	0.96	0.96	0.96	0.96	0.96	0.96
23	0.95	0.95	0.95	0.95	0.95	0.95	0.95	0.95	0.95	0.95	0.95	0.95	0.95	0.95	0.95	0.95	0.95	0.95	0.95	0.95	0.95
24	0.95	0.95	0.95	0.95	0.95	0.95	0.95	0.95	0.95	0.95	0.95	0.95	0.95	0.95	0.95	0.95	0.95	0.95	0.95	0.95	0.95
25	0.95	0.95	0.95	0.95	0.95	0.95	0.95	0.95	0.95	0.95	0.95	0.95	0.95	0.95	0.95	0.95	0.95	0.95	0.95	0.95	0.95
26	0.95	0.95	0.95	0.95	0.95	0.95	0.95	0.95	0.95	0.95	0.95	0.95	0.95	0.95	0.95	0.95	0.95	0.95	0.95	0.95	0.95
27	0.95	0.95	0.95	0.95	0.95	0.95	0.95	0.95	0.95	0.95	0.95	0.95	0.95	0.95	0.95	0.95	0.95	0.95	0.95	0.95	0.95
28	0.94	0.94	0.94	0.94	0.94	0.94	0.94	0.94	0.94	0.94	0.94	0.94	0.94	0.94	0.94	0.94	0.94	0.94	0.94	0.94	0.94
29	0.94	0.94	0.94	0.94	0.94	0.94	0.94	0.94	0.94	0.94	0.94	0.94	0.94	0.94	0.94	0.94	0.94	0.94	0.94	0.94	0.94
30	0.94	0.94	0.94	0.94	0.94	0.94	0.94	0.94	0.94	0.94	0.94	0.94	0.94	0.94	0.94	0.94	0.94	0.94	0.94	0.94	0.94
31	0.93	0.93	0.93	0.93	0.93	0.93	0.93	0.93	0.93	0.93	0.93	0.93	0.93	0.93	0.93	0.93	0.93	0.93	0.93	0.93	0.93
32	0.93	0.93	0.93	0.93	0.93	0.93	0.93	0.93	0.93	0.93	0.93	0.93	0.93	0.93	0.93	0.93	0.93	0.93	0.93	0.93	0.93
33	0.92	0.92	0.92	0.92	0.92	0.92	0.92	0.92	0.92	0.92	0.92	0.92	0.92	0.92	0.92	0.92	0.92	0.92	0.92	0.92	0.92
34	0.92	0.92	0.92	0.92	0.92	0.92	0.92	0.92	0.92	0.92	0.92	0.92	0.92	0.92	0.92	0.92	0.92	0.92	0.92	0.92	0.92
35	0.91	0.91	0.91	0.91	0.91	0.91	0.91	0.91	0.91	0.91	0.91	0.91	0.91	0.91	0.91	0.91	0.91	0.91	0.91	0.91	0.91
36	0.91	0.91	0.91	0.91	0.91	0.91	0.91	0.91	0.91	0.91	0.91	0.91	0.91	0.91	0.91	0.91	0.91	0.91	0.91	0.91	0.90
37	0.90	0.90	0.90	0.90	0.90	0.90	0.90	0.90	0.90	0.90	0.90	0.90	0.90	0.90	0.90	0.90	0.90	0.90	0.90	0.90	0.90
38	0.89	0.89	0.89	0.89	0.89	0.89	0.89	0.89	0.89	0.90	0.90	0.90	0.90	0.90	0.90	0.89	0.89	0.89	0.89	0.89	0.89
39	0.89	0.89	0.89	0.89	0.89	0.89	0.89	0.89	0.89	0.89	0.89	0.89	0.89	0.89	0.89	0.89	0.89	0.89	0.89	0.89	0.89
40	0.88	0.88	0.88	0.88	0.88	0.88	0.88	0.88	0.88	0.88	0.88	0.88	0.88	0.88	0.88	0.88	0.88	0.88	0.88	0.88	0.88
41	0.87	0.87	0.87	0.87	0.87	0.87	0.87	0.87	0.87	0.87	0.87	0.88	0.88	0.88	0.88	0.87	0.87	0.87	0.87	0.87	0.87
42	0.87	0.87	0.87	0.87	0.87	0.87	0.87	0.87	0.87	0.87	0.87	0.87	0.87	0.87	0.87	0.87	0.87	0.86	0.86	0.86	0.86
43	0.86	0.86	0.86	0.86	0.86	0.86	0.86	0.86	0.86	0.86	0.87	0.86	0.86	0.86	0.86	0.86	0.86	0.86	0.86	0.86	0.86
44	0.85	0.85	0.85	0.85	0.85	0.85	0.85	0.85	0.86	0.86	0.86	0.86	0.86	0.86	0.86	0.86	0.85	0.85	0.85	0.86	0.86
45	0.84	0.84	0.84	0.85	0.85	0.85	0.85	0.85	0.85	0.85	0.86	0.85	0.85	0.85	0.85	0.85	0.85	0.85	0.85	0.85	0.85
46	0.83	0.83	0.84	0.84	0.84	0.84	0.84	0.84	0.84	0.84	0.85	0.84	0.84	0.84	0.84	0.84	0.84	0.84	0.84	0.84	0.84
47	0.82	0.83	0.83	0.83	0.83	0.83	0.84	0.84	0.84	0.84	0.84	0.84	0.83	0.83	0.83	0.83	0.83	0.83	0.84	0.83	0.83
48	0.82	0.82	0.82	0.82	0.82	0.82	0.82	0.83	0.82	0.83	0.83	0.83	0.83	0.83	0.83	0.83	0.83	0.83	0.82	0.82	0.82
49	0.81	0.81	0.81	0.81	0.82	0.82	0.82	0.81	0.82	0.82	0.82	0.82	0.81	0.81	0.82	0.82	0.82	0.82	0.82	0.81	0.82
50	0.80	0.80	0.80	0.81	0.81	0.81	0.81	0.81	0.81	0.81	0.81	0.81	0.81	0.81	0.81	0.81	0.81	0.81	0.81	0.81	0.81
51	0.79	0.79	0.80	0.80	0.80	0.80	0.80	0.80	0.80	0.80	0.80	0.80	0.80	0.80	0.80	0.80	0.80	0.80	0.80	0.80	0.80
52	0.78	0.79	0.79	0.79	0.79	0.79	0.79	0.79	0.79	0.79	0.80	0.79	0.79	0.79	0.79	0.79	0.79	0.79	0.79	0.80	0.79
53	0.78	0.78	0.78	0.78	0.78	0.78	0.78	0.78	0.78	0.78	0.79	0.78	0.79	0.79	0.78	0.79	0.79	0.79	0.79	0.79	0.79
54	0.77	0.77	0.77	0.77	0.77	0.77	0.77	0.77	0.77	0.77	0.78	0.78	0.78	0.77	0.78	0.78	0.78	0.78	0.78	0.78	0.78
55	0.76	0.76	0.76	0.77	0.77	0.77	0.76	0.77	0.77	0.78	0.77	0.78	0.78	0.77	0.77	0.77	0.77	0.77	0.77	0.77	0.77
56	0.75	0.75	0.76	0.76	0.76	0.76	0.76	0.76	0.76	0.76	0.76	0.76	0.76	0.76	0.76	0.76	0.76	0.76	0.76	0.76	0.76
57	0.75	0.75	0.75	0.75	0.75	0.75	0.76	0.75	0.75	0.75	0.76	0.76	0.75	0.75	0.75	0.76	0.76	0.76	0.76	0.76	0.76
58	0.74	0.74	0.74	0.74	0.74	0.74	0.74	0.75	0.75	0.75	0.75	0.75	0.75	0.75	0.75	0.75	0.75	0.75	0.75	0.75	0.75
59	0.73	0.74	0.74	0.73	0.74	0.74	0.74	0.74	0.74	0.74	0.75	0.74	0.74	0.74	0.74	0.74	0.74	0.74	0.74	0.74	0.74
60	0.72	0.72	0.72	0.73	0.73	0.73	0.73	0.73	0.74	0.74	0.74	0.74	0.74	0.74	0.74	0.74	0.73	0.73	0.73	0.73	0.73

East or West facing window partial shading correction factors for fins, Ff

Latitude Fin Angle	45	46	47	48	49	50	51	52	53	54	55	56	57	58	59	60	61	62	63	64	65
0	1.00	1.00	1.00	1.00	1.00	1.00	1.00	1.00	1.00	1.00	1.00	1.00	1.00	1.00	1.00	1.00	1.00	1.00	1.00	1.00	1.00
1	1.00	1.00	1.00	1.00	1.00	1.00	1.00	1.00	1.00	1.00	1.00	1.00	1.00	1.00	1.00	1.00	1.00	1.00	1.00	1.00	1.00
2	0.99	0.99	0.99	0.99	0.99	0.99	0.99	0.99	0.99	0.99	0.99	0.99	0.99	0.99	0.99	0.99	0.99	0.99	0.99	0.99	0.99
3	0.99	0.99	0.99	0.99	0.99	0.99	0.99	0.99	0.99	0.99	0.99	0.99	0.99	0.99	0.99	0.99	0.99	0.99	0.99	0.99	0.99
4	0.99	0.99	0.99	0.99	0.99	0.99	0.99	0.99	0.99	0.99	0.99	0.99	0.99	0.99	0.99	0.98	0.98	0.98	0.98	0.98	0.98
5	0.99	0.99	0.99	0.98	0.98	0.98	0.98	0.98	0.98	0.98	0.98	0.98	0.98	0.98	0.98	0.98	0.98	0.98	0.98	0.98	0.98
6	0.98	0.98	0.98	0.98	0.98	0.98	0.98	0.98	0.98	0.98	0.98	0.98	0.98	0.98	0.98	0.98	0.98	0.98	0.98	0.98	0.98
7	0.98	0.98	0.98	0.98	0.98	0.98	0.98	0.98	0.98	0.98	0.98	0.98	0.98	0.98	0.97	0.97	0.97	0.97	0.97	0.97	0.97
8	0.98	0.98	0.98	0.97	0.97	0.97	0.97	0.97	0.98	0.97	0.97	0.97	0.97	0.97	0.97	0.97	0.97	0.97	0.97	0.97	0.97
9	0.98	0.97	0.97	0.97	0.97	0.97	0.97	0.97	0.97	0.97	0.97	0.97	0.97	0.97	0.97	0.97	0.97	0.97	0.97	0.97	0.97
10	0.97	0.97	0.97	0.97	0.97	0.97	0.97	0.97	0.97	0.97	0.97	0.97	0.97	0.97	0.97	0.97	0.96	0.96	0.96	0.96	0.96
11	0.97	0.97	0.97	0.97	0.97	0.97	0.97	0.97	0.97	0.97	0.96	0.96	0.96	0.96	0.96	0.96	0.96	0.96	0.96	0.96	0.96
12	0.97	0.96	0.96	0.96	0.96	0.96	0.96	0.96	0.96	0.96	0.96	0.96	0.96	0.96	0.96	0.96	0.96	0.96	0.96	0.96	0.96
13	0.96	0.96	0.96	0.96	0.96	0.96	0.96	0.96	0.96	0.96	0.96	0.96	0.96	0.96	0.96	0.96	0.96	0.95	0.95	0.95	0.95
14	0.96	0.96	0.96	0.96	0.96	0.96	0.96	0.96	0.96	0.96	0.96	0.95	0.95	0.95	0.95	0.95	0.95	0.95	0.95	0.95	0.95
15	0.96	0.96	0.95	0.95	0.95	0.95	0.95	0.95	0.95	0.95	0.95	0.95	0.95	0.95	0.95	0.95	0.95	0.95	0.95	0.95	0.95
16	0.95	0.95	0.95	0.95	0.95	0.95	0.95	0.95	0.95	0.95	0.95	0.95	0.95	0.95	0.95	0.95	0.95	0.95	0.94	0.94	0.94
17	0.95	0.95	0.95	0.95	0.95	0.95	0.95	0.94	0.94	0.94	0.94	0.94	0.94	0.94	0.94	0.94	0.94	0.94	0.94	0.94	0.94
18	0.95	0.94	0.94	0.94	0.94	0.94	0.94	0.94	0.94	0.94	0.94	0.94	0.94	0.94	0.94	0.94	0.94	0.94	0.94	0.94	0.94
19	0.94	0.94	0.94	0.94	0.94	0.94	0.94	0.94	0.94	0.94	0.94	0.94	0.94	0.94	0.94	0.94	0.94	0.94	0.93	0.93	0.93
20	0.94	0.94	0.94	0.94	0.94	0.94	0.94	0.94	0.94	0.93	0.93	0.93	0.93	0.93	0.93	0.93	0.93	0.93	0.93	0.93	0.93
21	0.94	0.94	0.94	0.93	0.93	0.93	0.93	0.93	0.93	0.93	0.93	0.93	0.93	0.93	0.93	0.93	0.93	0.93	0.93	0.93	0.93
22	0.94	0.93	0.93	0.93	0.93	0.93	0.93	0.93	0.93	0.93	0.93	0.93	0.93	0.93	0.93	0.93	0.93	0.93	0.92	0.92	0.92
23	0.93	0.93	0.93	0.93	0.93	0.93	0.93	0.93	0.93	0.93	0.93	0.93	0.93	0.92	0.92	0.92	0.92	0.92	0.92	0.92	0.92
24	0.93	0.93	0.93	0.93	0.93	0.93	0.92	0.92	0.92	0.92	0.92	0.92	0.92	0.92	0.92	0.92	0.92	0.92	0.92	0.92	0.92
25	0.93	0.93	0.92	0.92	0.92	0.92	0.92	0.92	0.92	0.92	0.92	0.92	0.92	0.92	0.92	0.92	0.92	0.92	0.92	0.92	0.91
26	0.92	0.92	0.92	0.92	0.92	0.92	0.92	0.92	0.92	0.92	0.92	0.92	0.92	0.92	0.92	0.91	0.91	0.91	0.91	0.91	0.91
27	0.92	0.92	0.92	0.92	0.92	0.92	0.92	0.92	0.92	0.91	0.91	0.91	0.91	0.91	0.91	0.91	0.91	0.91	0.91	0.91	0.91
28	0.93	0.92	0.92	0.92	0.92	0.92	0.91	0.91	0.91	0.91	0.91	0.91	0.91	0.91	0.91	0.91	0.91	0.91	0.91	0.91	0.91
29	0.92	0.92	0.92	0.91	0.91	0.92	0.91	0.91	0.91	0.91	0.91	0.91	0.91	0.91	0.91	0.91	0.91	0.91	0.90	0.90	0.90
30	0.92	0.92	0.92	0.91	0.92	0.91	0.91	0.91	0.91	0.91	0.91	0.91	0.91	0.91	0.91	0.91	0.90	0.90	0.90	0.90	0.90
31	0.91	0.91	0.91	0.91	0.91	0.90	0.90	0.90	0.91	0.90	0.90	0.90	0.90	0.90	0.90	0.90	0.90	0.90	0.90	0.90	0.89
32	0.91	0.91	0.91	0.91	0.91	0.90	0.90	0.90	0.90	0.90	0.90	0.90	0.90	0.90	0.90	0.90	0.89	0.89	0.89	0.89	0.89
33	0.90	0.90	0.90	0.90	0.90	0.90	0.90	0.90	0.90	0.89	0.89	0.89	0.89	0.89	0.89	0.89	0.89	0.89	0.89	0.89	0.88
34	0.90	0.90	0.90	0.89	0.89	0.89	0.89	0.89	0.89	0.89	0.89	0.89	0.89	0.89	0.88	0.88	0.88	0.88	0.88	0.88	0.88
35	0.89	0.89	0.89	0.89	0.89	0.89	0.89	0.89	0.89	0.88	0.88	0.88	0.88	0.88	0.88	0.88	0.88	0.88	0.88	0.88	0.88
36	0.89	0.89	0.88	0.88	0.88	0.88	0.88	0.88	0.88	0.88	0.88	0.88	0.88	0.87	0.87	0.87	0.87	0.87	0.87	0.87	0.87
37	0.88	0.88	0.88	0.88	0.88	0.88	0.87	0.88	0.87	0.87	0.87	0.87	0.87	0.87	0.87	0.87	0.87	0.87	0.86	0.86	0.86
38	0.88	0.88	0.88	0.87	0.87	0.87	0.87	0.87	0.87	0.87	0.86	0.86	0.86	0.86	0.86	0.86	0.86	0.86	0.86	0.86	0.86
39	0.87	0.87	0.87	0.87	0.87	0.86	0.87	0.86	0.86	0.86	0.86	0.86	0.86	0.86	0.86	0.86	0.85	0.85	0.86	0.86	0.85
40	0.87	0.87	0.86	0.86	0.86	0.86	0.86	0.86	0.86	0.86	0.86	0.86	0.86	0.86	0.86	0.85	0.85	0.85	0.85	0.85	0.85
41	0.86	0.86	0.86	0.86	0.85	0.85	0.85	0.85	0.85	0.85	0.85	0.85	0.85	0.85	0.85	0.85	0.85	0.85	0.85	0.85	0.85
42	0.86	0.85	0.85	0.85	0.85	0.85	0.85	0.85	0.85	0.85	0.85	0.84	0.84	0.84	0.84	0.84	0.84	0.84	0.84	0.84	0.84
43	0.85	0.85	0.85	0.84	0.84	0.84	0.84	0.84	0.84	0.84	0.84	0.84	0.84	0.84	0.84	0.84	0.84	0.84	0.84	0.84	0.84
44	0.85	0.84	0.84	0.84	0.84	0.84	0.84	0.84	0.84	0.84	0.84	0.84	0.84	0.84	0.84	0.84	0.84	0.83	0.84	0.83	0.83
45	0.84	0.84	0.84	0.83	0.83	0.83	0.83	0.83	0.83	0.83	0.83	0.83	0.83	0.83	0.83	0.83	0.83	0.83	0.83	0.83	0.82
46	0.83	0.83	0.83	0.83	0.83	0.83	0.83	0.82	0.83	0.82	0.82	0.82	0.82	0.82	0.82	0.82	0.82	0.82	0.82	0.82	0.81
47	0.83	0.83	0.83	0.82	0.82	0.82	0.82	0.82	0.82	0.82	0.82	0.82	0.82	0.82	0.81	0.81	0.81	0.81	0.81	0.81	0.81
48	0.82	0.82	0.82	0.82	0.82	0.81	0.81	0.81	0.81	0.81	0.81	0.81	0.81	0.81	0.81	0.81	0.81	0.80	0.80	0.80	0.80
49	0.82	0.82	0.81	0.81	0.81	0.81	0.81	0.81	0.81	0.81	0.80	0.80	0.80	0.80	0.80	0.80	0.80	0.80	0.80	0.80	0.80
50	0.81	0.81	0.81	0.81	0.80	0.81	0.80	0.80	0.80	0.80	0.80	0.80	0.80	0.80	0.80	0.80	0.80	0.79	0.80	0.79	0.79
51	0.80	0.80	0.80	0.80	0.80	0.80	0.79	0.79	0.79	0.79	0.79	0.79	0.79	0.79	0.79	0.79	0.79	0.79	0.79	0.79	0.79
52	0.80	0.80	0.80	0.79	0.79	0.79	0.79	0.79	0.79	0.79	0.79	0.79	0.79	0.79	0.78	0.78	0.78	0.78	0.78	0.78	0.78
53	0.79	0.79	0.79	0.79	0.78	0.78	0.78	0.78	0.78	0.78	0.78	0.78	0.78	0.78	0.78	0.78	0.78	0.78	0.78	0.77	0.77
54	0.79	0.78	0.78	0.78	0.78	0.78	0.78	0.78	0.78	0.78	0.78	0.77	0.77	0.77	0.77	0.77	0.77	0.77	0.77	0.77	0.77
55	0.78	0.78	0.78	0.78	0.78	0.77	0.77	0.77	0.77	0.77	0.77	0.77	0.77	0.77	0.77	0.77	0.77	0.76	0.76	0.76	0.76
56	0.77	0.77	0.77	0.77	0.77	0.77	0.77	0.77	0.77	0.77	0.77	0.77	0.77	0.76	0.76	0.76	0.76	0.76	0.76	0.76	0.76
57	0.77	0.77	0.77	0.76	0.76	0.76	0.76	0.76	0.76	0.76	0.76	0.76	0.76	0.76	0.76	0.76	0.76	0.76	0.75	0.75	0.75
58	0.76	0.76	0.76	0.76	0.76	0.76	0.76	0.76	0.76	0.76	0.75	0.75	0.75	0.75	0.75	0.75	0.75	0.75	0.75	0.75	0.75
59	0.76	0.76	0.76	0.75	0.76	0.75	0.75	0.75	0.75	0.75	0.75	0.75	0.75	0.75	0.75	0.75	0.74	0.74	0.74	0.74	0.74
60	0.75	0.75	0.75	0.75	0.75	0.75	0.75	0.75	0.75	0.75	0.75	0.75	0.75	0.74	0.74	0.74	0.74	0.74	0.74	0.73	0.73

North facing window partial shading correction factors for fins, Ff

Fin Angle	45	46	47	48	49	50	51	52	53	54	55	56	57	58	59	60	61	62	63	64	65
0	1.00	1.00	1.00	1.00	1.00	1.00	1.00	1.00	1.00	1.00	1.00	1.00	1.00	1.00	1.00	1.00	1.00	1.00	1.00	1.00	1.00
1	1.00	1.00	1.00	1.00	1.00	1.00	1.00	1.00	1.00	1.00	1.00	1.00	1.00	1.00	1.00	1.00	1.00	1.00	1.00	1.00	1.00
2	1.00	1.00	1.00	1.00	1.00	1.00	1.00	1.00	1.00	1.00	1.00	1.00	1.00	1.00	1.00	1.00	1.00	1.00	1.00	1.00	1.00
3	1.00	1.00	1.00	1.00	1.00	1.00	1.00	1.00	1.00	1.00	1.00	1.00	1.00	1.00	1.00	1.00	1.00	1.00	1.00	1.00	1.00
4	1.00	1.00	1.00	1.00	1.00	1.00	1.00	1.00	1.00	1.00	1.00	1.00	1.00	1.00	1.00	1.00	1.00	1.00	1.00	1.00	1.00
5	1.00	1.00	1.00	1.00	1.00	1.00	1.00	1.00	1.00	1.00	1.00	1.00	1.00	1.00	1.00	1.00	1.00	1.00	1.00	1.00	1.00
6	1.00	1.00	1.00	1.00	1.00	1.00	1.00	1.00	1.00	1.00	1.00	1.00	1.00	1.00	1.00	1.00	1.00	1.00	0.99	0.99	0.99
7	1.00	1.00	1.00	1.00	1.00	1.00	1.00	1.00	1.00	1.00	1.00	1.00	1.00	1.00	1.00	0.99	0.99	0.99	0.99	0.99	0.99
8	1.00	1.00	1.00	1.00	1.00	1.00	1.00	1.00	1.00	1.00	1.00	0.99	0.99	0.99	0.99	0.99	0.99	0.99	0.99	0.99	0.99
9	1.00	1.00	1.00	1.00	1.00	1.00	1.00	1.00	0.99	0.99	0.99	0.99	0.99	0.99	0.99	0.99	0.99	0.99	0.99	0.99	0.99
10	1.00	1.00	1.00	1.00	1.00	1.00	0.99	0.99	0.99	0.99	0.99	0.99	0.99	0.99	0.99	0.99	0.99	0.99	0.99	0.99	0.99
11	1.00	1.00	1.00	1.00	1.00	1.00	0.99	0.99	0.99	0.99	0.99	0.99	0.99	0.99	0.99	0.99	0.99	0.99	0.99	0.99	0.99
12	1.00	1.00	1.00	1.00	1.00	1.00	0.99	0.99	0.99	0.99	0.99	0.99	0.99	0.99	0.99	0.99	0.99	0.99	0.99	0.99	0.99
13	1.00	1.00	1.00	1.00	1.00	1.00	0.99	0.99	0.99	0.99	0.99	0.99	0.99	0.99	0.99	0.99	0.99	0.99	0.99	0.99	0.99
14	1.00	1.00	1.00	1.00	1.00	1.00	0.99	0.99	0.99	0.99	0.99	0.99	0.99	0.99	0.99	0.99	0.99	0.99	0.99	0.99	0.99
15	1.00	1.00	1.00	1.00	1.00	1.00	0.99	0.99	0.99	0.99	0.99	0.99	0.99	0.99	0.99	0.99	0.99	0.99	0.99	0.99	0.98
16	1.00	1.00	1.00	1.00	1.00	1.00	0.99	0.99	0.99	0.99	0.99	0.99	0.99	0.99	0.99	0.99	0.99	0.99	0.99	0.98	0.98
17	1.00	1.00	1.00	1.00	1.00	1.00	0.99	0.99	0.99	0.99	0.99	0.99	0.99	0.99	0.99	0.99	0.99	0.98	0.98	0.98	0.98
18	1.00	1.00	1.00	1.00	1.00	1.00	0.99	0.99	0.99	0.99	0.99	0.99	0.99	0.99	0.99	0.99	0.98	0.98	0.98	0.98	0.98
19	1.00	1.00	1.00	1.00	1.00	1.00	0.99	0.99	0.99	0.99	0.99	0.99	0.99	0.99	0.99	0.99	0.98	0.98	0.98	0.98	0.98
20	1.00	1.00	1.00	1.00	1.00	1.00	0.99	0.99	0.99	0.99	0.99	0.99	0.99	0.99	0.99	0.98	0.98	0.98	0.98	0.98	0.98
21	1.00	1.00	1.00	1.00	1.00	1.00	0.99	0.99	0.99	0.99	0.99	0.99	0.99	0.99	0.99	0.98	0.98	0.98	0.98	0.98	0.98
22	1.00	1.00	1.00	1.00	1.00	1.00	0.99	0.99	0.99	0.99	0.99	0.99	0.99	0.99	0.98	0.98	0.98	0.98	0.98	0.98	0.98
23	1.00	1.00	1.00	1.00	1.00	1.00	0.99	0.99	0.99	0.99	0.99	0.99	0.99	0.99	0.98	0.98	0.98	0.98	0.98	0.98	0.98
24	1.00	1.00	1.00	1.00	1.00	1.00	0.99	0.99	0.99	0.99	0.99	0.99	0.99	0.99	0.98	0.98	0.98	0.98	0.98	0.98	0.98
25	1.00	1.00	1.00	1.00	1.00	1.00	0.99	0.99	0.99	0.99	0.99	0.99	0.99	0.98	0.98	0.98	0.98	0.98	0.98	0.98	0.98
26	1.00	1.00	1.00	1.00	1.00	1.00	0.99	0.99	0.99	0.99	0.99	0.99	0.99	0.98	0.98	0.98	0.98	0.98	0.98	0.98	0.98
27	1.00	1.00	1.00	1.00	1.00	1.00	0.99	0.99	0.99	0.99	0.99	0.99	0.99	0.98	0.98	0.98	0.98	0.98	0.98	0.98	0.98
28	1.00	1.00	1.00	1.00	1.00	1.00	0.99	0.99	0.99	0.99	0.99	0.99	0.98	0.98	0.98	0.98	0.98	0.98	0.98	0.98	0.98
29	1.00	1.00	1.00	1.00	1.00	1.00	0.99	0.99	0.99	0.99	0.99	0.99	0.98	0.98	0.98	0.98	0.98	0.98	0.98	0.98	0.98
30	1.00	1.00	1.00	1.00	1.00	1.00	0.99	0.99	0.99	0.99	0.99	0.99	0.98	0.98	0.98	0.98	0.98	0.98	0.98	0.98	0.98
31	1.00	1.00	1.00	1.00	1.00	1.00	0.99	0.99	0.99	0.99	0.99	0.99	0.98	0.98	0.98	0.98	0.98	0.98	0.98	0.98	0.98
32	1.00	1.00	1.00	1.00	1.00	1.00	0.99	0.99	0.99	0.99	0.99	0.99	0.98	0.98	0.98	0.98	0.98	0.98	0.98	0.98	0.98
33	1.00	1.00	1.00	1.00	1.00	1.00	0.99	0.99	0.99	0.99	0.99	0.99	0.98	0.98	0.98	0.98	0.98	0.98	0.98	0.98	0.98
34	1.00	1.00	1.00	1.00	1.00	1.00	0.99	0.99	0.99	0.99	0.99	0.99	0.98	0.98	0.98	0.98	0.98	0.98	0.98	0.98	0.98
35	1.00	1.00	1.00	1.00	1.00	1.00	0.99	0.99	0.99	0.99	0.99	0.99	0.98	0.98	0.98	0.98	0.98	0.98	0.98	0.98	0.98
36	1.00	1.00	1.00	1.00	1.00	1.00	0.99	0.99	0.99	0.99	0.99	0.99	0.98	0.98	0.98	0.98	0.98	0.98	0.98	0.98	0.98
37	1.00	1.00	1.00	1.00	1.00	1.00	0.99	0.99	0.99	0.99	0.99	0.99	0.98	0.98	0.98	0.98	0.98	0.98	0.98	0.98	0.98
38	1.00	1.00	1.00	1.00	1.00	1.00	0.99	0.99	0.99	0.99	0.99	0.99	0.98	0.98	0.98	0.98	0.98	0.98	0.98	0.98	0.98
39	1.00	1.00	1.00	1.00	1.00	1.00	0.99	0.99	0.99	0.99	0.99	0.99	0.98	0.98	0.98	0.98	0.98	0.98	0.98	0.98	0.98
40	1.00	1.00	1.00	1.00	1.00	1.00	0.99	0.99	0.99	0.99	0.99	0.99	0.98	0.98	0.98	0.98	0.98	0.98	0.98	0.98	0.98
41	1.00	1.00	1.00	1.00	1.00	1.00	0.99	0.99	0.99	0.99	0.99	0.99	0.98	0.98	0.98	0.98	0.98	0.98	0.98	0.98	0.98
42	1.00	1.00	1.00	1.00	1.00	1.00	0.99	0.99	0.99	0.99	0.99	0.99	0.98	0.98	0.98	0.98	0.98	0.98	0.98	0.98	0.98
43	1.00	1.00	1.00	1.00	1.00	1.00	0.99	0.99	0.99	0.99	0.99	0.99	0.98	0.98	0.98	0.98	0.98	0.98	0.98	0.98	0.98
44	1.00	1.00	1.00	1.00	1.00	1.00	0.99	0.99	0.99	0.99	0.99	0.99	0.98	0.98	0.98	0.98	0.98	0.98	0.98	0.98	0.98
45	1.00	1.00	1.00	1.00	1.00	1.00	0.99	0.99	0.99	0.99	0.99	0.99	0.98	0.98	0.98	0.99	0.98	0.98	0.98	0.98	0.98
46	1.00	1.00	1.00	1.00	1.00	1.00	0.99	0.99	0.99	0.99	0.99	0.99	0.99	0.99	0.99	0.99	0.98	0.98	0.98	0.98	0.98
47	1.00	1.00	1.00	1.00	1.00	1.00	0.99	0.99	0.99	0.99	0.99	0.99	0.99	0.99	0.99	0.99	0.98	0.98	0.98	0.98	0.98
48	1.00	1.00	1.00	1.00	1.00	1.00	0.99	0.99	0.99	0.99	0.99	0.99	0.99	0.99	0.99	0.99	0.98	0.98	0.98	0.98	0.98
49	1.00	1.00	1.00	1.00	1.00	1.00	0.99	0.99	0.99	0.99	0.99	0.99	0.99	0.99	0.99	0.99	0.98	0.98	0.98	0.98	0.98
50	1.00	1.00	1.00	1.00	1.00	1.00	0.99	0.99	0.99	0.99	0.99	0.99	0.99	0.99	0.99	0.99	0.98	0.98	0.98	0.98	0.98
51	1.00	1.00	1.00	1.00	1.00	1.00	0.99	0.99	0.99	0.99	0.99	0.99	0.99	0.99	0.99	0.99	0.98	0.98	0.98	0.98	0.98
52	1.00	1.00	1.00	1.00	1.00	1.00	0.99	0.99	0.99	0.99	0.99	0.99	0.99	0.99	0.99	0.99	0.98	0.98	0.98	0.98	0.98
53	1.00	1.00	1.00	1.00	1.00	1.00	0.99	0.99	0.99	0.99	0.99	0.99	0.99	0.99	0.99	0.99	0.98	0.98	0.98	0.98	0.98
54	1.00	1.00	1.00	1.00	1.00	1.00	0.99	0.99	0.99	0.99	0.99	0.99	0.99	0.99	0.99	0.99	0.98	0.98	0.98	0.98	0.98
55	1.00	1.00	1.00	1.00	1.00	1.00	0.99	0.99	0.99	0.99	0.99	0.99	0.99	0.99	0.99	0.99	0.98	0.98	0.98	0.98	0.98
56	1.00	1.00	1.00	1.00	1.00	1.00	0.99	0.99	0.99	0.99	0.99	0.99	0.99	0.99	0.99	0.99	0.98	0.98	0.98	0.98	0.98
57	1.00	1.00	1.00	1.00	1.00	1.00	0.99	0.99	0.99	0.99	0.99	0.99	0.99	0.99	0.99	0.99	0.98	0.98	0.98	0.98	0.98
58	1.00	1.00	1.00	1.00	1.00	1.00	0.99	0.99	0.99	0.99	0.99	0.99	0.99	0.99	0.99	0.99	0.98	0.98	0.98	0.98	0.98
59	1.00	1.00	1.00	1.00	1.00	1.00	0.99	0.99	0.99	0.99	0.99	0.99	0.99	0.99	0.99	0.99	0.98	0.98	0.98	0.98	0.98
60	1.00	1.00	1.00	1.00	1.00	1.00	0.99	0.99	0.99	0.99	0.99	0.99	0.99	0.99	0.99	0.99	0.98	0.98	0.98	0.98	0.98

References

A comprehensive guide to managing asbestos in premises, HSE Books, 2002 (ISBN 0 7176 2381 5)

Accredited Construction details, Communities and Local Government (CLG) Publications, 2007 (Product Code 07BD04040)

Approved Document for Part F of the Building Regulations, CLG, 2006, available at www.planningportal.gov.uk/uploads/br/BR_PDF_ADF_2006.pdf

BRE Digest 498 selecting lighting controls (ISBN 1 86081 905 2)

BRE Good Building Guide 61, Lighting non-domestic, 2004 (ISBN 1 86081 719X)

BRE, A Technical Manual for SBEM, 24 October 2008, Communities and Local Government (current version available from BRE www.ncm.bre.co.uk)

BRE, A User Guide to iSBEM, 29 April 2009, Communities and Local Government (current version available from BRE www.ncm.bre.co.uk)

BSRIA The Illustrated Guide to Electrical Building Services, 2nd edition BG 5/2005 ISBN 0 86022 653 0 www.bsria.co.uk

CIBSE Guide F Energy Efficiency in Buildings, CIBSE, 2004 (ISBN 1903287340)

CIBSE Lighting Guide – The Industrial Environment, CIBSE, 1989 (ISBN 0 900953 38 1)

Conservation of fuel and power: Approved Document L2A 'Conservation of fuel and power in new buildings other than dwellings' (2006 edition) ODPM, available at www.planningportal.gov.uk/uploads/br/BR_PDF_ADL2A_2006.pdf

Conservation of fuel and power: Approved Document L2B 'Conservation of fuel and power in existing buildings other than dwellings' (2006 edition) OPDM, available at http://www.planningportal.gov.uk/uploads/br/BR_PDF_ADL2B_2006.pdf

DFP Technical Booklet F1: 2006 – *Conservation of fuel and power in dwellings*, available at www.dfpni.gov.uk/tb_f1_mp_v6.pdf

DFP Technical Booklet F2: 2006 – *Conservation of fuel and power in buildings other than dwellings*, available at www.dfpni.gov.uk/tb_f2_v15-2.pdf

Display lighting – Technology Guide, Carbon Trust CTG010, 2008

Environmental Design CIBSE Guide A, The Chartered Institution of Building Services Engineers (CIBSE), London, 2006 (ISBN 1 903287 66 9)

Five step risk assessment available at www.hse.gov.uk/risk/fivesteps.htm

Garratt, J. and Nowak, F., *Tackling Condensation: A guide to the causes of, an remedies for, surface condensation and mould in traditional housing*, Building Research Establishment, Watford, 1991 (ISBN 0 85125 444 6)

Green Value: Green buildings, growing assets, RICS, 2005

July 2008 CLG guidance – Improving the Energy Efficiency of Our Buildings – A Guide to Air-Conditioning Inspections For Buildings (2nd edition), CLG, July 2008 available at www.communities.gov.uk/documents/planningandbuilding/pdf/nondwellingsguidance.pdf

Improving the energy efficiency of our buildings – Local weights and measures guidance for Energy Certificates and air-conditioning inspections for buildings, CLG, October 2008

Installers' Guide to the Assessment of Energy Efficient Lighting Installations – Lighting Guide 007, Action Energy from the Carbon Trust 2004 (ref ILG 007)

Introduction to Asbestos Essentials, HSE Books, 2001 (ISBN 0 7176 1901 6)

Kyoto Protocol to the United Nations Framework Convention on Climate Change available at http://unfccc.int/resource/docs/convkp/kpeng.pdf

Lamp Guide 2001, Lighting Industry Federation Ltd available at www.lif.co.uk

Lighting – bright ideas for more efficient illumination Carbon Trust CTV 021 Technology Overview 2007

LZC Strategic Guide – Low or Zero Carbon Energy Sources: Strategic Guide (1st edition) May 2006 available at www.planningportal.gov.uk/uploads/br/BR_PDF_PTL_ZEROCARBONfinal.pdf

Management of Health and Safety at Work, Approved Code of Practice and Guidance L21 (2nd edition), HSE Books, 2000 (ISBN 0 7176 7488 9)

Minimum Requirements For Energy Assessors For Non-Dwellings, CLG, October 2007 available at www.communities.gov.uk/documents/planningandbuilding/pdf/nondwellings.pdf

National Occupational Standards for Building Energy Assessment (Non dwellings) on Construction, Sale or Rent, Final Version, February 2009, available at www.assetskills.org/nmsruntime/saveasdialog.asp?lID=972&sID=439

NHER, OCDEA Technical Bulletin Issue 19, National Energy Services, September 2008

Non-Domestic Heating, Cooling and Ventilation Compliance Guide (1st Edition), CLG, May 2006 available at www.planningportal.gov.uk/uploads/br/BR_PDF_PTL_NONDOMHEAT.pdf

Parnham, P. and Russen, L. *Domestic Energy Assessor's Handbook,* RICS, Coventry, 2008 (ISBN 9781842193734)

Requirements for energy performance certificates (EPCs) when marketing commercial (non-domestic) properties for sale or let, CLG, October 2009

RICS Code of Measuring Practice, RICS guidance note (6th Edition), RICS Books, Coventry, September 2007

Your Guide to important issues in lighting design, Lighting Industry Federation Ltd available at www.lif.co.uk

LEGISLATION

Acts
Building Act 1984
Contracts (Rights of Third Parties) Act 1999
Health and Safety at Work etc Act 1974
Occupiers' Liability Act 1957
Occupiers' Liability Act 1984
Office Shops and Railways Premises Act 1963
Thermal Insulation (Industrial Buildings) Act 1957

Directives
Directive 2002/91/EC of the European Parliament and of the Council of 16 December 2002 on the Energy Performance of Buildings http://eur-lex.europa.eu/LexUriServ/LexUriServ.do?uri=OJ:L:2003:001:0065:0071:EN:PDF

Statutory Instruments
Control of Asbestos Regulations 2006

The Energy Performance of Buildings (Certificates and Inspections) (England and Wales) Regulations 2007 (SI 2007/991) available at www.opsi.gov.uk/si/si2007/uksi_20070991_en_1

The Management of Health and Safety at Work Regulations 1999 (SI 1999/3242) available at www.opsi.gov.uk/si/si1999/19993242.htm

The Reporting of Injuries, Diseases and Occurrences Regulations 1995 (SI 1995/3163) available at www.opsi.gov.uk/SI/si1995/Uksi_19953163_en_1.htm

The Work at Height Regulations 2005 (SI 2005/735)

Notification of Cooling Towers and Evaporative Condensers Regulations 1992 (SI 1992/2225), available at www.opsi.gov.uk/SI/si1992/Uksi_19922225_en_1.htm

USEFUL WEBSITES

www.abbeqa.co.uk
www.assetskills.org
www.besca.org.uk;
www.bre.co.uk/accreditation;
www.chpqa.com
www.ciat.org.uk;
www.cibse.org;
www.cityandguilds.com
www.communities.gov.uk
www.eca.gov.uk/etl
www.ecmk.co.uk;
www.elmhurstenergy.co.uk;
www.energy-assessors.org.uk
www.hicertification.co.uk;
www.hse.gov.uk/falls/ladders.htm
www.hse.gov.uk/risk/fivesteps.htm
www.isurv.co.uk
www.knauf.co.uk;
www.mcrma.co.uk
www.napit.org.uk;
www.nfopp.co.uk
www.nher.co.uk;
www.northgate-ispublicservices.com;
www.planningportal.gov.uk
www.quidos.co.uk;
www.rics.org/hips;
www.sbsa.gov.uk
www.stroma.com.

Index

accreditation schemes 6–7
Accredited Building Details 35
air conditioning *see also* heating, ventilation and air conditioning (HVAC) systems; ventilation and air conditioning
    inspections 5
asbestos 12–13
    awareness training 14
    diseases 15
    emergencies 15
    management in non-domestic premises 15
    properties 14
    risk 14
    typical locations for most common asbestos-containing materials 13
assessments *see* commercial energy assessments; training and assessment
asset rating 3
    energy performance certificate *see* energy performance certificates (EPCs)
auditing *see* energy performance certificates (EPCs); quality and auditing
audit trail 133
    'King's Lynn protocol' 133, 134

boilers *see* heating, ventilation and air conditioning (HVAC) systems
building construction *see* construction; energy in construction
Building Regulations 3–4, 35–36
    Approved Documents (ADs) 4
    lighting 115
    new build construction 27–28
        Building Regulations in other parts of UK 28–29
building systems 48–50

ceilings 47
central heating *see* heating, ventilation and air conditioning (HVAC) systems
cladding panels 42
client instructions 135
client report 165
climate change
    commercial buildings contribution 1
commercial buildings
    contribution to climate change 1
commercial energy assessment
    absence of fixed services that provide conditioning 17–18
    building, definition 17
    complaints procedure 22
    conditioning, definition 17
    contribution of commercial buildings to climate change 1
    data gatherers (DGs) 22–23
    different levels 3
        existing buildings 18
    enforcement 19
    European context 1–2
    exemptions 18–19, 167–172
    historical development of energy issues 1
    inspection equipment 23–26

    level 3, case studies 167–267
    life-long learning 22
    new build construction
        Building Regulations 27–28
        Building Regulations in other parts of UK 28–29
    non-domestic energy assessors 1
    professional indemnity insurance (PII) 22
    quality management system (QMS) 19–22
    RICS Code of Measuring Practice 26–27
    running/organising professional practice as a business 22
Communities and Local Government (CLG) 2
    Minimum Requirements for Energy Assessors for Non-Dwellings, CLG, October 2007 7–8
complaints procedure 22
conflicts of interest 133
conservation areas 36–37
construction *see also* energy in construction
    date of 37–38
    forms of 38
    framed buildings 39–40
        controlling dimensions 41
    identification 37
    traditional methods 38–39
    types of external envelope *see* external envelope
contract law *see* Law of Tort
core knowledge and recognition
    energy and construction 31–35
    heating and hot water services 51–56
        heating sources 65–73
        heating systems 56–65
        hot water generation 75–80
        HVAC system controls 73–75
    lighting 113
        artificial systems and their recognition 117–126
        controls 129–131
        damage caused by light 132
        display lighting 131
        safe inspection of lamps 131–132
        Simplified Building Energy Model (SBEM) 127–131
        theory and practice 113–114
    planning and Building Regulations issues 35–37
    specific construction issues 37–41
    types of external envelope 41–49
    ventilation and air conditioning 83
        air conditioning systems 95–109
        renewable energy systems 109–111
        ventilation 83–86
        ventilation and air conditioning equipment 86–94
curtain walling 42–43

data entry into program 143
    main SBEM tabs 143
        'geometry' *see* 'geometry'
        'project database' *see* 'project database'
data gatherers (DGs) 22–23
display energy certificates (DECs) 2
Domestic Energy Assessor's Handbook (2008) 38
doors 46

data entry
    construction for doors 144
    external doors 153, 289
    glazed doors 153
    high usage entrance doors 153
    'U values' of vehicular entrance doors 153

energy assessment *see* commercial energy assessment
energy in construction 31
    Accredited Building Details 35
    effective thermal capacity (K_m value) 31–34
    thermal bridges 34–35
    thermal transmittance (U value) 31
energy performance certificates (EPCs) 2, 3, 158
    audit 20–21, 163
        recording comments on assessment program 164
    graphic rating 158
Energy Performance of Buildings (Certificates and Inspections) (England and Wales) Regulations 2007 (SI 2007/991) 2
    reg. 38(1) 19
Energy Performance of Buildings Directive (EPBD) 1–2
    UK response 2–5
European legislation
    Energy Performance of Buildings Directive (EPBD) 1–2
        UK response 2–5
enquiries
    initial 133
    post-inspection 142
envelopes *see* external envelope; 'geometry'
existing buildings
    commercial energy assessment 18
    inspection and reflection methodologies 133
        after inspection 142
        first contact with client 133–135
        pre-inspection practice 135–136
        site inspection – suggested procedure 136–142
external envelope
    building systems 48–49
    cladding panels 42
    curtain walling 42–43
    doors 46
    external walls 41–42
    floors and ceilings 46–47
    profiled cladding systems 43
    retaining structures 43–44
    roofs 46
    windows 44–45

fees 133–134
fins and overhangs 153
    transmission factor correction 293–300
floors 46–47
framed buildings 39–40
    controlling dimensions 41
    types of external envelope *see* external envelope

'geometry'
    activity 144–146
        choosing 148
    building types and activities
        case study 147–148
        example 146–147
    conditioning
        indirect 152
        specific issues relating to 150–152
    entrance doors
        high usage 153
        vehicular, 'U values' 153
    envelopes 149–150
        measurement conventions 152–153
    external doors and windows 153, 289
    fins and overhangs 153, 293–300
    glazed doors 153
    global building infiltration 144
    thermal capacities within merged areas 153
    zones and zoning 148–149
        adjoining conditions 150, 151
        zone height and corners 149, 150
glazing *see also* windows
    doors 153

hazards 12
health and safety
    Approved Codes of Practice (ACOP) 11–12
    asbestos 12–15
    hazards 12
        simple risk estimator 13
    legislation 11
    risk assessment 12
heating, ventilation and air conditioning (HVAC) systems 51
    controls 73
        central time control 73
        local temperature control 74
        local time control 73
        optimum start/stop control 74–75
        weather compensation temperature control 74
    drawing symbols and abbreviations 281–287
    efficiency 54–55
    essential task 51–52
    fuels and emission factors 55–56
    heating sources 65
        boiler flues 66–67
        boiler heat sources 65
        combined heat and power (CHP) 70–71
        district heating 71–72
        gas and oil fuelled boilers 67–68
        heat pumps 72–73
        medium and high temperature boilers 70
        mineral fuel and biomass boilers 69
    heating systems 56–57
        central heating using air 60–61
        destratification fans 65
        forced convection heating 62, 63
        local room heaters 62–65
        radiant heating 61–62
        radiators and convectors 57–59
        underfloor heating 59–60
    hot water generation 75
        hot water demands 75
        hot water energy losses 75–78
        hot water generators 78–80
    level 4 assessment, case study 269–277
    organisation 52–54
    pipe and duct identification 279–280
    power factor 56
    sources of information 52
    systems not represented in SBEM 56
heat recovery *see* ventilation and air conditioning
hot water generation *see* heating, ventilation and air conditioning (HVAC) systems

initial enquiry 133
inspection and reflection methodologies 133
    existing buildings
        after inspection 142
        first contact with client 133–135
        pre-inspection practice 135–136
        site inspection – suggested procedure 136–142
inspection equipment 23–26

Kappa value 31–34
Kyoto Protocol 1

lamps *see* lighting
Law of Tort 9–10
    case study 10
    law of contract 10
    privity of contract and third party rights 10–11
liability 8
    health and safety *see* health and safety
    occupiers 9
lighting 113
    artificial systems and their recognition
        Carbon Trust lamp identification information 123–126
        compact fluorescent discharge lamps – (CFLs) 120
        fluorescent control gear and ballasts, identification 121–122
        fluorescent 'strip' discharge lamps – (MCF) 118–119
        high pressure mercury discharge lamps (MBF) 121
        high pressure sodium discharge lamps – (SON) 120
        induction discharge lamps – (QL) 121
        lamp identification – case study 122–123
        light emitting diodes – (LEDs) 121
        low pressure sodium discharge lamps – (SOX) 120
        metal halide discharge lamps – (MBI or HPI or MBI/MH or CDM) 120
        tungsten filament (GLS) lamps – incandescent 117–118
        tungsten halogen (TH) lamps – incandescent 118
        type of lamp and recognition 117
    controls 129–130
        Simplified Building Energy Model (SBEM) 130–131
    damage caused by light 132
    display lighting 131
    safe inspection of lamps 131–132
    Simplified Building Energy Model (SBEM)
        air extracting luminaries 129
        defining lamp type 127
        efficiencies and efficacies of different lamps 127–129
        'full lighting design carried out' 129
        light controls 129–130
        practical issues 131
        zone lighting energy 127
    theory and practice 113–114
        Building Regulations 115
        colour rendering 115
        colour temperature 114
        glossary 115–117
        identification of lamps 115
        levels of lighting 115
        nature of light 113–114
listed buildings 36–37

Minimum Requirements for Energy Assessors for Non-Dwellings, CLG, October 2007 7–8
modular/portable buildings 48–49

national calculation methodology (NCM) 2
    elements 2
national legislation
    Energy Performance of Buildings (Certificates and Inspections) (England and Wales) Regulations 2007 (SI 2007/991) 2
National Occupational Standards (NOS)
    asset skills 5
new build construction
    Building Regulations 27–28
        in other parts of UK 28–29
non-domestic energy assessors (CEAs) 1
    guidance available 4–5

occupiers' liability 9
operational rating 3

photovoltaic systems 110
post-inspection enquiries 142

pre-inspection practice 135
    desk duty 135–136
    seller's questionnaire 135
professional indemnity insurance (PII) 22
profiled cladding systems 43
'project database' 143–144
    construction for
        doors 144
        floors 144
        glazing 144
        roofs 144
        walls 144

quality and auditing
    accreditation schemes 6–7
    energy performance certificates (EPCs) 20–21
    liability 8
    Minimum Requirements for Energy Assessors for Non-Dwellings, CLG, October 2007 7–8
quality management system (QMS) 19–22

recommendations 159
    how to use results page
        developing own recommendations 163
        recommendations generated by SBEM 163
        recommendations tab 161–163
renewable energy systems 109
    photovoltaic systems 110
    solar energy systems 109–110
    wind turbines 111
reporting to client 165
    client report 165
results page
    different building types 157
    example 155–157
    how to use 159
        audit checks after calculation 160–161
        audit checks before calculation 159–160
        developing own recommendations 163
        recommendations generated by SBEM 163
        recommendations tab 161–163
        reflection on EPC and results generally 159
    interpretation of results 157–158
    other SBEM descriptions 157
retaining structures 43–44
risk assessment 12
roofs 46
    area ratio covered by array of rooflights 291
    construction, data entry 144

Simplified Building Energy Model (SBEM) 2
    Building Regulations 3–4
    documents 4
    lighting
        air extracting luminaries 129
        defining lamp type 127
        efficiencies and efficacies of different lamps 127–129
        'full lighting design carried out' 129
        light controls 129–130
        practical issues 131
        zone lighting energy 127
    other program options 2–3
    results *see* results page
site inspection
    suggested procedure 136
        arrival 136–137
        contents of plan 138
        extent of inspection – 'invasive' inspections 140
        information from client 137
        order of inspection 138–140
        photographs 140

plan of property 137–138
reflection notes 140–141
two different possible data collection methodologies 141–142
zoning 141
solar energy systems 109–110

thermal bridges 34–35
   repeating 35
thermal capacity (K$_m$ value) 31–34
thermal transmittance 31
training and assessment
   assessment of knowledge, understanding and competence 5–6
   end test 6
   National Occupational Standards (NOS) – asset skills 5
   qualification 6
   training 5

'U value' 31
   vehicular entrance doors 153

ventilation and air conditioning 83
   air conditioning systems 95–96
      air-water systems 95, 100–103
      all-air systems 95
      constant air volume systems 98–100
      cooling sources 106–109
      refrigerant systems 95, 103–105
      variable air volume (VAV) systems 96–98
   heat recovery 93
      heat pipe devices 94
      other heat recovery systems 94–95
      plate heat exchangers (recuporator) 93
      run-around coils 94
      thermal wheels 93–94
HVAC systems *see* heating, ventilation and air conditioning (HVAC) systems
renewable energy systems 109
   photovoltaic systems 110
   solar energy systems 109–110
   wind turbines 111
ventilation 83
   central air conditioning system ventilation 85
   local extract 85–86
   mechanical ventilation 85
   mixed-mode ventilation 86
   natural ventilation 84
   ventilation openings 83–84
ventilation and air conditioning equipment 86–87
   air handling units 91–92
   air heat exchangers 89–91
   dampers 88
   fans 87
   filters 88

walls
   cladding panels 42
   construction, data entry 144
   external 41–42
water heaters *see* heating, ventilation and air conditioning (HVAC) systems
windows 44–45
   data entry 144
      external windows 153, 289
      fins and overhangs 153, 293–300
wind turbines 111